Network Cabling Illuminated

Robert J. Shimonski, RSNetworks.net
Richard Steiner, Leviton Manufacturing, Voice and Data Division
Sean Sheedy, Industry Consultant

JONES AND BARTLETT PUBLISHERS
Sudbury, Massachusetts
BOSTON   TORONTO   LONDON   SINGAPORE

World Headquarters

Jones and Bartlett Publishers
40 Tall Pine Drive
Sudbury, MA 01776
978-443-5000
info@jbpub.com
www.jbpub.com

Jones and Bartlett Publishers
Canada
6339 Ormindale Way
Mississauga, ON L5V 1J2
CANADA

Jones and Bartlett Publishers
International
Barb House, Barb Mews
London W6 7PA
UK

Jones and Bartlett's books and products are available through most bookstores and online booksellers. To contact Jones and Bartlett Publishers directly, call 800-832-0034, fax 978-443-8000, or visit our website www.jbpub.com.

Substantial discounts on bulk quantities of Jones and Bartlett's publications are available to corporations, professional associations, and other qualified organizations. For details and specific discount information, contact the special sales department at Jones and Bartlett via the above contact information or send an email to specialsales@jbpub.com.

Library of Congress Cataloging-in-Publication Data
Shimonski, Robert.
Network cabling illuminated / by Robert J. Shimonski, Richard
T. Steiner, and Sean M. Sheedy.
p. cm.
ISBN 0-7637-3393-8
1. Telecommunication cables. 2. Telecommunication wiring.
3. Computer networks. I. Steiner, Richard T. II. Sheedy, Sean
M. III. Title.
TK5103.12.S55 2005
621.382—dc22
2005007397

Production Credits
Acquisitions Editor: Tim Anderson
Production Director: Amy Rose
Production Assistant: Alison Meier
Editorial Assistant: Kate Koch
Marketing Manager: Andrea DeFronzo
V.P. of Manufacturing: Therese Connell
Cover Design: Kristin E. Ohlin
Composition: Northeast Compositors
Printing and Binding: Malloy Inc.
Cover Printing: Malloy Inc.

Printed in the United States of America
09 08 07 06 05   10 9 8 7 6 5 4 3 2 1

Dedications

This book is dedicated to you, the *reader*, the most important person in my mind while operating within the book creation process. From an author's perspective, it's hard to judge how a book may represent itself to a reader. I hope that this book will serve as what the authors wanted it to be—a needed gateway into the cabling profession, providing you with the essentials and fundamentals that will take you to new levels in your career. This textbook should also serve as a reference to revisit when you need to make a cable run, as an example. Voice and data cabling is a professional skill that should be viewed, approached, and mastered just like any other profession such as art, music, English, auto mechanics, or environmental science. There are a lot of things to know and a lot of hands-on skills that need to be acquired in the process of doing the work. I hope that this book serves that purpose to you, to bring closure to a possible hole in your knowledge base or enlighten you in areas where you may not have a lot of experience. This book is dedicated to the readers, who hopefully have taken away what they needed in this book.

– *Robert Shimonski*

I would like to dedicate this to my wife for her patience and understanding while pursuing my career and to my son, without whom I wouldn't have found the courage and energy to do some of the things I've done.

– *Richard Steiner*

Preface

Purpose of this book

The concepts of electricity, waves, and signaling, in addition to standards and codes, are essential to understanding and working safely with data cabling. A basic cabling course should cover theory and foundational material related to cabling in addition to practical, hands-on procedures, and coverage of standards and safety practices. Having worked in the information technology and security fields for over 40 years combined, the authors of this book have written from their extensive experience to present cabling topics using a no-nonsense approach. This text uses an easy-to-understand, practical format, making it not only more interesting to the student but easier for the instructor to explain and hold the attention of the students. With pertinent lab exercises, strong real-world scenarios, and instruction on the use of common, popular tools and practices, and recommended strategies and implementations, this book provides coverage of all necessary topics for individuals interested in learning basic cabling concepts and procedures.

Structure

Chapter 1, "Voice and Data Cabling Basics," covers the history and basic elements of cabling, such as cable types, transmission and interference characteristics, and emerging cable technologies. Chapter 2, "Cabling Standards and Specifications," focuses on national (that is, United States) and international standards and codes that apply to data cabling. Chapter 3, "Understanding Signals," covers the basics of conducted media, optical signals, and wireless and radiated media. Chapter 4, "Data Network Signals," covers analog and digital signals, encoding, circuit communication, and broadband networking.

The next three chapters focus on popular types of cabling. Chapter 5, "Copper Media: Twisted-Pair Cabling," covers twisted-pair, Chapter 6, "Copper Media: Coaxial Cabling," covers coax, and Chapter 7, "Fiber-Optic Media," discusses various aspects of fiber-optic cabling.

Chapter 8, "Cabling System Connections and Termination," covers the most commonly used cabling connectors (such as RJ-45, F connectors, SC connectors, etc.) and cabling termination, including punchdown blocks. Chapter 9, "Safety Considerations," covers safety codes and standards applicable to cabling and electricity in general, and personal and workplace safety tips and procedures. Chapter 10, "Electrical Protection Systems," explains general concepts and terminology involved with electrical exposure, protectors, and grounding.

Chapter 11, "SOHO and Residential Infrastructure Technology," discusses common cabling technologies in use in small office/home office environments. Chapter 12, "Structured Cabling," focuses on backbone and horizontal cabling, raceway types, and structured cabling types. Chapter 13, "Building Your Cabling Toolkit," discusses many of the most useful and popular tools you should acquire or have access to when working on commercial cabling projects. Chapter 14, "Planning and Implementation of Premise Wiring Installations," describes in detail what you need to know to properly plan for a commercial cabling project. And finally, Chapter 15, "Cabling Installation and Testing," walks you through actual cable installation and testing procedures. The appendices offer tips on obtaining and keeping employment in the cabling industry, outlooks on the future of cabling, references and resources for further learning, and a glossary of terms used throughout the book.

Learning Features

The writing style is conversational. Each chapter begins with a statement of learning objectives. Step-by-step examples of cabling-related concepts and procedures are presented throughout the text. Illustrations are used both to clarify the material and to vary the presentation. The text is sprinkled with Notes, Tips, and Warnings meant to alert the reader to additional and helpful information related to the subject being discussed. End-of-chapter elements include sections titled Chapter Summary, Key Terms, Challenge Exercises, and Challenge Scenarios. The "Chapter Summary" section ties together important topics discussed in detail within each chapter. The "Key Terms" section is comprised of a glossary of significant terms used throughout the chapter. The "Challenge Exercises" section provides students the opportunity to gain hands-on experience with topics discussed in the chapter body, whereas the "Challenge Scenarios" section provides real-world applications of the chapter material.

Chapter summaries are included in the text to provide a rapid review or preview of the material and to help students understand the relative importance of the concepts presented.

Audience

The material is suitable for freshman or sophomore computer science majors or information science majors, or students at a two-year technical college or community college, with a basic technical background.

Acknowledgments

The authors would like to thank Jones and Bartlett for the opportunity to write a cabling textbook that combines the foundational concepts with the practical aspects of the topic. We appreciate Jones and Bartlett project management—Tim Anderson, Acquisitions Editor, and Lesley Chiller, Editorial Assistant, who both took an active and patient role in seeing this book to completion. Amy Rose and Alison Meier, our production editors, did a fantastic job tracking the chapters throughout the author review and page proofs phases. In our collective opinion, editorial staff members rarely get the credit they deserve for seeing a book from rough manuscript to the finished product you're reading now. Many thanks to Amy, Alison, the copyeditor, and proofreader for the hours of work you put into this book.

The authors would also like to thank Lloyd Bardell, our technical reviewer, and Kim Lindros, the authors' project manager. Lloyd pointed out our shortcomings (and outright mistakes, on occasion), and offered pertinent suggestions for changes and additions to make the book more readable and useful. Kim managed the project on our behalf, reviewing manuscript, watching the schedule, and much more, which resulted in a smoother writing process.

—*Robert Shimonski, Richard Steiner, and Sean Sheedy*

I would personally like to thank *everyone* who has helped me along the way—the number of peers I have accumulated in the past 5 years alone is unbelievable to me, and I could probably put out a book on just "who to thank." In any profession, working as a team is always better; it helps forge teamwork, which is where the real magic happens. You can't run cable without a team. A sincere *mega*-thank-you goes to Kim Lindros for being a great project manager to work with, and thanks to Jones and Bartlett for putting out a much-needed title. I also thank the other authors and the team assembled to create this book.

This book was written at the right time—a perfect example of a real-world cabling job was underway at the same time as the writing of this book. What could be better than reading a book written by actual cable installers performing a cable installation that covers the scope of this book? This installation helped to give this book a real-world feel because the recent facility move required the installation of all new fiber and copper cabling that had to be designed, implemented, and tested. Nearly every aspect of a real-world cable installation is found within this book, to include planning, safety, and the actual technical aspects of cutting and crimping cable, testing it, and so on. The following is a snapshot of the actual installation.

As you can see, a cable installer's job is a physical one—you *will* sweat and get dirty. I would like to sincerely thank all those who helped to perform this installation, as it was truly amazing how quickly and efficiently it was done. I especially want to thank Siew Tan, Todd Fitch, and Stu Dobson for the amazing job they did.

—*Robert Shimonski*

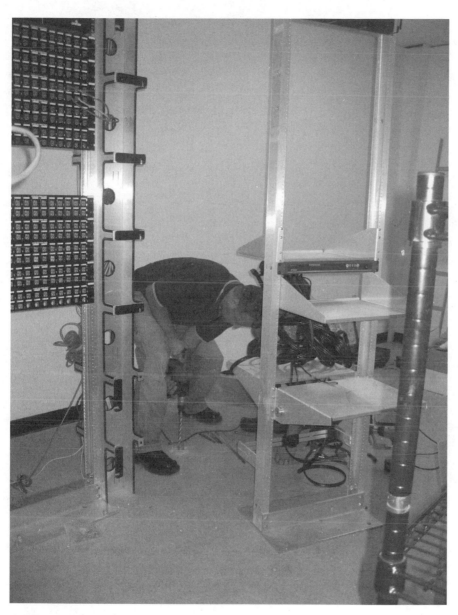

Rob Shimonski and Todd Fitch drilling and installing racks

About the Authors

Robert J. Shimonski is a well-noted industry specialist, speaker, trainer, author, and editor of over 50 published books and thousands of magazine articles and certification study guides. Robert works within today's most challenging technological and business environments in a global setting. When writing about information technology, Robert brings proven frontline industry knowledge to the reader in every page, following the discipline that a reader needs—a writer deeply rooted in the heat of the battle to be the reporter. Robert is constantly on top of the latest trends and reporting the state of the IT industry from a real-world perspective. Robert is there mentoring, leading, planning, designing, implementing, reading, and writing. Robert also works very hard to mix and balance college, certification, and experience into a useful skill set defined only by the ever-changing and ever-interfusing fields of technology and business.

Blessed with years of frontline experience in a myriad of positions and companies, Robert has implemented technology in over 250+ different companies worldwide. Robert is also a 4-year college graduate who holds many information technology credentials, which include TruSecure TICSA, Cisco CCDP, CCNP, Nortel NNCSS, Microsoft MCSE, Novell Master CNE, Prosoft Master CIW, SANS GSEC, GCIH, CompTIA HTI+, Security+, Server+, Network+, i-Net+, A+, e-Biz+, Symantec SPS, and NAI Sniffer SCP, among many others.

Robert is author of many network and security-related articles and published books including the *Sniffer Pro Network Optimization and Troubleshooting Handbook* from Syngress Media Inc. Robert is also a lead writer at www.WindowsNetworking.com, a source for many Windows-based networking white papers and solutions.

When Robert is not busy traveling and working as an ubergeek, he is spending time with Erika, playing music with his band, or plotting his next business startup. You can contact Robert at *www.RSNetworks.net.*

Richard Steiner has over 13 years of experience in the telecommunications industry. His experience includes installing and troubleshooting, estimating, designing, and project management of telecommunication cabling networks. Among his industry certifications, he is a BICSI Registered Communications Distribution Designer (since 2000) and a certified Fiber Optics Installer through the Electronic Technician's Association. He also holds a Bachelor of Science degree in Information Technology. He is currently an applications engineer for a major telecommunications manufacturing company.

Sean Sheedy has over 20 years of experience in the telecommunications industry, specializing in fiber optics. During his years in the field, he has worked as a technician/installer, troubleshooter, lead tech, project manager, emergency restoration technician, and independent consultant. Some of the projects he has work on have included clients such as Microsoft, federal and state prisons, divisions of the military, and other government agencies. He holds over 30 industry certifications.

In 1997, Sean started teaching a 5-week fiber-optic course and continues to teach fiber-optic and copper installation and troubleshooting courses at Edmonds Community College in Edmonds, Washington. He is a certified AMP instructor and 3M instructor, and is certified through Cisco to teach its "Fundamentals of Voice, Data, and Cabling" course. Currently, Sean still serves as an industry consultant and a fiber-optics engineer.

Contributing author: Mark Mirrotto

About the Technical Editor

Lloyd Bardell, CCNA, CCA, MCP, A+, Network+, i-Net+, holds a Bachelor of Arts degree (summa cum laude) from Jacksonville State University and a Masters of Science degree in Information Architecture, Capitol College and a Master of Arts, Jacksonville State University. He was a New York University Faculty Summer Scholar, the recipient of the 1991 Lilly Foundation Research Grant, and a two-time recipient of the Pelham Scholarship. He has taught at Talladega College, Southwestern Community College, and Bangkok University, and currently teaches part-time at Foundation College.

Lloyd has worked in various facets of the IT industry since 1992. He currently owns and operates a consulting firm, Lloyd Technical (*http://www.lloyd-technical.com*), which specializes in network design and evaluation of existing networking systems. His clients include SBC, ARC, and TAC Worldwide. Lloyd lives in San Diego, California.

Contents

CHAPTER 1

Voice and Data Cabling Basics

Learning Objectives

After reading this chapter you will be able to:

- Understand wired media and its primary purpose

- Grasp cable basics

- Discuss the evolution of cabling in voice and data communications

- Discuss telephony cabling and how it differs from data cabling

- Recognize important data cabling types, such as coaxial, twisted-pair copper cabling, and fiber-optic

- Identify emerging technologies, such as the convergence of voice and data

Networks are an essential part of modern life, and most networks require cabling. Whether we use networks in our professional or private lives, or both, they now are the basis of most modern communications. Most modern households have at least one computer—and many have home networks—largely due to the exponential growth and popularity of the Internet. Virtually every home in the developed world depends on network cabling for telephones and home entertainment.

Network cabling supports worldwide electronic communications (such as by computer, phone, or fax), allowing organizations, corporations, and individuals to exchange information by both voice and data, including multimedia. With cabling costs decreasing while quality and performance increase, and with the continued growth of networks including the Internet, the demand for premise wiring and the number of cabling jobs will also continue to increase over time.

Along with this increased demand, there is increased responsibility when working with cabling. There are a number of logistical, technical, and legal guidelines relating to safety and other issues. Those in the cabling profession must have both technical knowledge and the ability to handle those parts of the job that require manual labor.

The quality of a network's performance is a combination of the characteristics of the cable used and the skill with which the cable is installed. Significant effort has been made in establishing standards that govern how each type of cabling should be implemented to ensure that the cabling provides years of service for clients and creates a network infrastructure that will support several generations of hardware. When doing an installation, one should always plan for future bandwidth, throughput, and technological needs.

1.1   Conducted (Wired) Media

Communication standards and principles rest on the basic principal that there are four basic components to communication: information, a sender, a receiver, and a medium for the transfer of information between the sender and receiver. On an electronic network, the **medium** can be one of two kinds: radiated (or wireless) and conducted (or wired). Both kinds of media have been in use throughout the history of electronic communications.

Wired media entails transferring information by means of a physical medium or cable. **Wireless media** allows information to be sent independently of physical cables. Wired media arrived on the scene more than half a century before wireless media, when Samuel Morse invented the telegraph in 1837.

1.2   The History of Cable Communications

The first communications network of modern times as well as the first network to be powered by electricity was the **telegraph**. Telegraphy was also the first technology to use cabling as the medium for communication. The telegraph was developed to support another network, the U.S. railway system. Telegraph lines followed routes originally created for America's railroads.

Telegraphy was the first communication system to use a special language or protocol. Data traveled along telegraph lines in the form of **Morse code**, named for its creator, Samuel Morse. Morse code, which was adopted as the standard code for communication, is a system of long and short electronic signals interspersed with intervals of silence and is still used today. Some historians might view Samuel Morse as the father of digital communication, but machines could not interpret or process the code because of the variation in the signals used to represent each alphabetic character. It wasn't until the development of a code with consistent signal length that digital communication became possible. Emil Baudot, often considered the pioneer of digital communication, developed that code. Baudot's code was known as the International Telegraph Alphabet number 1 and used five signals to represent each character, which meant that communications could go beyond basic alphanumeric code. These signal states were sent by a keyboard having five keys with each key representing a different bit of the five-state signal. Baudot's code evolved with the help of Donald Murray and Western Union into what is known now as Baudot's Code or International Telegraph Alphabet number 2 (ITA2), and is still used in telecommunication devices for the deaf (such as telecommunication devices for the deaf [TDDs]) and in some ham radio applications.

Telegraph networks used wire as the physical medium for communication until Guglielmo Marconi developed the wireless telegraph in 1890. The first automatic telegraph equipment arrived in 1910, along with the development of the postal telegraph system. Many newspapers and news organizations still use these teleprinter, or "wire," services today.

With the invention and standardization of telephone equipment, the use of the telegraph declined drastically, as people enjoyed the luxury of being able to talk to someone several miles away. The telephone was a great advance over the telegraph. It was user-friendly, and an individual could use the device directly, without having to have an operator translate messages into Morse code, although telegraph messages or "cablegrams" were still used for some purposes. Telephone communication operated by converting spoken sound into a signal through modulation, expanding on the old "tin can on a string" principle by using electricity to amplify and contain the modulated signal that carried voice traffic. (We will discuss modulation in the next section.) Telegraph networks were much simpler than those for the telephone, as they used specific stations for access to telegraph cables and equipment. The widespread demand for the easier-to-use phone resulted in an expansion of the telecommunications infrastructure of the time, which would allow businesses and eventually homes to connect directly to the telephone network using on-site equipment.

1.2.1   Telephony Cabling

Telephony, the basis of telephone communications, allows a device to convert sound into an electrical signal, transmit the represented sound over electrical cabling, and convert the signal back into sound. The primary patents of telephony are attributed to Alexander Graham Bell, who was able to transmit the spoken word through a series of vibrations that captured and reproduced the spoken information. The process of modulation was used to convert the captured vibrations and render them as intelligible signals. **Modulation** establishes a continuous, variable wave, which represents a signal that contains sound information. The signal, called an analog signal, varies along its axis, similar to human speech.

When a person spoke into a telephone in the early days of the technology, the sound was transformed into an analog signal that was sent to a telephone switching office. At the switching station, an operator would answer and take the call request, then physically "patch" the call through to the connection to the switching office belonging to the person being called. This was a connection-oriented form of communication with a circuit lasting for the duration of the call.

When the patch was established, the phone belonging to the person being called would ring, notifying that person that a call request was being made.

The concept of the ringing phone is attributed to Alexander Graham Bell's assistant, Thomas Watson, who experimented with different ways of notifying someone that they were being called. His first idea was to use a button activated by a special tone that would cause a "tick-tock" sound on the intended recipient's telephone that would continue until the receiver was picked up or the operator stopped signaling. Eventually, it was decided that ringing worked better than ticking.

If someone were to make a non-local call, the operator would have to patch the caller through an interoffice trunk connection to another switching office, where another operator would connect the caller to whomever they were calling.

Telephony technologies evolved and automatic electronic switches replaced friendly or not so friendly operators, the human equivalent of switches. This technology allowed a caller to dial special numbers that told the switch to whom the call would be connecting. Operators became unnecessary for direct-call routing.

The underpinning of any telephony network is its electronic devices. As the telephony network grew, more cabling was necessary to allow more people to communicate. Larger networks first developed in metropolitan areas, expanding in the 20th century to include virtually every home in the developed world.

The telephone network, referred to as **plain old telephone service (POTS)**, began humbly enough, but quickly was involved in great competition and controversy with the need for ever-larger telephone networks, connections with the telegraph networks, and the use of patented technology. As telephone and telegraph networks were used more frequently to communicate across the United States, the federal government began to regulate both networks under the auspices of the Interstate Commerce Commission (ICC). The Federal Communications Commission (FCC) was formed in 1934, and the ICC delegated the responsibility of regulating telephone traffic between the states to the FCC. Much of today's telecommunications industry is still under FCC supervision.

Initially, the utility companies controlled their own infrastructures, which led to companies becoming centralized and monopolistic. This was the case in nearly every country that implemented telephone networks. The first U.S. telephone exchange was built in New Haven, Connecticut, in

1878; however, the growth of telephone service exploded and the National Telephone Company was born. The National Telephone Company later became **American Telephone and Telegraph Corporation (AT&T)**. Known as "Ma Bell" to most North American telephone users, AT&T would become the dominant telephone company (particularly in terms of control of the infrastructure) and would be a monopoly until the 1980s, when anti-trust regulations resulted in its breakup into **regional Bell operating carriers (RBOCs)**. Other countries would deal with monopolies either by use of anti-trust regulations similar to those in the United States or by their own government's control of the infrastructure.

Whether public trusts or private enterprises controlled the infrastructure, the issue of cable installation and management remained constant throughout the evolution of these networks. The job of the cable installer was born with the implementation of telegraphy, and with the evolution of telephony, the profession continued to grow steadily. The steady growth became a sudden boom with the increasing demand for data networks in the late 20th century. The demand for cable installers and managers persists to the present day, as does the demand for standards for skill sets and cabling certifications.

1.2.2   Wireless Media

In the last two decades, telephony networks have grown beyond the traditional cable-based network, or *land lines*. Wireless networks, such as those used by cellular and satellite telephones, use radio wave extensions to receive and transmit telephone calls within a coverage area. Various forms of point-to-point communication (walkie-talkies, CBs, and ham radios) as well as other forms of radio and television broadcasting also used wireless networks. Bluetooth and the Institute of Electrical and Electronics Engineers (IEEE) 802.11 are examples of wireless standards for computer networks and their peripherals.

1.3   Data Networks and Cabling

Early computer networks worked along a centralized computing concept, in which all processing took place and all information was stored at a centrally located mainframe or minicomputer. Users would connect to the computers via asynchronous terminals, or *dumb terminals*. The communication did not require numerous protocols because the terminals sent and received user input and output only in ASCII (American Standard Code for Information Exchange), a text-based format. Special character code sets such as EBCDIC (Extended Binary-Coded Decimal Interchange Code) could be used over IBM networks. A significant difference between EBCDIC and ASCII was the number of bits upon which the protocols were based for the transmission of data—seven for the former and eight for the latter. Many of these terminal-based environments connected terminals to mainframes using the same cabling as that used for the telephone networks, making life simpler for cable installers.

Early data networks used modems to communicate over voice-grade analog lines in existing telephony networks. Data transfer rates were dramatically slower than today's modem standards, but capacity demands were also much lower in the early days. In 1958, the U.S. Department of Defense put the first data communications network into place: SAGE (Semi-Automatic Ground Environment). In 1964, the first major corporate network was created; a joint venture between IBM and American Airlines, the Sabre network handled commercial airline reservations.

Data networking wasn't a normal part of everyday business until the early 1970s, when network providers and hardware manufacturers began developing standardized networks. The Xerox Corporation published the standards for Ethernet, its proprietary network, in 1972. With the spread of Ethernet and its counterparts, newer standards for computer and digital networking evolved that led to the development of new kinds of cabling. Ethernet also provided the foundation for the development of local area networks (LANs).

The history of networking using today's conventional methods is fairly young in comparison to the history of telecommunications in general. Computer networks are data communication networks, but they owe their development to the voice, wire, and broadcasting telecommunications networks.

Table 1.1 chronologically lists many prominent dates in the history of telephony and data cabling.

TABLE 1.1 Important Events in the Development of Communications

Date	Event
1837	Samuel Morse invents the electromagnetic telegraph
1844	Morse dispatches the first U.S. inter-city telegraph message over a public line
1861	Coast-to-coast telegraph communication begins in the United States
1876	Alexander Graham Bell invents the telephone
1877	The telephone is used in a private home for the first time
1884	First long-distance telephone line is established, connecting Boston to New York City
1886	AT&T offers private line service to customers
1885–89	Heinrich Rudolf Hertz first broadcasts and receives radio wave transmissions
1890–96	Guglielmo Marconi develops the first telegraph that transmits radio waves
1899–1900	First use of a wireless telephone
1902	The human voice is first heard via radio transmission
1910	Automated telegraph wire service is commercially available
1920	Sound broadcasting begins
1927–28	Philo T. Farnsworth develops broadcast television
1928	Teletype invented
1929	Coaxial cable invented
1930	First two-way video telephone call takes place
1946	ENIAC, the first totally electronic digital computer, is developed at the University of Pennsylvania
1947	The cellular telephone is developed by Bell Labs
1951	Direct-dial long distance telephone service is available
1957	Sputnik-1, Earth's first artificial satellite, is launched
1960	First electronic switch developed
1968	The FCC makes a landmark decision (Carterfone Decision) that allows customers to use their own equipment to access networks
1970	Packet switching is available
1971	Intel develops the first microprocessor, named the 4004
1971	Ray Tomlinson develops e-mail

TABLE 1.1   (continued)

Date	Event
1972	Xerox develops Ethernet
1975	MITS introduces the first personal computer, ALTAIR
1975	IBM develops the 5100 personal computer
1976	Apple Computer introduces the Apple I
1981	The IBM PC is developed
1984	Introduction of cellular telephones to replace radio phones
1990	The first national wireless network, ARDIS, is available
1991	Digital subscriber line (DSL) becomes available
1992	Release of the first standard for Integrated Services Digital Network (ISDN)
1993	The first digital mobile network is established in the United States

Source: Evolution of Networks (*http://www.mnetworks.net*)

1.4   What's in a Cable?

Cabling is made up of a conductive material covered with one or more sheathing and/or insulation elements. The conductive material is the primary medium for signal transmission. Cabling can use a single strand of conductive material or multiple strands contained under a single universal sheathing and/or insulation medium. There are three major types of cabling—coaxial, twisted-pair, and fiber-optic—each of which we will look at in the next section.

When installed properly, all cables interface with hardware components that serve as the sender or the receiver of information—or as both simultaneously. Cables connect to these components by means of an interface that helps pass signals to the sending/receiving components, which are called interface connectors. Throughout this book, we will discuss different types of interface connectors along with different types of cabling.

1.5   Well-Known Cable Types and Varieties

Most wired media use three main types of cabling: twisted-pair, coaxial, and fiber-optic. Voice and data networks use different cabling types depending on the kind and amount of traffic the cabling is to carry. Signaling standards and protocols are also important in determining what kind of cabling to use. Electricity is the signal medium over coaxial and twisted-pair cabling; fiber-optic cabling uses coherent optics (light). All cabling varies widely in characteristics such as size, capacity, and cost. Regarding the latter, the more difficult the network cabling is to install, the more expensive it is to install.

1.5.1   Twisted-Pair Cabling

The most widely used network cabling type is **twisted-pair cabling**. In the 1970s and 1980s, twisted-pair cabling was used for voice communications. In the 1980s, data networks also started using twisted-pair cabling because it offered a simple, inexpensive, and modular basis for local area networking. Multiple bundled cables allow for duplex communication (telephone conversations), for separate paths for communications signaling, and setup and for broadband communications. The twisted wires in a twisted-pair cable controls signal degradation caused by electromagnetic interference (EMI) and radio frequency interference (RFI). Different numbers of twists in each of the wire pairs control problems such as interference and cross-talk between the pairs in the cable (discussed in Chapters 3 and 5 in this book). Interference is potentially more serious in data cabling, in which cross-talk can severely damage the integrity of data communications. However, cross-talk occurs more often in voice networks, where at times users can sometimes hear conversations taking place over other wires within the same bundle. Often, however, cross-talk occurs at non-audible levels.

Unshielded Twisted-Pair (UTP)

Unshielded twisted-pair (UTP) cable is the most common and widely available network cabling today. Business phone systems, LANs, and wide area networks (WANs), and more recently home networks with high-speed Internet access, use UTP. UTP cable is composed of eight copper wires coupled in four pairs, with each wire wrapped with its own insulation. The twisting of the wires conjoins each pair. Twisting also protects the wire pairs through the cancellation effect, produced by the twisted-wire pairs, which limits signal degradation caused by EMI and RFI. The number of twists in the wire pairs varies to reduce cross-talk between the pairs. Specifications for UTP cabling govern bandwidth, color schemes, length, and how many twists or braids are allowed per meter of cable.

There are differences between the UTP cabling used for data networking and that used in voice networking. UTP used as a data networking medium has electrical property differences as well as other differences in physical character from that used in voice networking. Because of its small size and pliability, UTP is one of the easiest types of cabling to install. It is easy to bundle multiple runs of UTP cabling, which makes it a flexible choice for installations in either old or new buildings. In addition, nearly every major data networking architecture has options for running over UTP, which is one of the reasons it is so popular. Finally, UTP costs less per meter than any other type of data network cabling.

However, there are some disadvantages to using UTP. It lacks shielding, which makes it more susceptible to EMI and RFI than most other kinds of network media. Careful consideration must be given to the placement of cable runs containing UTP. There are several standards that assist in UTP premise wiring.

UTP also has distance limitations, primarily as a consequence of the twisting of the wire pairs. Although the cancellation effect is beneficial in trying to eliminate noise, it contributes to early attenuation (signal degradation). Early in its history, UTP was slower at transmitting data than other cable types, but this is no longer the case. In fact, UTP now is considered the fastest copper-based network medium, and it is used in digital phone systems and the popular Ethernet local area network standard. Both the 10BaseT and 100BaseT variants of Ethernet use UTP.

Shielded Twisted-Pair (STP)

Shielded twisted-pair (STP) cabling is in some ways an improvement over UTP. It combines UTP's wire twisting techniques and cancellation with additional shielding. Each of the four pairs of wires is wrapped in metallic foil and the set of four pairs is wrapped in a larger metallic foil or braid. Whereas UTP cable is usually 100 Ohms, STP cabling is usually 150 Ohms. Token Ring networks and Ethernet networks that demand the extra reduction of electrical noise and other interference use STP. The trade-offs in using STP rather than UTP are increased costs, reduced pliability, and the need for proper grounding, which can make installation more difficult.

Newer Twisted-Pair Standards

A newer, more cost-effective solution as an alternative to UTP is a type of twisted pair cabling called screened twisted-pair (ScTP). ScTP is also known as foiled twisted-pair (FTP). ScTP cabling wraps regular UTP cabling with a metallic foil shield, or *screen*. ScTP is usually 100- or 120-Ohm cable, because it has been designed to mesh readily with existing UTP premise wiring.

The major downside to using this metallic shielding with both STP and ScTP is that STP and ScTP can suffer serious noise problems if the cable is improperly grounded. The solution is to install and terminate the cable properly. Although it might be cumbersome and difficult at times—and is always rather tedious—good workmanship is essential. Most individuals and businesses use UTP rather than STP cabling because STP is more cumbersome to handle, as well as being tedious to ground and terminate properly.

Another disadvantage to using STP and ScTP is that they have shorter distance limitations—and thus require smaller catchment areas (areas of coverage)—than other types of network cabling. Attenuation is greater, which can mean that *repeaters* must be used to re-amplify the signals in longer cable runs. The shielding materials make terminations more difficult.

The various types of twisted-pair cabling are discussed in detail in Chapter 5, "Copper Media: Twisted-Pair Cabling."

1.5.2   Coaxial Cable

Coaxial cabling is found in most homes, thanks to the advent of cable television, VHF antennae, satellite television, VCR/DVD systems, and broadband Internet access. Coaxial cabling is also used in businesses that must manage and support hybrid communications such as voice, video, and data traffic. Even though coaxial cabling has been replaced over the years with cheaper twisted-pair and faster, more reliable fiber-optic cabling in data-exclusive networks, coaxial cabling installations are still common.

Coaxial cabling was first developed in the late 1920s to expedite a rapidly growing transcontinental telephone network. The demand for multiple simultaneous calls between major metropolitan areas accelerated implementation of coaxial telephone trunks. In 1936, the first long-distance coaxial cable was installed between New York and Philadelphia, and was made available for multi-channel telephone communications and testing.

Coaxial cabling is composed of multiple conductors. At the cable's center is an inner wire core comprised of a copper wire surrounded by insulation and a braid or foil covering that acts as a shield to the inner conductor. A cable jacket surrounded by a cylindrical insulator covers the inner wire set.

You might not see much coaxial cabling in voice networks. However, the trunks carrying multiple voice traffic channels switched between offices are coaxial cable runs because the coaxial design allows the cable to carry greater bands of traffic. This is the basis for the term "broadband cabling," which is often used as a synonym for coaxial cabling. Coaxial cable can serve as an excellent backbone cabling for setting up computer networks, because the additional shielding allows it to be run for long distances without using repeaters. Coaxial cable is also cheaper than fiber-optic cabling.

Coaxial cabling is a broad category compared to twisted-pair and fiber-optic cabling because it includes the widest variations in size and capacity. Thicker coaxial cabling is used for outdoor lines that require greater carrying capacity. For indoor networks, thinner coaxial cable is used because it is easier to route through conduits and other tight areas. The largest diameter of coaxial cable specified for use with Ethernet was 10Base5, also referred to as Thicknet. Because of its thickness, this kind of cable can be too rigid to install easily in certain environments. Thicknet cabling is rarely used today, except for special-purpose installations that might involve certain specialized networks.

Smaller diameter coaxial cabling (.35 cm) was used with Ethernet as a less expensive Thicknet alternative (affectionately referred to as "cheapernet"). This smaller, Thinnet cable became very popular with growing use of the personal computer (PC) in the 1980s. Thinnet Ethernet was easier—and hence less expensive—to install. Thinnet was more pliable than Thicknet, making it suitable for cable runs that had to make many bends and turns. Thinnet, like Thicknet, required proper connections and solid terminations because of its reliance on proper electrical grounding.

Coaxial cable is more expensive to install than twisted-pair cable. It also relies heavily on proper grounding, requiring a solid termination at both ends. It is imperative that there are strong and solid electrical connections with all connectors that interface with the cabling. This was, and still is, an issue when coaxial cabling is improperly installed. As a result, coaxial cabling is not considered cost-effective. Coaxial cabling is discussed in more detail in Chapter 6, "Copper Media: Coaxial Cabling."

1.5.3   Fiber-Optic Cable

Fiber-optic cabling is designed to carry coherent pulses or a continuous modulated stream of coherent light. Because pure light has a far lower degree of signal resistance than electricity, it is much faster and can carry far more data than other types of cabling. More importantly, fiber-optic cabling is exempt from the problems of electromagnetic and radio-frequency interference. In fiber cabling, the signal is controlled optically and guided through a solid, nonconductive medium. For that reason, radiated or wireless traffic will not interfere with the signal. The trade-off is that fiber-optic cabling is much more expensive than twisted-pair cabling and coaxial cabling.

Experiments dating back to the 19th century have been conducted using fiber-optics. Patents were first issued for fiber-optic technology in the 1950s; however, the first functional fiber-optic networks did not appear until the late 1970s. The challenge for two decades was being able to create solid-state light sources and high-quality, pure-perfect glasses, as well as the equipment for testing, fusing, crimping, and repairing the delicate cables. Telephone companies first implemented large-scale fiber-optic networks. Many people will remember the Sprint television commercials of the early 1980s that discussed the wonders of fiber-optic communications and the benefits for long-distance telephone communication. Soon after, data networks were developed with high-speed backbones that made possible the growth of high-capacity data networks such as the National Science Foundation Network (NSFNet) and the other networks that make up the Internet.

There are many different types of fiber-optic cabling, along with different applications for it. In recent years, less expensive glasses—and in some cases even glass–quality plastics—has led to the rapid expansion of fiber in many areas, including the novelty and entertainment industries. A number of holiday trees and decorations demonstrate an aesthetic use of fiber-optic technology.

Fiber-optic cabling used in voice and data networking consists of two separate fibers encased in separate sheaths. Fiber-optic cabling is most often used to expedite bi-directional communication. A separate cable is used for transmitting and for receiving. Each cable of fiber is very tiny. The largest fiber-optic core fiber has the diameter of a human hair, about 63 microns. Many layers of protection, usually made of Kevlar®, and an outer insulation jacket surround each thin wire. (Kevlar is also used for bulletproof vests and airplane skins.) The outer insulation provides elemental and fire protection for the entire cable. When fiber is being buried outdoors, it will have an additional metal wire for added strength and support. It will also be contained in a special pipe to help protect against errant shovels or back hoes.

Fiber-optics use the principles of reflection and refraction. Light left undisturbed will travel in straight lines. Even when bent by lenses and deflected by mirrors, light will travel in a straight line between optical devices. Fiber-optics technology grew out of the desire to control the path of light flow. In 1880, William Wheeler, a contemporary and competitor of Thomas Edison, patented a scheme for piping light through buildings because he did not believe that Edison's incandescent light bulb would be practical. He wanted to use light from a bright electric arc to light rooms from a distance and increase the area covered by artificial lighting. Unfortunately, it was his invention that proved impractical and the incandescent bulb can now be purchased for as little as 25 cents. Wheeler's original ideas did not go to waste; they would eventually be the basis of fiber-optic cabling.

All fibers in fiber-optics have the same fundamental structure. The fiber located in the center of the cable is referred to as the core and has a higher refractive index than the glass (or plastic) cladding that surrounds it. This difference in the refractive index causes the *total internal reflection* that guides light along the core. The size of the core and the size of the cladding vary depending on the type of cabling.

In a communication system based on fiber-optics, an optical signal originates from a sender, or "light source." The light source feeds a modulated signal into a fiber, which delivers the signal to a receiver at almost the speed of light. The receiver demodulates or decodes the light signal and converts it into a direct electrical signal for internal processing or for retransmission across an electrically-conducted communications network. We discuss fiber-optics in more detail in Chapter 7, "Fiber-Optic Media," and briefly in other chapters in this book, in the context of communications.

1.6   Cable Transmission Characteristics

One of the major considerations in transmitting via cabling is whether the signal's energy is to be electrically generated or optically generated. Theoretically, both travel at the speed of light, less the resistance of the material in the medium.

Once a signal is generated, it is carried in the form of a wave. There are a number of waveforms, including square (pulse, discrete) for digital signals and sine (continuous) for analog signals. We discuss signals, waves, analog, and digital signals in detail in Chapter 3, "Understanding Signals."

1.7   Cable Interference Characteristics

As we briefly mentioned previously, many elements in the decision-making process help determine what type of cabling to install, including the type of **interference** to which the cabling is susceptible; the characteristics of that interference; and how, if at all or to what degree, the interference affects the type of cabling under consideration. Attenuation and interference are the principal physical constraints on cabling performance. Attenuation is the reduction of the original signal's amplitude (loss of strength) with little, if any, distortion. Problems with attenuation require consideration of the distance from the originating device. Distance limitations on cable runs, which help prevent attenuation problems, are established after the cabling has been surveyed. For example, the standard for UTP cabling from a computer to any other network device is 100 meters. This does not necessarily mean that once a cable reaches 101 meters, the signal will automatically disappear; however, the standards body cannot certify that the cable meeting those physical properties will be able to exceed that limit without any attenuation occurring. In determining and placing distance limitations, standards bodies often err on the side of caution. We discuss standards bodies in detail in Chapter 2, "Cabling Standards and Specifications."

Interference can come in a number of forms, the most common type being noise. Noise affects cable selection indirectly because some types of signals can tolerate noise better than others. For example, using STP or fiber-optic cabling might be unnecessary if you are selecting cabling for a voice-grade line. However, the integrity of the entire transmission can become compromised when sending data or fax transmissions over voice-grade lines (circuits), because voice-grade lines are engineered to tolerate only a certain amount of noise. One of the major differences between data-grade and voice-grade communication lines is how the lines have been engineered to handle noise and other types of interference.

The two most common types of interference are EMI and RFI. EMI disrupts communication over cabling located in the electromagnetic field of some other electricity-generating device. RFI occurs when the frequency spectrum on a network falls into the same radio frequency (RF) range of another signal generated outside the network. The major difference between the two types of interference is the interference's source. If the source is another electronic or electrical device, it is attributed to EMI. If another traveling or wireless signal is the source of interference, it is considered RFI.

1.8 Voice Communications

The first voice cable consisted of a single wire that contained its own grounding. The two-wire concept would come in 1881, with Alexander Bell's patent for two-wire technology. The two-wire system allowed for the creation of a complete electrical circuit called a local loop, connecting a user's phone with a switching office. The two wires would be twisted a few years later to prevent crosstalk, which was the first implementation of twisted-pair cabling. Individual local loops were combined at junction points to maximize the efficiency of cable runs, especially in larger metropolitan areas. Multiple lines could be combined into larger trunk lines (often using coaxial or fiber-optic cabling). Multiplexing methods would combine all the signals into a single larger signal that carried all calls simultaneously.

The service most people use for regular voice and fax communication today is POTS. Most areas of North America support POTS as a legacy service; the majority of the telephony network in North America has migrated to the PSTN (public switched telephone network), which carries both digital and analog telephony services.

1.9 Data Communications

Data communications have grown in a similar manner to telephony networks as their quality and reliability improves. The idea of multiplexing involves grouping traffic channels to carry multiple data streams or signals to reduce traffic on the network. Data networks employ a more complex, layered model than telephony networks, and operate with different standards and protocols. It could be claimed that data communications industries can still learn from the simplicity and efficiency of the telecommunications industry.

Prior to the late 1990s, accessing your local area network or the Internet from your home was limited to using the telephony network. High-speed access lines were viewed as being cost-effective only for mid- to large-sized businesses. The Integrated Services Digital Network (ISDN), which provided digital access to the telephone company, paved the way for emerging digital technologies such as digital subscriber line (xDSL). These digital technologies provided greater bandwidth at a pure digital-to-digital transmission level. We discuss many of these technologies in Chapter 3, "Understanding Signals."

1.10   Emerging Cabling Technologies

There is a growing demand for faster connections to accommodate increasing network traffic. Along with this demand, manufacturers are continuing to improve cabling standards. A prime example is the ever-evolving standards for UTP. In recent years, UTP has gone from a weak, if inexpensive, option to a cabling option that can carry data 100 times faster than it could 10 years ago—while remaining inexpensive. Modern twisted-pair cables, such as Category 5, 5e, and 6, have four pairs of wires with more twists than types that came before them. We are beginning to see the convergence of voice and data networks, which makes it feasible to use a single cabling system for both phones and computers.

Using a single cabling scheme to handle voice, data, and other services such as video is called structured cabling. **Structured cabling** includes all the devices through which the cable runs, such as the cable terminating equipment and the end devices that connect users, computers, and other resources. Today, previously separate technologies are being combined into one communications system, resulting in more manageable cabling systems, less expensive maintenance, increased flexibility, and more scalability.

Technology in the cabling industry doubles about every 12 to 18 months. As a cabling technician, you could be expected to learn new technologies at least every two to three years. It is critical that all entry-level cable installers are as up to date as possible on current and legacy cabling technologies, while remaining in touch with emerging standards and technologies. Most new cabling technologies focus on increasing bandwidth and data capacity. Some also focus on new media or making expensive media more cost-effective and accessible.

1.10.1   Internet Telephony

Internet telephony began to expand exponentially at the end of the 20th century. Telephony is defined as the electronic transmission of voice over electrical media. Computer telephony is the transfer of voice over traditional data networks. By making simple modifications to their computers, users make a voice call or send a fax transmission via the Internet. This has resulted in an uproar about the current lack of regulation. We need to decide how this technology will be regulated and how existing POTS will handle it.

1.10.2   Wireless Networks

Whether for convenience or in situations in which wireless is the only option, the last few years have seen a strong increase in the use of the airwaves to communicate data. Airwaves do not require a physical medium to send signals. Some wireless data networks use **spread spectrum** technologies to send signals, whereas others use infrared and cellular waves.

Some of the most important ideas in modern communications today involve using radio or terrestrial microwave signals for data or voice communication. Cellular telephones are part of a wireless network, and other technologies under development also use cellular technology. A wireless computer network can be installed using products built to the IEEE 802.11 family of standards. Other network systems use microwave signals from central towers or satellites. Light, in the form of infrared or laser, has been adapted for communications. Cabling installation technicians can expect to work with one or more of these technologies in the near future.

1.11   Chapter Summary

- The use of cabling as a medium for communication is more than 150 years old. The first large-scale implementation of communications cabling, the 19th century telegraph networks, were used by railroad companies to dispatch and manage traffic. It was at first a luxury to be able to afford to send cables (cablegrams). Before long, however, the telegraph was in widespread use.

- The telephone networks followed telegraph networks. Voice-network cabling evolved from a single wire in the late 19th century to the twisted-pair variant in use today.

- Cables consist primarily of a conductive wire over which a signal is generated by electrical voltage or optical light. The wire is either protected with shielding or simply insulated, depending on the type of cabling. Cabling design, performance, and physical properties vary greatly.

- Twisted pair-cabling comes in two major varieties: unshielded twisted-pair (UTP) cabling and shielded twisted-pair cabling (STP).

- Coaxial cabling is made up of multiple conductors. The center of the cable is an inner wire comprised of a copper wire surrounded by insulation and a braid or foil covering that acts as a shield. A cable jacket covers the inner wire set.

- Fiber-optic cabling used in voice and data networking is made up of two fibers encased in separate sheaths. Fiber-optic cabling is most often used for bidirectional communication. A separate cable is used for each communication direction.

- All three types of cabling are or have been used in both voice and data communications networks. All vary in their degree of susceptibility to internal and external factors such as attenuation, interference, and noise.

- The cabling industry is similar to the rest of the information technology industry in that technology doubles in complexity or evolves into something new and twice as good about every 12 to 18 months. As a potential cabling technician, you can expect to have to learn new technologies at least every two to three years. It is critical that cable technicians stay current with all existing cabling technologies, while keeping up with emerging standards and technologies.

1.12   Key Terms

American Telephone and Telegraph (AT&T) Corporation: Also known as "Ma Bell," it was the telecommunications provider that dominated the telephony industry until the 1980s.

coaxial cabling: A type of cable, often called "coax," made up of multiple conductors. Coaxial cabling is found in both voice and data networks.

fiber-optic cabling: A type of cabling designed to carry coherent light as either pulses or a continuous modulated stream. All fiber is made of the same inner core of pure glass.

interference: The external elements that can impede the transmission of information across a cable.

medium: The manner in which information is sent within a communication system.

modulation: The process of translating information into a continuous signal.

Morse code: The first communications protocol, or language, to be used across a cabled network.

plain old telephone service (POTS): The legacy phone service designed to carry primarily voice traffic.

regional Bell operating carriers (RBOCs): The seven companies that were formed as the result of the 1983 breakup of AT&T.

signal: The force of energy transmitted across a communication medium; it represents the information being communicated.

spread spectrum: A type of wireless communication in which the frequency of the transmitted signal is varied, resulting in a greater bandwidth than if the signal was not varied. Spread spectrum signals are usually specified in megahertz (MHz) or gigahertz (GHz).

telegraph: The first communications device, patented in 1837. The telegraph allowed the transmission of messages over electronic cables.

telephony: The process of converting sound or data into an electrical signal and transmitting the signal over electrical cabling.

twisted-pair cabling: The most widely used type of network cabling, using pairs of twisted wires. The twisting eliminates cross-talk and other degradation.

wired media: Also known as cabled media, in which information travels through a contained physical path or circuit.

wireless media: Also known as radiated media, in which information travels freely through the air or the vacuum of space.

1.13   Challenge Questions

1.1 Cabling has allowed people across the world to communicate with one another via _____.

   a. fax

   b. phone

   c. computer

   d. All of the above

1.2 There are many technical, logistical, and legal guidelines surrounding safety and standards when working with _____.

1.3 Communication requires at least the following components:

   a. Sender and receiver

   b. Sender, receiver, medium, and information

   c. Sender, cable, receiver, and information

   d. Sender, node, medium, and receiver

   e. None of the above

1.4 All types of cabling consist of a conductive _____ material protected by one or more sheathing and or insulation components.

   a. clad

   b. core

   c. fiber

   d. copper

1.5 The three major types of cabling include _____, _____, and _____.

   a. fiber-optic, coaxial, twinaxial

   b. fiber-optic, unshielded twisted-pair, shielded twisted-pair

   c. coaxial, twisted-pair, fiber-optic

   d. Thinnet, Thicknet, twisted-pair

1.6 The first device to use cabling for communications was the _____.

a. telegraph

b. telephone

c. teletype

d. television

1.7 The first communications protocol to be used over a cabled network was _____.

a. ASCII

b. Baudot Code

c. EBCDIC

d. Morse code

1.8 The telephone communicated voice by translating the sound directly into a signal through _____.

a. transcending

b. modulation

c. shifting

d. decoding

1.9 _____ is the process of a device converting direct imaging, sound, and/or data into an electrical signal and then transmitting that sound over electrical cabling.

a. Telegraphy

b. Cabling

c. Telephony

d. Teleprompting

1.10 The primary patents of telephony were established by the communications pioneer, _____.

a. Bell

b. Baudot

c. Marconi

d. Edison

1.11 In 1934, the FCC was formed and the responsibility for regulating communication was transferred to the FCC from the

_____.

a. American National Standards Institute (ANSI)

b. Interstate Commerce Commission (ICC)

c. Federal Bureau of Investigation (FBI)

d. Building Industry Consulting Services International (BICSI)

e. None of the above

1.12 For an entire century, one company dominated the U.S. telecommunications industry. It started as the National Telephone Company and would later become _____.

a. Sprint

b. AT&T

c. MCI

d. BTI

1.13 The first telephone _____ was built in New Haven, Connecticut, in 1878.

a. cable

b. exchange

c. trunk

d. device

1.14 In 1983, AT&T was divided up into seven _____ as the result of antitrust regulations.

a. local exchanges

b. private branches

c. trusts

d. RBOCs

1.15 Wireless networks now make it possible to communicate via telephone calls anywhere within coverage areas through which technologies?

    a. Cellular

    b. Satellite

    c. Radio waves

    d. All of the above

    e. None of the above

1.16 In 1958, the U.S. Department of Defense put the first _____ communications network into place.

    a. telephone

    b. data

    c. aerial

    d. video

1.17 The more complex the network cabling is, the more _____ it is to install.

1.18 Which of the following are common types of twisted-pair cabling?

    a. Unshielded twisted-pair cabling

    b. Shielded twisted-pair cabling

    c. Screened twisted-pair cabling

    d. All of the above

    e. Both a and b

1.19 Coaxial cabling is found in the following types of networks except _____.

    a. cable television

    b. broadband Internet access

    c. VHF/UHF antenna connections

    d. residential telephone networks

1.20 Coaxial cabling is used on _____ networks.

    a. voice

    b. data

    c. video

    d. All of the above

1.21 One of the primary disadvantages of fiber-optic cabling is
      _____.

      a. cost

      b. resistance to weather

      c. interference

      d. flexibility

1.22 Fiber-optic cabling is exempt from the problems of
      _____ and _____ interference.

1.23 Fiber-optics work off the principles of _____ and
      _____.

      a. reflection, electricity

      b. fusion, reflection

      c. refraction, electricity

      d. reflection, refraction

1.24 Cable transmissions result from the generation of signals using
      either optics or _____.

      a. kinetic energy

      b. electrical voltage

      c. silence

      d. pure energy

1.25 The most common types of interference are _____
      and _____.

1.14 Challenge Exercises

Challenge Exercise 1.1

The development of Ethernet was one of the historical topics mentioned in
this chapter. For over 30 years, Ethernet has been a fixture in computer data
networking. In this exercise, you search the Internet for information about
Ethernet standards. To complete this exercise, you need a computer with a
Web browser and Internet access.

1.1   Log on to your computer, open your Web browser, and access the Internet. In your Web browser, type:

http://www.ethermanage.com/ethernet/ethernet.html

1.2   Using the links found on the resulting Web page, read about Ethernet standards, and answer the following questions:

    a. How many different standards exist for Ethernet?

    b. Which standards are public and which standards are private?

    c. Of the three major categories of data cabling (coaxial, twisted-pair, and fiber-optics), which have corresponding standards for Ethernet and what are they?

Challenge Exercise 1.2

The Building Industry Consulting Services International, Inc. (BICSI) is considered the industry standard for cable installers. Navigate to the BICSI Web site at *http://www.bicsi.org* and research the BICSI Level I certification for cable installations. Find out the following information:

    a. What parts make up the BICSI Level 1 Exam?

    b. What are the major objectives and categories required for the exam?

    c. What materials are recommended to use for passing the BICSI exam?

    d. How many different levels can you achieve with on-the-job training?

1.15   Challenge Scenarios

Challenge Scenario 1.1

A small nonprofit organization that gathers and provides books to soldiers overseas recently received its first computer as a donation. The director has asked you to help the organization set up Internet access from the computer. You must research all available options and provide answers to these questions for the director in the form of a table:

    a. What types of media are used for Internet access in your area?

b. Survey the Internet service providers in your area. Find out the various types of services they offer and determine how they differ in cabling types (if they do at all).

c. Which types of cabling are likely to be more costly and which ones are more reliable based on physical characteristics?

d. Which media offer the fastest speeds? Does cabling play a factor in this?

CHAPTER 2

Cabling Standards and Specifications

Learning Objectives

After reading this chapter you will be able to:

- Understand standards, including cabling standards, and how they form the basis of rules and guidelines and implementation of technology

- Understand the development of private standards

- Discuss the formation of public standards through consortiums

- Associate specifications and standards for cabling with the standards bodies responsible for their development

- Identify the dominant standards organizations in the United States and North America

- Identify the dominant standards bodies in other developed countries, such as Europe and Japan

- Understand international standards organizations and how they help span the gaps among divergent national standards

- Recognize the naming conventions used by the more widespread standards bodies

For some time, companies in the technology industry, including telecommunications, established their own rules and guidelines for design and implementation. These guidelines are usually referred to as **standards** or protocols. *Standard* usually refers more properly to a particular technology's layout, operational, and/or physical characteristics. In the cabling industry, standards directly govern both informal and structured cabling installations. *Protocol* more often refers to a description of a communications process or model. For example, protocols used in network communications are developed so that they conform to cabling and other physical-layer standards.

2.1   Standards

Standards permit differing technologies to interface. They might be specified and controlled by a single **vendor** or a consortium of vendors with similar interests in common. Standards are often born out of experience, especially negative experiences. Standards are intended to serve as benchmarks that guarantee quality and stability. This is particularly true in regard to network cabling. The focus of this book is on standards for networking media and structured cabling practices; however, we will consider all important **standards bodies** that relate to network media, either directly or indirectly.

As with many other new industries, the computer and telecommunications industries saw a need for standards as soon as intense competition developed. Such competition often warrants the development of **de jure** standards, that is, standards controlled by an independent judicial body or organization. Prior to the development of such standards, existing standards are **de facto**. De facto standards are set by a dominant vendor within an industry. However, certain areas of the industry are still volatile and do not yet have universally adopted standards. Many vendors view a common standard as an obstruction to their competitive edge.

A need for open systems developed, which are systems in which network components could be combined on any network. Standards bodies cooperated to ensure that consistent standards were available worldwide, creating many standards that overlap. A number of technologies have various identifications that correlate with more than one standards body. This can be somewhat confusing when examining the various technologies and standards, especially standards governing cabling and media.

2.1.1 Private (De Facto) Standards

De facto standards are informal. De facto means "existing in fact," and de facto standards develop either with the inordinate success of a single company or through a consortium of private corporate partners. De facto standards often become de jure standards depending on the success of the developer when they are accepted by a substantial segment of the industry. What is often referred to as an "industry standard" is one that has been developed in the field rather than undergoing an official standards-setting process.

One popular de facto standard is the Ethernet local area network (LAN) technology. Bob Metcalf introduced it in 1972 originally as a graduate project. Metcalf, who eventually went to work for Xerox, helped to develop the Ethernet standard further. A modified standard was established later, often known as Ethernet II. It was referred to as DIX Ethernet in honor of the consortium of corporations that developed it: Digital (DEC), Intel, and Xerox.

The IBM personal computer (PC) is also an example of a de facto standard. IBM set the standard for the personal computer early on with the IBM PC and the XT. However, IBM's market share declined with the development of the AT standard by a consortium of corporations who developed competing standards for PC hardware. More recently, Microsoft's Windows operating system has become a de facto standard for PC operating systems. Alternatives to Microsoft Windows, such as the Linux operating system, must conform to the de facto standards in order to be competitive.

2.1.2 Public (De Jure) Standards

De jure means literally "according to law," and refers to an official standard, usually legislated by a standards body or organization with no ties to (or bias against) a particular company. De jure standards are meant to ensure compatibility among technologies, but are often used to bar monopolies or to give the users a choice of systems. Many of these standards bodies are large scale with a wide variety of interests. Some technologies have kept up with rapid growth in their industry through forums, which are special interest groups concerned with a specific topic. Asynchronous Transfer

Mode (ATM) and Frame Relay are popular examples of standards that emerged from forums.

Every facet of networking, from physical media to user applications, uses de facto standards, some of which include:

- **TIA/EIA-RS-232:** An interface standard for serial connections

- **Transmission Control Protocol/Internet Protocol (TCP/IP):** The protocol suite used for internal intranets and the World Wide Web

- **X.25:** An international packet-switching standard

- **802.x:** A standard series for network architectures and data formats

- **Open Systems Interconnection (OSI) reference model:** An international protocol stack and reference model used in education and protocol development

- **V.90:** A recent standard for 56 Kbps modems that ended the debate over x2 and Flex

Ethernet originally emerged as a private standard, but is now available as a public standard called Carrier Sense Multiple Access/Collision Detection (CSMA/CD). It is more commonly known as IEEE 802.3.

2.1.3   The Purpose of Standards Bodies

Standards bodies provide the impetus for the development of de jure and industry standards. As with other governing bodies, there is an established process for proposing, developing, and ratifying standards. These organizations, along with the vendors and users, play a vital role in determining whether a standard exists for a technology. Often, standards bodies establish the middle ground and arbitrate technical conflicts between the users and the vendors.

2.1.4   Understanding Cabling Specifications and Standards

Cabling and media currently fall under a number of public and private standards. Cabling specifications and standards help to establish comprehensive guidelines for network cable planning, design, and implementation. These guidelines cover cabling elements such as wiring, connections, terminations, supporting equipment, insulation, distributions centers, and installation methods. The combined specifications and standards are called **structured cabling systems**.

A structured cabling system includes the following:

- Description of the media (wiring) and its layout for the work area, horizontal cabling, and backbone cabling

- Standards for cable connectors and other modular interfaces

- A clear and concise design plan

- Procedures for testing and certifying cable installation

- Troubleshooting and maintenance procedures

All of these elements are governed by various standards that ensure uniformity, safety, and compliance of the design plan and finished installation. You must be aware of all local, state, national, and international standards and codes that affect a particular cabling project.

A network that grows without a structured cabling system plan is ultimately extremely expensive to expand and maintain. Structured cabling systems and applicable standards organizations are covered in detail in Chapter 12, "Structured Cabling Guidelines."

2.1.5   International versus National

The goal of international standards bodies is the development of universally accepted standards. Many of these standards bodies adopt an international standard and cross-reference it in their catalog of standards. Often, they choose to maintain a local standard rather than adopt an international one. In addition, only specific portions of a standard are often adopted rather than the standard as a whole.

International standards bodies include the **International Organization for Standardization (ISO)** and the International Telecommunication Union (ITU). The international bodies often involve professionals from industries and governments of many different nations. National standards bodies include the **American National Standards Institute (ANSI)** and the **Institute of Electronics and Electrical Engineers (IEEE)**.

The basic rule of thumb regarding the establishment of standards is this: Something that is called a standard is usually voluntary. At the point at which the standard becomes a "code," it is taken as an enforceable regulation. Standards implemented at the international and national level are considered voluntary standards. At state, municipal, and other local levels, the same standards are often codes or regulations, possibly incurring civil fines or criminal penalties for violations. However, many standards are not a part of any regulatory codes and remain voluntary.

ANSI and ITU are the two most common standards bodies governing telecommunications and other computer technologies. Although the former is a national standards body and the latter international, industry demands have compelled them to work together on many standards. They often share the same standards under different identifiers.

International Telecommunication Union Telecommunication Standardization Sector (ITU-T)

The **International Telecommunication Union-Telecommunication Standardization Sector (ITU-T)** was created in 1993, replacing the International Telegraph and Telephone Consultative Committee (known by its French acronym of CCITT). The CCITT was originally founded in 1865 as the International Telegraph Union. The group reformed in 1948 with U.N. backing to deal with the original goals of recommendations concerning telegraph, telephone, and tariff issues. The U.S. voting member is a representative of the U.S. State Department.

The ITU-T approved and published recommendations once every four years until 1988, after which, all standards were published upon completion. These standards are organized into series or categories, identified with a letter prefix corresponding to the standard's series. These series are listed in Table 2.1.

When an ITU-T standard needs revision, the ITU-T revises the standard using a French suffix rather than assigning a new code. The term *-bis* denotes a second version; *-ter* denotes a third version.

Selected ITU-T Standards

The two most common series of ITU-T standards are the V Series and the X Series. V Series standards cover data communication over the telephone network. Table 2.2 includes a partial list of the V Series standards. X Series standards cover services in facilities over data networks. Table 2.3 includes a partial list the X Series.

ANSI

The American National Standards Institute (ANSI) body was established in 1984 and sets the standards for most computer software and hardware development in North America that are not covered by an international standard. ANSI often brings together other organizations and members as

TABLE 2.1 ITU-T Letter Series

Series Recommendations	Description
A	Organization of the work of the ITU-T
B	Means of expression (definitions, symbols, and classification)
C	General telecommunication statistics
D	General tariff principles
E	Overall network operation, telephone service, service operation, and human factors
F	Telecommunication services other than telephone
G	Transmission systems and media, and digital systems and networks
H	Audiovisual and multimedia systems
I	Integrated Services Digital Network (ISDN)
J	Transmission of sound programs, television, and other multimedia signals
K	Protection against interference
L	Construction, installation, and protection of cable and other elements of outside plant
M	Maintenance: transmission systems, telephone circuits, telegraphy, facsimile, etc.
N	Maintenance: international sound program and television transmission circuits
O	Specifications of measuring equipment
P	Telephone transmission quality, telephone installations, local line networks
Q	Switching and signaling
R	Telegraph transmission
S	Telegraph services terminal equipment
T	Terminals for telematic services

TABLE 2.1 **(continued)**

Series Recommendations	Description
U	Telegraph switching
V	Data communication over the telephone network
X	Data networks and open system communication
Y	Global information infrastructure and Internet protocol aspects
Z	Programming languages

Source: ITU-T (*http://www.itu.int/ITU-T/*)

TABLE 2.2 **Partial List of ITU-T V Series Standards in Force (V Series)**

Series	Description
V.8 bis (09/98)	Procedures for the identification and selection of common modes of operation between data circuit-terminating equipment (DCE) and between data terminal equipment (DTE) over the public switched telephone
V.24 (10/96)	List of definitions for interchange circuits between DCE and DTE
V.34 (02/98)	A modem operating at data-signaling rates of up to 33.6 Kbps for use on the general switched telephone network and on leased point-to-point, two-wire telephone-type circuits
V.38 (10/96)	A 48-, 56-, or 64-Kbps DCE standardized for use on digital point-to-point leased circuits
V.42 bis (01/90)	Data compression procedures for DCE using error-correction procedures
V.90 (09/98)	A digital modem and analog modem pair for use on the public switched telephone network (PSTN) at data-signaling rates of up to 56 Kbps downstream and up to 33.6 Kbps upstream
V.130 (08/95)	ISDN terminal adapter framework
V.140 (02/98)	Procedures for establishing communication between two multiprotocol audiovisual terminals using digital channels at a multiple of 64 Kbps or 56 Kbps

Source: ITU-T (*http://www.itu.int/ITU-T/*)

TABLE 2.3 Partial List of ITU-T X-Series Standards in Force (X Series)

Series	Description
X.3 (3/93)	Packet assembly/disassembly (PAD) facility in a public data network
X.21 bis (11/88)	Use on public data networks of DTE which is designed for interfacing to synchronous V-Series modems
X.25 (10/96)	Interface between DTE and DCE for terminals operating in the packet mode and connected to public data networks by dedicated circuit
X.29 (12/97)	Procedures for the exchange of control information and user data between a PAD facility and a packet mode DTE or another PAD
X.121 (10/96)	International numbering plan for public data networks
X.150 (11/88)	Principles of maintenance testing for public data networks using DTE and DCE test loops
X.500 (8/97)	Information technology—Open Systems Interconnection—The Directory: Overview of concepts, models, and services
X.509 (8/97)	Information technology—Open Systems Interconnection—The Directory: Authentication framework

Source: ITU-T (*http://wwww.itu.int/ITU-T/*)

delegates to ensure impartiality and defines software, hardware, electronic, and industrial standards for the international as well national community when working on joint projects.

The T-1 committee (not related to the high-speed link) of ANSI deals with the development of telecommunications standards. One of the primary duties of the T-1 committee is to coordinate closely with the ITU and address specific North American concerns.

ANSI is the official U.S. representative to the International Accreditation Forum (IAF), the ISO, and the International Electrotechnical Commission (IEC). It is also the U.S. member of the Pacific Area Standards Congress (PASC) and the Pan American Standards Commission (COPANT).

ANSI creates and publishes standards for programming languages, communications, and networking. Table 2.4 contains a partial list of some of the ANSI standards. Table 2.5 is a partial list of ANSI LAN standards.

ANSI's X3 committee establishes standards for transmission media.

TABLE 2.4   Selected ANSI Standards

Standard	Description
T1.102	Electrical interfaces for the digital hierarchy
T1.103	Format specifications for synchronous DS3 lines in the digital hierarchy
T1.113 SSN7	ISDN User Part
T1.116 SSN7	Operations, maintenance, and administration part (OMAP)
T1.219	ISDN Management: Overview and Principles
X3.1	Synchronous data transmission signaling rates (similar to ITU V.6 and V.7)
X3.4	Seven-bit ASCII character set
X3.36	Signaling rates for high-speed communications between DTE and DCE
X3.44	Determining the performance of data communications systems
X3.66	Advanced data communication and control procedures (ADCCP)
X3.92	Data encryption algorithm (DEA)
X3.105	Data link encryption

Source: ANSI (*http://www.ansi.org*)

TABLE 2.5   ANSI LAN Standards

Standard	Description
X3T9.5	Fiber Distributed Data Interface (FDDI) standards
X3T9.3	High-Performance Parallel Interface (HPPI)
X3.2	High-Speed Serial Interface (HSSI)
X3.230	Fiber Channel Physical and Signaling Interface (FCPSI)
X3.92	A privacy and security encryption algorithm
X12	Electronic Data Interchange (EDI)

Source: ANSI (*http://www.ansi.org*)

Telecommunications Industry Association (TIA) and Electronic Industries Alliance (EIA)

The Telecommunications Industry Association (TIA) and Electronic Industries Alliance (EIA) are mainly North American organizations that mandate and oversee many of the standards for voice and data cabling, industries that emerged after the deregulation of the U.S. telecommunications industry in the 1980s. The industry standards evolved after the deregulation of the U.S. telephone industry in 1984, which made the building owner responsible for on-premises cabling. Before deregulation, AT&T used its own standards.

The development of standards for twisted-pair cabling, fiber optics, customer premises equipment, network termination equipment, and wireless and satellite networking are also the responsibility of TIA/EIA. The TIA and EIA standards are designed to be used over a period of years, and TIA/EIA rarely completely revise a standard. Instead, they publish appendices or, in the case of the TIA, Telecommunications Systems Bulletins (TSB). Bulletin TSB-95 is a commonly referred to standard that provides guidelines to the minimum expected level of performance for existing Category 5 cable installations, and in light of newer categories of cabling, new performance levels have been devised (categories 5e and 6, for example). The bulletin also outlines expected performance criteria for new networking transmission methods such as Gigabit Ethernet. We will discuss cabling Category 5, 5e, and 6 in Chapter 4, "Data Network Signals."

TIA/EIA Standards

Of the many **Telecommunications Industry Association/Electronic Industries Alliance (TIA/EIA)** standards and supplements, those listed in Table 2.6 are used most frequently by cable installers.

National Electrical Code (NEC)

The NEC is, first and foremost, a safety code that was established to protect people and property from electrical hazards. NEC is developed through a consortium of standards bodies and safety organizations, including ANSI

TABLE 2.6   Selected TIA/EIA Standards

Standard	Title	Description
TIA/EIA-568-A	Commercial Building Standard for Telecommunications Wiring	Specifies the minimum requirements for telecommunications cabling, topology, and distance limits; media and connecting hardware performance specifications; and connector and pin assignments.
TIA/EIA-568-B (T568B)	Commercial Building Telecommunications Cabling Standard	This standard specifies the component and transmission requirements for media for multiproduct and multivendor environments similar to T568A. Contains many addenda and is updated frequently. (T568A is rarely updated).
TIA/EIA-569-A (T569A)	Commercial Building Standards for Telecommunications Pathways and Spaces	This standard specifies design and construction practices within and between buildings that support telecommunications media and equipment.
TIA/EIA-570-A	Residential Telecommunications Cabling Standard	The cabling infrastructure specifications within this standard are intended to include support for security, audio, television, sensors, alarms, and intercom systems.
TIA/EIA-606-A	Administration Standard for the Telecommunications Infrastructure of Commercial Buildings	This standard specifies that each hardware termination unit have a unique identifier.
TIA/EIA-607	Commercial Building Grounding and Bonding Requirements for Telecommunications	This standard supports a multivendor, multiproduct environment, as well as the grounding practices for various systems that may be installed on customer premises.

and the **National Fire Protection Association (NFPA),** primarily for adoption by other standards bodies and governments to adopt. A prime example is the **Occupational Safety and Health Administration (OSHA).**

NEC type codes are listed in cable and cable supply catalogs, and classify explicit product categories for specific uses. NEC Code Section 310-11, for example, states that all conductors and cables must be marked in according to their American Wire Gauge (AWG) sizes, at intervals of not more than 24 inches. NEC Code Section 250 deals with grounding standards. Article 250 stipulates that all 120- and 240-volt systems have a grounding system. Code Section 110 deals with wire standards and their relationship with connectors, screws, and so on.

National Fire Protection Association (NFPA)

In addition to being familiar with standards and codes that govern occupational, consumer, and fire safety, it is important to understand environmental regulations that relate to cabling.

The NFPA establishes guidelines meant to prevent fire. There are more than 300 NFPA standards, and NFPA is one of the most influential organizations in instituting local fire and building codes.

Underwriters Laboratories Inc. (UL)

Underwriters Laboratories Inc. (UL) is an independent, not-for-profit organization that has tested and certified products for public safety for more than a century. Each year, more than 17 billion UL marks are applied to products worldwide. UL was founded in 1894 and is the leading organization in the U.S. for product safety and certification. Their services today include helping corporations gain global acceptance for their products, whether that product is an electrical device, a programmable system, or the company's quality process.

The UL LAN Certification Program addresses performance as well as safety. Companies that manufacture cable that earn a UL mark (e.g., Level I, LVL I, or LEV I) display them on the outer product jacket or packaging. Fire, building, and insurance regulators use UL, as they do the NEC. Here are some common UL codes:

- **UL 96A:** Defines guidelines for the certification of systems materials and components.

- **UL 444:** Defines the standard for safety for communications cable.

- **UL 13:** Defines the standard for safety for power-limited circuit cables.

UL tests and evaluates components and determines whether they should be given a UL listing. UL conducts follow-up inspections and tests on products to which they have given the listing. You no doubt can find five items with UL ratings in your immediate vicinity right now.

Occupational Safety and Health Administration (OSHA)

In the United States, OSHA is the organization that is primarily responsible for the safety of American workers. OSHA, which was formed in 1971, operates under the auspices of the U.S. Department of Labor. Since its inception, OSHA has drastically reduced the number of injuries and fatalities in workplaces.

OSHA has the responsibility of protecting workers by enforcing U.S. labor laws. Although OSHA as an agency does not deal with building codes or permits, they have the authority to impose heavy fines and/or shut down a job site when they find serious safety violations.

OSHA's budget for fiscal year 2005 was $468.1 million. The agency has a staff of 2,220, which includes about 1,100 compliance safety and health officers. At this time, 26 states run their own OSHA programs. OSHA inspected 39,167 workplaces during fiscal year 2003. State OSHA programs conducted an additional 57,866 inspections.

State and Local Building Codes

Many structured cabling projects are incorporated into construction projects, and they will require many permits. Cabling jobs can require permits for new construction and, in some areas, for modifications to existing structures or retrofits. Since these building codes vary from state to state and from county to county, we recommend that you check your local municipal district office or other zoning department for information about permit requirements. To obtain local or state building codes, contact the building official for the local jurisdiction. You can buy a listing of all of the basic building codes that are in effect in the United States and internationally from the International Code Council (ICC).

Federal Communications Commission (FCC) Rules

The **Federal Communications Commission (FCC)** is an independent United States government agency that answers directly to Congress. The Communications Act of 1934 established the FCC, which is charged with the regulation of interstate and international radio, television, wire, satellite, and cable communications, including regulation of the airwaves we use. The FCC has jurisdiction in the 50 states, the District of Columbia, and U.S. possessions. All communication devices and cabling must be in compliance with FCC regulations and must make FCC-compliance statements. All computer devices, components, connectors, and the like are required to carry a specific FCC identification number. The FCC ID number of any device manufactured in the last 20 years indicates its place of origin.

International Organization for Standardization (ISO) Standards

The ISO is a nongovernmental international standards organization that provides not only standards, but also reference models for member nations and organizations. The ISO was formed in 1947 and is made up of standards bodies from over 140 member nations. ANSI is a member of ISO. The work of the ISO produces international agreements, which are published as international standards. ISO regulates everything from physical communications all the way to organizational practices (i.e., ISO 9000). The ISO is best known for the **Open Systems Interconnection (OSI) reference model**, developed in the early 1980s.

Institute of Electrical and Electronics Engineers (IEEE)

The IEEE is a non-profit, technical professional association. Through its members, the IEEE is a leading authority in technical areas ranging from computer engineering, biomedical technology, and telecommunications, to electric power, aerospace, and consumer electronics, among others. Through its technical publishing, conferences, and consensus-based standards activities, the IEEE produces 30 percent of the world's published literature in electrical engineering, computers, and control technology, holds annually more than 300 major conferences, and has nearly 900 active standards with an additional 700 under development.

IEEE 802 Standards   The IEEE is well known for its **802** family of standards, which refers to standards that define LAN and metropolitan area network (MAN) protocols. The IEEE 802 committee has been at the forefront in the development of LAN and MAN standards. The IEEE 802 family of standards defines rules for three layers (Physical, Data Link, and Network) and two sublayers (Logical Link Control and Media Access Control). Table 2.7 lists the IEEE subcommittees and their responsibilities.

TABLE 2.7   IEEE 802 Networking Standards

Subcommittee	Protocol Standard
802.1	High Level Interface
802.2	Logical Link Control (Sub-layer of Data Link)
802.3	Carrier Sense Multiple Access/Collision Detect (CSMA/CD) Networks (Ethernet)
802.3u	100 Mbps Ethernet (Fast Ethernet)
802.3z	Gigabit Ethernet
802.4	Token Bus Networks
802.5	Token Ring Networks
802.6	MANs
802.7	Broadband Technical Advisory Group
802.8	Fiber Optic Technical Advisory Group
802.9	Integrated Voice and Data LAN Working Group
802.10	LAN Security Working Group
802.11	Wireless Networking Working Group
802.12	Demand Priority Access Methods
802.14	Cable Modems

2.1.6   Forums

Some organizations, referred to as forums, are dedicated to standardizing a specific protocol or family of protocols. Forums existed long before official standards arose and many provided the foundation for drafting standards that would eventually be adopted by organizations such as ANSI and ISO.

ATM Forum

The goal of the ATM Forum, formed in 1991, is expediting use of ATM protocol and services by rapidly merging interoperability specifications. The ATM Forum was formed in 1991. It comprises some 150 member companies and remains open to any organization interested in hastening the availability of ATM-based solutions.

Frame Relay Forum

The Frame Relay Forum is an association of vendors, carriers, users, and consultants with the goal of implementing Frame Relay in conformity to national and international standards. Like the ATM Forum, this group was formed in 1991. The forum works to improve existing Frame Relay standards that are necessary, but not sufficient. The forum also creates implementation agreements (IAs), which are agreements by all members of the Frame Relay community as to the specific way in which standards are to be applied. The Frame Relay Forum's marketing committees are charged with the development of worldwide markets through user education.

2.1.7   Canadian Standards

Countries other than the United States have standards bodies to develop and manage standards that are unique to a particular nation. Most of the time, these standards bodies address safety concerns, providing standards designed to protect public safety, general workplace safety, compliance with environmental regulations, and proper disposal of hazardous waste. They also address worker protection (unions) and establish guidelines for laboratory safety. Centre for Occupational Health and Safety (CCOHS) is the Canadian organization that parallels OSHA. CCOHS promotes safe working environments by providing information and advice about occupational health and safety.

Canadian Standards Association (CSA)

The **Canadian Standards Association (CSA)**, formed in 1919, developed the CSA Canadian Electrical Standard that is generally adopted as the Canadian Electrical Code (CEC), and is enforceable by law. CEC is similar to the NEC. It establishes the standard for electrical wiring in Canada and forms the basis for all provincial codes. CSA was the first Canadian organization designed to develop industrial standards. The association provides an open forum in which the public, government, and business community can agree on the criteria that best meets community interests for materials, products, structures, and services in a wide variety of fields. The CSA has more than 1,500 standards.

2.1.8   European Commission (EC) Standard

The European Commission (EC) established the European Union Directives, which require certain categories of goods sold in Europe to conform to a single document in order to satisfy the legal requirements of the European community. Products that meet the directives carry the Conformité Européene (CE) logo; the program is referred to as the CE mark. A CE marking indicates conformity to the legal requirements of the EU directives rather than conformity to a standard. For more information, visit *http://europa.eu.int/comm/enterprise/faq/ce-mark.htm*.

European Committee for Electrotechnical Standardization (CENELEC)

Comité Européen de Normalisation Electrotechnique (CENELEC) is known in English as the European Committee for Electrotechnical Standardization. CENELEC was set up under Belgian law in 1973 as a nonprofit organization and develops electrotechnical standards for most of Europe. The organization works with 35,000 technical experts from 19 European countries to develop and publish standards for the European market. In Directive 83/189/EEC, the CE officially recognized CENELEC as the European standards organization. Many CENELEC cabling standards mirror ISO cabling standards, with minor changes.

Although CENELEC and the International Electrotechnical Commission (IEC) operate at two different levels, they are the most important standardization bodies in the electrotechnical field in Europe. Cooperation between CENELEC and the IEC is described in the Dresden Agreement, approved and signed by both partners in that German city in 1996.

2.1.9   Japanese Industrial Standards Committee (JISC)

The Japanese Industrial Standards Committee (JISC) develops the Japanese Industrial Standards (JIS). JIS are voluntary standards, so it is necessary to get the consensus of all the companies involved to ensure effective application. JISC formulates a long-range plan every five years for promoting industrial standardization. JISC participates in ISO and IEC technical committees.

Japanese Standards Association (JSA)

The Japanese Standards Association (JSA), established in 1945, has the objective of public education about the standardization and unification of industrial standards, thus contributing to the improvement of technology and heightening production efficiency. The JSA conducts surveys and researches standardization in a number of fields, including networking. JSA also develops and publishes JIS. The JSA actively participates on ISO and IEC committees.

2.1.10   Obtaining Standards

Public standards can be found in various locations on the Internet, usually in their original comprehensive format directly from the standards body responsible for them. (See Appendix C, "Cabling Resources and Information," for a detailed list of Web sites for organizations discussed in this chapter.) The following is a list of addresses at which you can contact the standards bodies directly:

> American National Standards Institute (ANSI)
> 11 West 42nd Street, 13th Floor
> New York, NY 10036
> (212) 642-4900
> (202) 639-4090 (Washington, D.C. office)
> *http://www.ansi.org*

> Canadian Standards Association (CSA)
> 178 Rexdale Boulevard
> Rexdale, Ontario M9W 1R3
> Canada
> (416) 747-4000
> *http://www.csa.com*

> European Computer Manufacturers Association (ECMA)
> 114 Rue de Rhone
> CH-1204 Geneva
> Switzerland

> Electronic Industries Alliance (EIA)
> 2001 Eye Street, NW
> Washington, D.C. 20006
> (202) 457-4966

> Federal Information Processing Standard (FIPS)
> U.S. Department of Commerce
> National Technical Information Service
> 5285 Port Royal Road
> Springfield, VA 22161

Institute of Electrical and Electronics Engineers (IEEE)
345 East 47th Street
New York, NY 10017
(212) 705-7900

International Organization for Standardization (ISO)
Central Secretariat
1, Rue de Varembe
CH-1204 Geneva
Switzerland

ISO (U.S. Office)
c/o American National Standards Institute (ANSI)
11 West 42nd Street, 13th Floor
New York, NY 10036
(212) 642-4900
(202) 639-4090 (Washington, D.C. office)

International Telecommunication Union (ITU)
General Secretariat
Place des Nations
CH-1211 Geneva 20
Switzerland

ITU (U.S. Office)
c/o U.S. Department of Commerce
National Technical Information Service
5285 Port Royal Road
Springfield, VA 22161
(703) 487-4650
http://www.itu.int/home/index.html

2.2 Chapter Summary

- As companies develop various technologies, the need for consistency, universal design approaches, and best practices become obvious. In response, companies and organizations will devise standards that allow technologies to interface with one another. Standards can be de facto (resulting from a private standard that gains market acceptance) or de jure (established by an independent body or organization). Standards bodies cooperate to ensure network components and other technologies continue to be implemented worldwide.

- Standards bodies define cabling specifications and implementation guidelines for structured cabling systems. This is important because a network without a structured cabling system plan will be difficult and expensive to expand and maintain.

- International standards bodies are committed to developing universal standards. They will often have members from national standards bodies around the globe. ANSI and the IEEE are examples of national standards bodies. International standards bodies include organizations such as the ISO and the ITU. These bodies standardize all major aspects of communications technologies.

- Organizations concerned with direct standardization of network cabling include ANSI, TIA, and EIA. Cabling installers use many ANSI TIA/EIA standards and cabling technicians have to comply with safety regulations as well as building and fire codes, which typically vary by local jurisdiction. Some of these codes are NEC codes, UL codes, NFPA codes, and OSHA regulations. OSHA is a federal agency that enforces all U.S. labor safety laws. The FCC (another federal agency) enforces laws governing communications across network cabling.

- Other countries have their own governing standards and regulations relating to communications and cabling. In Canada, the CSA is the Canadian universal electrical code. In Europe, the European Commission establishes requirements or directives for communication and safety. Standards can be obtained (often for a fee) by writing to the organization or ordering them online at the agency's Web site.

2.3 Key Terms

802: A standard series for network architectures and data formats.

American National Standards Institute (ANSI): The North American standards body for most software and hardware areas of the computer industry.

Canadian Standards Association (CSA): Canadian standards body responsible for developing the Canadian electric standard.

de facto: A standard that originated privately and emerged on a widespread basis through market acceptance.

de jure: A legislated or official standard, usually the product of an independent standards body or organization.

Federal Communications Commission (FCC): An independent U.S. government agency that answers directly to Congress, and is responsible for regulating interstate and international radio, television, wire, satellite, and cable communications.

Institute of Electrical and Electronics Engineers (IEEE): A nonprofit standards union that is a leading authority in technical areas ranging from computer engineering, biomedical technologies, and communications, to electrical power, aerospace, and consumer electronics.

International Organization for Standardization (ISO): An organization responsible for developing standards and reference models for member nations and organizations for communication and other commercial areas.

International Telecommunication Union-Telecommunication Standardization Sector (ITU-T): A U.N.-governed organization that is responsible for international communications standards.

National Fire Protection Association (NFPA): Responsible for the establishment of guidelines to prevent fire.

Occupational Safety and Health Administration (OSHA): A U.S. Department of Labor organization responsible for regulating workplace safety.

Open Systems Interconnection (OSI) reference model: An international protocol stack and reference model used in education and protocol development.

standard: Guidelines representing the layout, operation, and/or the physical characteristics of a particular technology or the best practices for implementing a technology.

standards bodies: Organizations formed to establish and maintain rules and guidelines.

structured cabling system: Guidelines for such things as wiring, connections, terminations, supporting equipment, insulation, distribution centers, and cable installation methods.

Transmission Control Protocol/Internet Protocol (TCP/IP): The protocol suite used for the Internet and for internal intranets.

Telecommunications Industry Association/Electronic Industries Alliance (TIA/EIA): North American organizations that establish and govern most of the standards for voice and data cabling.

TIA/EIA-568: Commercial building standards for telecommunications wiring.

TIA/EIA RS-232: An interface standard for serial connections.

Underwriters Laboratories Inc. (UL): An independent, not-for-profit organization that tests and certified product safety.

vendor: A provider or manufacturer of a technical product or service.

2.4   Challenge Questions

2.1 Standardizing the telephone system in the United States brought about what results?

2.2 After the deregulation of AT&T in 1983, how did standards play a role in maintaining stability?

2.3 Like many other industries, the computer and telecommunications industries soon saw a need for standards once _____ arose.

   a. monopolies

   b. competition

   c. bankruptcy

   d. recessions

2.4 A standard that originates through a private vendor that becomes adopted through market acceptance is referred to as a _____ standard.

   a. de jure

   b. de facto

   c. ad hoc

   d. ad hominum

2.5 Which of the following is an example of a de facto standard?

   a. Ethernet

   b. TCP/IP

c. X.500

d. 802.3

2.6 When a standard is legislated or established through an independent consortium, it is referred to as a _____ standard.

a. de facto

b. de jure

c. ad hoc

d. ad hominum

2.7 One common suite of standards is _____, which is an entire protocol suite dedicated for use on the worldwide Internet as well as for heterogeneous enterprise and local area networks.

a. 802.3

b. TCP/IP

c. IPX

d. FDDI

2.8 The _____ model is an international protocol stack and reference model used for education and protocol development.

a. TCP/IP

b. ANSI

c. CSMA/CD

d. OSI

2.9 What is the de facto IEEE parallel standard to Ethernet?

a. 802.5

b. X.500

c. 802.3

d. TCP/IP

2.10 A structured cabling system includes which specifications?

a. Standards for cable connectors and other modular interfaces used with the cabling

    b. Procedures and methodologies for troubleshooting and maintenance

    c. Procedures for testing and certification of the cable installation

    d. Description of the media and its layout descriptions

    e. All of the above

2.11 Which of the following is a U.S. standards bodies?

    a. ISO

    b. ITU

    c. ANSI

    d. CEC

2.12 Why is it disadvantageous for cable installers not to keep up to date on new technologies?

2.13 Which U.N.-backed group formed originally in 1865 as the International Telegraph Union?

    a. ANSI

    b. IEEE

    c. ITU-T

    d. ISO

2.14 ITU-T recommendations are divided into categories labeled by _____, which preface every standard identifier.

    a. numerical series

    b. letter series

    c. date code

    d. None of the above

2.15 What do the ITU Series X recommendations deal with?

    a. Switching and signaling

    b. Maintenance, transmission systems, and circuits

    c. Data communication over the telephone network

    d. Data networks and open system communications

2.16 _____ standards define software, hardware, electronic, and industrial standards for the American community.

    a. ISO

    b. ITU

    c. ANSI

    d. X.500

2.17 The _____ and the _____ are organizations that establish and govern a significant amount of standards relating to voice and data cabling.

2.18 Which TIA/EIA series of standards details commercial building telecommunication wiring standards?

    a. 802

    b. X.500

    c. 568

    d. TCP/IP

2.19 The _____ is a U.S.-based body responsible for protecting people and property from electrical hazards through type codes.

    a. NEC

    b. NFPA

    c. EIA

    d. TIA

2.20 What is the leading organization in U.S. product safety and certification?

    a. TIA

    b. UL

    c. NEC

    d. ISO

2.21 In the United States, what organization is responsible for workplace safety?

    a. OSHA

    b. HEPA

    c. UL

    d. NEC

2.22 The IEEE is known for its _____ family of networking standards.

    a. 1010

    b. 568

    c. 802

    d. X-dot

2.23 Similar to the NEC, the _____ is the standard for electrical wiring in Canada upon which all provincial codes are based.

    a. CSA

    b. CEC

    c. NEC

    d. CE

2.24 Which organization developed the Japanese Industrial Standards?

    a. JISC

    b. ANSI

    c. ISO

    d. ITU

2.5 Challenge Exercises

Challenge Exercise 2.1

In this exercise, you research standards on the Internet. To complete this exercise, you need a computer with a Web browser and Internet access.

    2.1 Log on to your computer, open your Web browser, and access the Internet. In your Web browser, type:
http://global.ihs.com/industry_stds.cfm

2.2 Click the **Construction** link in the Popular Industry Standards section Which standards are listed as the most requested standards documents? Repeat this step for the **Electrical/Electronics** link.

2.3 Do you recognize any of the acronyms as being discussed in Chapter 2? If so, what are they? Is a fee associated with these documents?

Challenge Exercise 2.2

In this exercise, you research general regulations and codes devised by a local body on the Internet. To complete the exercise, you need a computer with a Web browser and Internet access.

2.1 Locate a fire extinguisher within your residence, school, or business. Notice that it is marked with quite a bit of specific information. For a fire extinguisher to properly put out a fire, it must be charged with the appropriate fire suppressant.

2.2 In your Web browser, enter the following address: *http://www.lafd.org/eqfirex.htm*

2.3 Using the information on this Web site and the information on the fire extinguisher, answer the following questions:

a. Which category is the fire approved to extinguish (Type A, B, or C)?

b. What's the definition of a Type A fire?

c. What's the definition of a Type B fire?

d. What's the definition of a Type C fire?

e. What size extinguisher is being rated?

f. Is the fire extinguisher fully charged?

Challenge Exercise 2.3

In this exercise, you research OSHA and ANSI standards on the Internet. You need a computer with a Web browser and Internet access.

2.1 In your Web browser, enter the following address: *http://public.ansi.org/ansionline/Documents/News%20and%20Pub lications/Other%20Documents/OSHA_Standards.xls*

2.2 Click **Save** to save the Excel file to your desktop. The file requires Microsoft Excel, or Excel Viewer, which is free from Microsoft.

Excel Viewer 2003 is available from *http://www.microsoft.com/downloads/.*

2.3 Observe the following ANSI standards relating to ladder safety listed below. As mentioned in this chapter, standards bodies are often used by regulating agencies to justify codes and other rules. Using the following standards IDs, determine what each one does and list how many times each (according to the spreadsheet) has been cited by OSHA for inclusion in their rules and regulations.

ANSI A14.1 _____

ANSI A14.2 _____

ANSI A14.3 _____

ANSI A14.4 _____

ANSI A14.5 _____

ANSI A14.6 _____

ANSI A14.7 _____

ANSI A14.8 _____

ANSI A14.9 _____

ANSI A14.10 _____

ANSI A14.11 _____

Challenge Exercise 2.4

In this exercise, you research computer components on the FCC Web site. To complete the exercise, you must have access to at least three different computer components (i.e., system boards, computer expansion cards, or communication devices) and be able to read their FCC identification numbers. You also need a computer with a Web browser and Internet access.

2.1 Find at least three computer devices, preferably for which you do not know the manufacturer. Locate the FCC identification numbers on each component.

2.2 In your Web browser, enter the following address: *http://www.fcc.gov/oet/fccid/*

2.3 The FCC ID Search Page opens. This Web site is useful for finding manufacturers and support information for unknown devices.

Using this site and the FCC identifications numbers, try to identify each of these devices.

2.6   Challenge Scenarios

Challenge Scenario 2.1

Two large financial institutions are planning a corporate merger. One of these companies is located in North America and the other is in Europe. Upon completion of the merger, one of the goals will be to establish a communications model that will allow the use of cross-compatible standards that will provide the new multinational organization ease of communication across both continents. You have been charged with selecting software and protocols that conform to these standards using the international reference model for communication—the OSI model—devised by the ISO.

You must be able to find at least one group of protocols that conform to and represent all seven layers of the OSI model while maintaining interoperability on an international basis. You want users to be able to facilitate communication with common physical interfaces as well as via software protocols. Answer the following questions in search of this potential solution:

2.1   Which standards organization would likely have all of these series of devised standards and protocols?

2.2   Using an Internet search engine, research the OSI model and find out if the ISO developed any protocols to correspond to the seven reference layers of the OSI model. If so, what are they?

2.3   Are there any ITU-T letter series standards that would also allow you to facilitate communication at all seven layers of the OSI model?

Challenge Scenario 2.2

For most network cabling projects, installers will find themselves working in tandem with other construction projects or immediately following the completion of other construction projects. As with most construction projects, a permit is required to ensure that the work is being done properly. Contact your local zoning departments for information on permit requirements in your area. Inquire for local or state building codes by contacting the building official for your local jurisdiction.

CHAPTER 3

Understanding Signals

Learning Objectives

After reading this chapter you will be able to:

- Distinguish signals and various kinds of signal transmission over media

- Comprehend the differences between analog and digital signals

- Discuss the basics of conducted media (wired/cabling) and radiated media (wireless)

- Distinguish electrical from optical signals

- Understand various signal characteristics and properties, including noise, distortion, attenuation, degradation, impedance, cross-talk, and interference

- Understand the basics of electricity and the electrical characteristics of cabling including voltage, current, amperage, AC/DC, and Ohm's law

- Understand cable construction and the use of insulation, grounding, and other protection in ensuring signal safety and integrity

- Discuss the basics of optical signals and the transmission of light waves as coherent signals by way of refraction and reflection

- Understand the transmission of signals through radiated waves such as radio, microwaves, and satellite

In Chapter 2, we introduced you to the concept of standards and standards organizations, including the International Organization for Standardization (ISO), which is responsible for the development of the Open Systems Interconnection (OSI) protocols and the OSI reference model. This is a seven-layer model used to break down the process of communication over data networks for the purpose of modularization and interoperability. Each layer represents a peer component that exists at both ends of an exchange of information. The seven layers of the OSI reference model are:

- **Application (Layer 7):** Provides support for a user's network applications. There is some standardization of Application layer protocols.

- **Presentation (Layer 6):** Involved with message syntax and semantics, and data conversion services, such as code conversion, data compression, and data encryption. This layer presents disparate end systems with the means to display data in an intelligible format.

- **Session (Layer 5):** Allows network users to establish sessions between machines and provides session management services. For instance, the Session layer can maintain virtual circuits for the duration of a connection-oriented data or voice connection.

- **Transport (Layer 4):** Allows data to pass through a network and provides users with services such as end-to-end packet reliability. This layer's end-to-end communication path is reliable as a rule. The Transport layer must ask the originator for acknowledgement in accounting for the packets sent. This layer and the layers above it deal primarily with end-station software, giving rise to the adage, "Layer 4 and up are for users."

- **Network (Layer 3):** Establishes connections between nodes, routes data packets, and accounts for the packets. At this level, there might be network congestion because a lot of packets are waiting to be routed and some packets might be lost. This layer is involved in path-selection and information-queuing algorithms.

- **Data Link (Layer 2):** Receives bits from the Physical layer (Layer 1) and converts them to frames. This layer also corrects errors that might have occurred during transmission across a link and provides an error-free transmission channel to the Network layer (Layer 3). The Data Link layer has two important sub-layers: the

MAC or Media Access Control layer and the LLC or Logical Link Control layer. The MAC provides an interface between the Physical layer's media and the network interface card of the Data Link layer. The LLC provides an interface between the hardware addressing of the Data Link layer and the logical addressing of the Network layer.

- **Physical (Layer 1):** Handles encoding/decoding and modulation/demodulation schemes, shift-key modulation, pulse-code modulation, signaling standards, and other signaling types. The Physical layer is the layer of the media of transmission such as UTP, RF, fiber-optic cabling, STP, and all other media types.

Our primary focus in this chapter is on OSI Layer 1, the Physical layer. We first discuss the concept of a signal, which is the format in which information is transmitted at this layer.

3.1   Signal Transmission

Before a signal can be transmitted across any medium, it must have a source or transmitter. The generator, or oscillator, uses energy to transmit a signal through a solid conductive medium, optical glass, or the air or vacuum of space. Whatever medium is used, communication requires energy.

Signals are electrical or optical energy sent from one device to another. In the case of wired media, the energy represents digital bits, voice, and/or video, traveling as either a series of voltages or coherent light patterns. The signal is converted back to its original format once it reaches the receiver.

Signals can be transmitted in the following ways:

- **Electrically:** Signals are transmitted through electrical pulses on copper wire.

- **Optically:** Signals are transmitted by the conversion of an electrical signal into a laser-guided signal or light pulses.

- **Radiated (wireless):** Signals are transmitted via infrared, microwave, or radio waves.

3.2   Conducted Media

Electrical signals are sent over cabling referred to as conducted media. Signals flow through conducted media by means of complex actions of atoms and charges, which make up electrical pulses by which most electronic devices send and receive information. Wires or cabling interconnecting these devices provide pathways for the electrical signals.

3.2.1   Electricity

All matter is made up of atoms, which consist of electrons, protons, neutrons, and other subatomic particles. The center of each is called the nucleus and contains the positively charged protons and the neutrons, which have no charge. Electrons are negatively charged and orbit the nucleus very quickly. They stay in orbit because of the magnetic force between the electrons and protons. Because protons and electrons have opposite polarities, the system blocks outside interference. This balance of polarity means that the atom has no net charge, as shown in Figure 3.1.

Electricity arises between two different substances, made up of different atoms, when they meet with additional energy from outside forces (such as friction), and electrons can move from one substance to the other. An atom with more or less than the normal number of electrons is called an **ion**. An atom that has more electrons than normal becomes a negatively charged ion, and an atom with fewer than normal electrons becomes a positively charged ion (Figure 3.2).

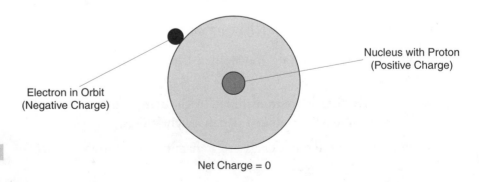

Figure 3.1

Atoms have no net charge

Electron in Orbit
(Negative Charge)

Nucleus with Proton
(Positive Charge)

Net Charge = 0

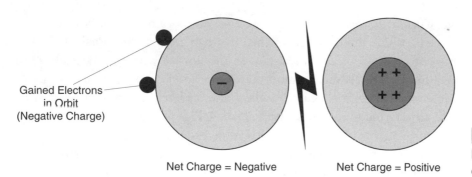

Net Charge = Negative   Net Charge = Positive

Figure 3.2
Negatively and positively charged ions

Voltage

Most reasons for net charges are physical, such as scuffing feet on a carpeted surface in a dry atmosphere, and chemical, such as the action inside a battery. Charges of a similar force repel each other; charges of different forces attract each other. If you have experimented with the traditional U-shaped magnets—marked N (north) and S (south)—then you have seen how similar magnetic poles attract and opposite poles repel. Electricity works the same way. The closer two charges are to each other, the stronger the forces of attraction and repulsion. The effort to separate opposing charges or to bring together similar charges produces energy, which can be stored for later use. We call this potential difference in energy **voltage**.

Voltage never occurs on a single point, which is why birds perched on power lines are not electrocuted. They are safe because the voltage cannot exist on a single point. On the other hand, if a bird were to touch two wires, it would meet with an untimely demise because there would be voltage between the two wires.

! WARNING

The "bird-on-the-wire" situation is one reason network technicians should always be careful when installing cable. A cable that is improperly terminated might not be able to pass energy through to the next cable. A cable can inadvertently act as an antenna if it is too close to electrical or radio interference.

Amperage and Current

Current is not voltage, nor does it always flow when voltage is present. There must be a conductive path, called a conductor, for there to be current. The path can be copper, metal, human flesh, water, wood, paper, or air. Current is measured in units called **amps** (amperage), which is the force by which the energy is pushed through a conductor. **Resistance** is a measure of how a path is resistant to the flow of current. Materials with very low resistance, such as most metals, are good conductors of electricity. Materials such as rock, cement, glass, and air have a very high resistance, and are referred to as insulators. A small voltage produces a current in conductors, whereas it takes a very high voltage to produce a current in insulators. Metal, usually copper, is used to conduct electricity, and a surrounding insulating substance is used to restrict the current to the metal. A conductor wrapped in an insulator is a wire. One or more wires bundled into the same jacket form a cable.

Resistance is measured in **ohms** (Ω). The resistance of conductors is approximately 1 ohm for every 10 feet of material. An insulator can have a resistance of millions of ohms.

Ohm's Law

Ohm's Law describes the relationship among **volts** (units of measurement for voltage), ohms (units of measurement for resistance), and amps (units of measurement for current), or voltage (V), resistance (R), and current (I). Ohm's Law states that $V = R \times I$, or voltage is equal to the amount of current multiplied by the resistance (Figure 3.3).

Anyone who works with electricity needs to properly understand Ohm's Law.

Current Containment

There must be a solid, consistent current traveling a clear conductive path in order to provide power to an electronic device. A current that is too

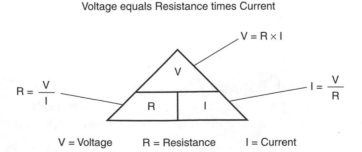

Figure 3.3

Ohm's Law

strong can produce a power surge or spike that damages the device. A current that is too weak can produce a power sag and the device will not function properly.

Alternating Current (AC)

Utilities in the United States and most other countries provide **alternating current (AC)**, which changes direction at intervals called hertz (Hz). Hertz is measured in cycles per second and most AC power operates at 50 or 60 Hz, managed by way of a grid of power lines that provide power to consumers.

There are a number of reasons for using AC. AC is not restricted to power line frequencies. Radio frequency currents also are AC. In addition, AC voltage can be changed by means of a **transformer**, which power stations use to collect and increase voltage. It is much easier and more efficient to move high-voltage power over long distances. At the destination point, transformers can collect high-voltage electricity and reduce it to 120 or 240 volts before it is carried to electrical outlets.

AC Line Noise   A major disadvantage for the cable technician in using AC power is AC line **noise**, which is always present. AC line noise can come from a device such as a computer peripheral and can interfere with network transmissions. Poor ground connections will exacerbate this problem.

AC Skin Effect   The AC **skin effect** is defined as the higher the frequency, the more the current traveling through the wire moves toward the outer edge. At the highest frequencies, the effect can be so pronounced that wires are not even used; the energy moves through tubes called waveguides (Figure 3.4).

From the cable technician's viewpoint, the AC skin effect is one of many reasons to avoid jerking or kinking cables when pulling them. Rough handling of cable can cause surface damage that reduces the cable's ability to pass signals.

Direct Current (DC)

Direct current flows continuously in only one direction. All batteries and generators produce direct current. Electronic and computer hardware use DC via an adaptor that converts an AC signal to a DC signal.

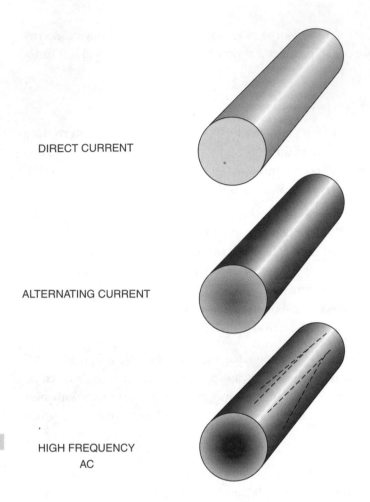

DIRECT CURRENT

ALTERNATING CURRENT

HIGH FREQUENCY
AC

Figure 3.4

Skin effect and wave-guides

In a DC circuit, electrons move from the negative pole to the positive pole. The general direction of movement is always the same, although current intensity can vary over time. There are two ways to transmit DC signals. One is simply turning the signal on or off. Morse code and the telegraph use this method.

The second method is more complicated. Today, **analog** systems and signaling use DC modulation, which is also called pulsating DC or varying DC signals.

Static Electricity

Static electricity is another common example of a charge. Most of us first learn about static electricity when we get a shock after we walk across a carpet on a cold, dry day and touch a doorknob or some other metal object.

Static electricity, which is caused by friction, is actually a collection of electrical phenomena in which the positive and negative electric charges within an object are not exactly equal. A static electricity discharge in the electronic environment of cables and networking is called **electrostatic discharge (ESD)**, and can damage or ruin sensitive electronic components. A sensitive apparatus should be handled using a ground strap. Once grounded, ESD charges quickly equalize and lose strength.

3.2.2   Cable Construction

Several factors go into the manufacture of cable. Wire thickness determines how much current the cable can carry (and consequently how dangerous the cable can be). **Capacitance** determines what kinds of signals the cable can carry. Uniformity ensures that there are no changes in the core diameter or imperfections in the construction of a cable. A lack of uniformity can cause impedance issues. Finally, the insulator's chemical components directly influence the amount of voltage the cable can carry before the insulation succumbs to a short circuit.

3.2.3   Grounding

Ground in the context of electricity refers to a connection between a circuit or other electrical equipment and the earth—literally, the ground. Virtually all safety codes stipulate that electrical equipment has to be connected to the ground or a grounded device, an important precaution designed to protect people and equipment. Figure 3.5 represents the symbol for an earth ground.

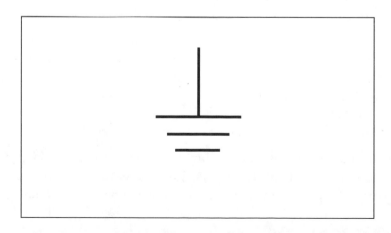

Figure 3.5
Earth ground symbol

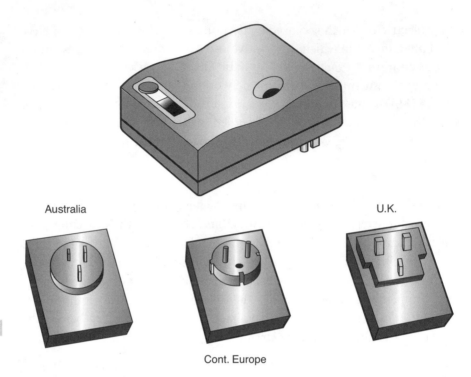

Australia

U.K.

Cont. Europe

Figure 3.6

**Ground prongs on
electrical plugs**

The chassis of most electrical equipment (such as computers) is grounded to reduce the necessity for return wires from the components, such as those you find in ceiling fans and lighting fixtures. Most internal voltages are measured to the chassis ground.

The power plugs attached to most electrical equipment provide grounding. Plugs vary from standard to standard and country to country, but most 120–240 volt connections (such as the plugs shown in Figure 3.6) use a three-prong plug, with the center prong serving as a ground.

Shielding wire is grounded in telecommunication cable to ensure that electrical signals traveling along the cable do not suffer interference from the electromagnetic radiation originating with other transmission lines and electrical equipment.

Telecommunications Grounding System

All power systems within a building have a grounding circuit that commonly connects all electrical devices to a terminal, or bus bar, in the main panel. The bus bar is connected to a metal pipe and a grounding rod driven into the earth. This method is also called the earth ground system. National Electrical Code (NEC) specifications stipulate that the grounding rod must go into the soil at least 8 feet. This method is also referred to as the earth ground system. A building's grounding system should be able to support the grounds for all telecommunications equipment and circuits within NEC-prescribed limits. (See Chapter 10 for details.)

The next section discusses optical signals and their properties.

3.3 Optical Signals

Optical transmission, used for years by long-distance telecommunications providers, is becoming the favored choice of highly reliable, high-capacity information links. As fiber becomes more affordable for the business desktop, this choice will be more common. Optical signals can travel in one of two formats:

- **Optical fiber:** Signals move through glass wires called fiber optics.

- **Optical free-space:** Lasers or light beams are used for point-to-point transmission of data. Optical free-space is a very expensive method of transmission and rarely used. On the other hand, infrared is also optical-based, popular, and inexpensive.

3.3.1 Photons and Light Waves

An electrical signal can be transformed into coherent pulses of light and transmitted over fiber-optic cables, a very different method from that used in transmissions over copper wire. The fundamental particle of a light wave is the **photon**. Photons travel in waves at frequencies higher than radio and microwaves, but lower than gamma and x-rays. The frequency of visible light is referred to as color, with a range from 430 trillion Hz to 750 trillion Hz. This is the range of the colors of the prism —red, orange, yellow, green, blue, indigo, violet—as illustrated in Figure 3.7.

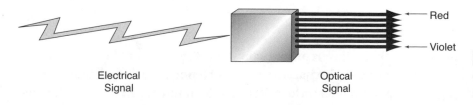

Electrical Signal   Optical Signal   Red

Violet

Figure 3.7

Conversion of an electrical signal to an optical signal

3.3.2   Reflection and Refraction

Reflection is the change in direction of a propagating wave, be it sound or light, after encountering (bouncing off of) a surface. Refraction is the bending of light, and takes place when light encounters substances such as air, water, and glass. The **refractive index** is a property of optical materials that relates to the velocity of light in the material. To calculate the refractive index, you divide light velocity through a particular material by light velocity in free space. Each kind of material has a different refractive index.

3.3.3   Free-Space Optical Communication

Free-space communication is a method of communication that uses light or laser to transmit information through the atmosphere or the vacuum of space. There must be a direct line of sight to use this method. Each transmission device has an optical transmitter and an optical receiver.

These systems often operate within the **infrared** spectrum. Infrared frequencies are below the visible light spectrum. Other versions of free-space communication operate in the blue-green spectrum. At present, free-space systems are reserved for confidential communications, such as military, medical, and finance. An obvious advantage of this technology is that it requires no licensing and emits no radiation that interferes with other transmissions. The U.S. military uses free-space technology for communicating securely with submerged submarines and between aircraft.

3.4   Wireless/Radiated Media

Signals travel through free space as waves of varying wavelengths. Before transmission, the waves are modulated—encoded with the information to be transmitted—by making slight changes in the waves, which are then transmitted over a specific frequency. For each frequency, there is a series of dimensions at which energy can easily move out of the line of transmission and into the surrounding space. This is called the *launch effect* and it typically occurs at half-wavelengths of the frequency.

3.4.1   Frequencies

Strictly speaking, the term **frequency** refers to the number of complete cycles per second in alternating current direction for an oscillating or varying electrical current. The hertz (Hz) is the unit of measurement for frequency. A frequency of 1 Hz means the current completes one cycle per second; 60 Hz (60 cycles per second) is the standard alternating current utility frequency.

Frequency is often difficult to understand. Suppose you are driving along an interstate highway and discover that you get along better with drivers who are traveling at the same speed as you. When there are cars traveling slower or faster than you, the result can be chaos, accidents, and road rage. The same principle is true for frequency. With waves, unless endpoints are communicating on the same frequency, all that is heard is incoherent noise, giving rise to the phrase, "not on the same wavelength."

Larger frequency measurements include: the kilohertz (KHz), referring to thousands of cycles per second; the megahertz (MHz), which refers to millions of cycles per second; and the gigahertz (GHz), referring to billions of cycles per second. You might occasionally hear terahertz (THz) used, which refers to trillions of cycles per second.

Frequency Spectrum

When speaking of the frequency spectrum, we are in fact talking about the electromagnetic radiation spectrum, which is the complete range of wavelengths of electromagnetic radiation. This spectrum starts with the longest audible radio waves and extends through visible light to radioactive gamma rays, which is the shortest wavelength in the spectrum. The audio frequency range is essentially 20 Hz to 20,000 Hz. *Frequency spectrum* is sometimes used to refer to the distribution of these frequencies. Bass-heavy sounds, for example, contain a large frequency content in the low end of the spectrum (20 Hz to 200 Hz).

3.4.2   Broadcasting Networks

U.S. broadcast television networks are designed to be public networks that provide news, education, and entertainment. These networks are free to the public as long as consumers have equipment that can receive the transmissions. The signals go out over public airwaves, and the Federal Communications Commission (FCC) regulates the content and the transmission method of these signals. Other countries also have state agencies to regulate public broadcast networks.

The term *broadcasting* means sending information from one endpoint to multiple endpoints simultaneously via a simplex method. The most common use of simplex communication is in radio and television broadcasting. TV and radio receive the signals on different frequencies (channels).

Broadcast networks use bands, which are a section of the frequency spectrum. Different bands are available depending on classification, such as bands allocated for commercial broadcast, AM broadcast (an MF band), and FM broadcast (a VHF band). The following is a list of the different bands of frequencies:

- **VLF (Very Low Frequency):** The part of the radio spectrum from 10 KHz to 30 KHz, or 10,000 to 30,000 cycles/second, respectively.

- **LF (Low Frequency):** The part of the radio spectrum from 30 KHz to 300 KHz, or 30,000 to 300,000 cycles/second, respectively.

- **MF (Medium Frequency):** The part of the radio spectrum from 300 KHz to 3 MHz, or 300,000 to 3 million cycles/second, respectively. The 160-meter, 1.8–2.0 MHz amateur band is an MF band, as is the commercial AM broadcast band 540–1600 KHz.

- **HF (High Frequency):** The part of the radio spectrum from 3 MHz to 30 MHz, or 3 million to 30 million cycles/second, respectively.

- **VHF (Very High Frequency):** The part of the radio spectrum from 30 MHz to 300 MHz, or 30 million to 300 million cycles/second, respectively.

- **UHF (Ultra High Frequency):** The part of the radio spectrum from 300 MHz to 3 GHz, or 300 million to 3 billion cycles/second, respectively.

- **SHF (Super High Frequency):** The part of the radio spectrum from 3 GHz to 30 GHz, or 3 billion to 30 billion cycles/second, respectively.

- **EHF (Extremely High Frequency):** The part of the frequency spectrum from 30 GHz to 300 GHz, or 30 billion to 300 billion cycles/second, respectively.

3.4.3   Two-Way Radio Networks

Two-way radio networks are used within locally contained areas of varying frequencies, which are used as channels similar to TV and radio networks, and often share the same bands. These networks and broadcast networks differ chiefly in that two-way radio networks are interactive communication and use a half-duplex transmission method in which one station cannot "talk" unless the channel is clear.

A band with a wavelength of approximately 11 meters is allocated for people using citizens band (CB) radios. A band wavelength of roughly 34 centimeters is designated for half-duplex cellular networks. In addition, bands have been set apart for police, fire, and ambulance service networks.

3.4.4 Spread Spectrum

It is possible to transmit either analog or digital data using an analog signal, which can be represented as a sine wave, over a spread spectrum transmission system. However, only an intended receiver with the same type of transmission system can receive and decode the transmissions over this kind of system. Spread spectrum systems bounce a signal around on what appear to be random frequencies rather than transmit the signal on a fixed frequency. Eavesdroppers will be stymied because transmission frequencies change constantly.

Spread spectrum technology includes wireless local area networks (WLANs), which use radio or infrared signals to connect computers and other electronic devices. Usually, such networks are extensions of a wired LAN, but they can also be freestanding.

Narrowband technology uses a specific radio frequency, requiring a license from the FCC, to transmit signals on LANs. The most commonly used method uses spread spectrum technology, which requires a wide bandwidth, but produces a louder, easier to detect signal.

The Infrared Data Association (IrDA) monitors and sets standards for infrared signals on LANs. The IrDA, a nonprofit organization, was started in 1993 with 50 companies, and has grown to hundreds of companies today.

3.4.5 IEEE 802.11

The Institute of Electrical and Electronics Engineers (IEEE) standards for wireless LANs are the 802.11a, 802.11b, and 802.11g standards for the Physical layer of communication. IEEE 802.11b states that a wireless LAN should transmit at 2.4 GHz and send data at speeds up to 11 Mbps, using direct sequence spread spectrum technology. IEEE 802.11a states that the

LAN should use 5 GHz at 54 Mbps, by means of orthogonal frequency division multiplexing. 802.11a is the newer and more powerful standard, but specifies a range of only 60 feet, compared to 300 feet with 802.11b. The 802.11a standard can be used with systems requiring high bandwidth, in which users are in close proximity. It also avoids 802.11b's 2.4 GHz band, which is now widely used by wireless phones. At present, the two standards are incompatible on the same network. The 802.11g standard operates in the same frequency range as the 802.11b standard with data rates approaching or equaling those of the 802.11a standard. Some WiFi equipment has been developed that handle both the 802.11b and 802.11g standards.

3.4.6   Cellular

The term *cellular* came from a blueprint for cellular technology in which each geographic region of coverage was divided into cells. The first cellular services operated at 800 MHz and used an analog signal. When a cellular phone customer turns on his or her phone, a signal is sent that identifies the caller as a paying customer, and then searches for a free channel to fit the call.

3.4.7   PCS Cellular

Personal communications services (PCS) operate at 1,850 MHz, using digital, rather than analog, technology. **Digital** technology samples pieces of the wave, chops it up, and sends it in strands of data. Digital technology encodes the voice into bit streams, making it a more suitable—and more secure—way of transmitting data. PCS benefits include better use of **bandwidth** (the power of the frequency) and a lower likelihood of receiving a corrupted call.

3.4.8   Antennas

Antennas are end points required for wireless communication. Free space is essentially chaotic, so antennas must be tuned to a matching frequency and mounted with the same orientation. This is called polarization (from the north and south poles). Normally, a horizontal antenna should receive a signal transmitted from a horizontally polarized antenna; likewise, vertical antennas should receive a signal from one vertically polarized. Antennas also can be omni-directional or directional:

- **Omni-directional antennas:** Transmit signals of equal strength in all directions, allowing every potential device within a specific radius to communicate with the antenna. The disadvantage is that with omni-directional antennas, anyone within the specified area can receive the signal and the signal can interfere with other transmissions. With unwanted signal reception, security issues are also problems.

- **Directional antennas:** Limit the scope of the signal and provide more secure transmissions. Terrestrial microwave communications use directional antennas for point-to-point communications.

3.5   Chapter Summary

- Communication requires energy that is used to generate a signal. A signal can make use of optical energy, electrical energy, or airwaves.

- Signals carry information through different forms of media, including optical fiber, conducted copper, air, or the vacuum of space.

- Signal formats can be either continuous (analog) or discrete (digital).

- Network cable technicians need to be aware of various characteristics and properties of signals including noise, distortion, attenuation, degradation, impedance, and interference.

- Network cable technicians should understand the different properties of optical and electrical signals and the different ways in which they operate. Electricity is the result of differences in charged atoms creating voltage and current. Optics work based on the principles of light refraction and reflection.

- Careful practices are required to protect the integrity of signals traveling over optical fiber and conducted cabling. Insulation, grounding, and proper termination help ensure the integrity of the connections.

3.6   Key Terms

alternating current (AC): A type of current that switches back and forth at a rate of 50 to 60 Hz (cycles per second).

amps: A unit used as a measurement for current.

analog: A type of signal that changes shape continuously even though it has a steady source.

attenuation: A loss of signal strength measured in decibels.

bandwidth: The data capacity of a link; specifically, the width of a band of electromagnetic frequencies.

capacitance: The limit of the amount of signal a cable can carry. This often depends on the proximity of the connectors.

cladding: A type of shielding used in cables.

current: The speed or flow rate of electricity on a circuit.

decibels (dB): In regard to signals, a measurement of attenuation.

digital: A signal type that is discrete (changes from one state to another [on/off]).

direct current (DC): Current that flows in only one direction (negative to positive).

dispersion: The scattering of light and signal on a fiber cable.

electrostatic discharge (ESD): The release of static electricity when two objects come in contact.

frequency: *1.* The number of complete cycles per second in alternating current direction for an oscillating or varying electrical current. The hertz is the unit of measurement for frequency. *2.* A particular range of the radio spectrum.

ground: A wire that sends stray current to the earth, creating a path to earth.

hertz: A unit of measurement that represents cycles per second.

impedance: A type of resistance that is intrinsic to the cable itself. It is the sum of inductive reactance and capacitive reactance.

infrared: A radiated technology that uses infrared light for point-to-point connections.

ion: An atom with more or fewer electrons than normal.

noise: Unwanted electrical or radio signals on a cable.

ohm: A unit of measurement for resistance.

Ohm's Law: A property that states the mathematical relationship between electrical voltage, resistance, and current: $V = IR$.

optical free-space: The atmosphere or the vacuum of space where wireless signals cross.

photon: A particle of light.

power: The measurement of rate at which work can get done using the electricity available.

radio spectrum: The complete spectrum of electromagnetic frequencies used for communications.

reactance: Opposition to the flow of alternating current due to capacitance or inductance.

refractive index: The property of optical materials that relates to the velocity of light in the material.

resistance: The opposition of the flow of electricity; represented as R.

radio frequency interference (RFI): Electromagnetic interference at radio frequencies, often occurring at a specific frequency or within a specific range of frequencies.

skin effect: The phenomenon that occurs when a current migrates out to the insulation (skin) of a conductor, rather than traveling through the core.

transformer: A mechanism that converts and reduces power on a power line.

volt: A unit of measurement for voltage.

voltage: Electromagnetic force or pressure; represented as V or E.

watt: A unit of measurement for power.

3.7 Challenge Questions

3.1 Define the following terms and provide the abbreviations for each:

    a. Alternating current

    b. Ampere

    c. Current

    d. Direct current

    e. Impedance

    f. Ohm

    g. Power

    h. Resistance

    i. Voltage

3.2 What are the three types of signal transmission? Provide a brief definition of each.

3.3 What are electromagnetic interference and radio frequency interference?

3.4 Define the following terms:

a. Reflection

b. Refraction

3.5 What is the current if resistance is 150 ohms and voltage is 150 volts?

3.6 Is it true that if a wire is positioned too close to sources of electrical noise or radio noise it may act as an antenna and will introduce stray signals?

3.7 How is attenuation measured?

3.8 Draw a picture of an analog signal.

3.9 What are four examples of insulators (electrons flow poorly)?

3.10 What are four examples of conductors (electrons flow well)?

3.11 What is electricity?

3.12 Do negative charges attract positive or negative charges?

3.13 Signals consist of electrical or optical _____ that are sent from one device to another.

3.14 _____ comprise all matter in the universe.

3.15 What are the seven layers of the OSI model (from the top down, in order)?

3.16 A potential difference in energy is referred to as _____.

a. amperage

b. voltage

c. current

d. wattage

3.17 The voltage of AC can be changed through the use of a _____.

a. conductor

b. CMOS

    c. cable

    d. transformer

3.18 Is AC restricted to power line frequencies?

3.19 The opposite of AC is _____.

3.20 _____ electricity is a type of charge that is often uninvited.

3.21 Grounding can occur with the aid of which of the following?

    a. Chassis

    b. Power plug

    c. Grounding rod

    d. All of the above

3.22 The bending of light is called _____.

    a. reflection

    b. refraction

    c. transformation

    d. dispersion

3.23 The _____ is the complete range of wavelengths of electromagnetic radiation.

3.24 A _____ is a section of frequencies reserved for a specific type of transmission.

    a. wave

    b. amp

    c. band

    d. phase

3.25 _____ are the end points that are required for wireless communications.

    a. Nodes

    b. Photons

    c. Antennae

    d. Rods

3.8   Challenge Exercises

Challenge Exercise 3.1

In this exercise, you use a digital multimeter to measure voltages. You need the following items:

- Five different batteries (preferably 1 AA, 1 C, 1 D, 1 9-volt, and 1 lantern battery)
- A digital multimeter
- Two connecting wires
- A power supply
- Some common materials (pencil, pen, desk, piece of wood, cup of water, eraser, sink, wool, and stapler)

3.1.1 Set the multimeter to read volts (V) if the multimeter measures both amperage and voltage. You will use the multimeter to measure the voltage of each battery by connecting the multimeter to the cells in the various ways as prescribed by your instructor. The goal is to measure positive voltage. If any of the readings display negative results, simply reverse the wires on the connection.

3.1.2 Test which of the following substances are conductors using either the power supply or the lantern battery as a source of energy and a multimeter as a sensitive current detector.

NOTE   If you are using a power supply, make sure it is set to a voltage of 10 volts or less. Verify the power supply's voltage with a multimeter prior to use.

3.1.3 Write your results next to each option:

Your skin _____

Pencil _____

Pen _____

Desk _____

Wood _____

Water _____

Eraser _____

Sink _____

Wool _____

Stapler _____

3.1.4 Answer the following questions:

    a. What is an insulator?

    b. What is a conductor?

    c. What type of solid conducts electricity best?

Challenge Exercise 3.2

In this exercise, you measure electricity in your home, a business, or your school. Electricity is sold in units called kilowatt-hours (kWh), which is a unit of energy. It is the amount of energy used by a 1-kilowatt appliance in 1 hour.

3.2.1 Locate a power meter at your residence, a neighbor's residence, a place of business, or your school.

3.2.2 Identify the meter type: _____ .

3.2.3 Read the meter and record the readings three times per day over a four-day period in the following table:

Date	Meter Readings		
	Morning	Afternoon	Evening

3.2.4 Calculate and record the total amounts of kilowatt-hours for all days per time period:

    a. Morning _____

    b. Afternoon _____

    c. Evening _____

3.2.5 How could performing this exercise become beneficial to the average family?

3.9   Challenge Scenarios

Challenge Scenario 3.1

You are involved in a discussion with a professional musician regarding the differences between analog and digital sound. The musician prefers analog technology to digital. One of his reasons is that he claims to actually hear the quantization (tiny details and elements) error in compact disc recordings as well as other digital media. Is this possible? If so, how have companies tried to remedy this?

Challenge Scenario 3.2

Have you ever noticed that when you drink through a straw, it takes more pressure to drink through a very narrow straw than it does through a short, wide one? How does this relate to electricity? Which electrical entities are analogous to the following:

- Water pressure
- Water flow
- Diameter of the straw

Challenge Scenario 3.3

You have just been hired as a covert agent for an unnamed intelligence agency. You will be traveling the world performing various operations of an electronic nature. You will need to be prepared and must obtain power adapters and sources for all of the areas of the world in which you will be traveling. Using the Internet and any other resources you have available to you, complete the following chart.

Country	Voltage	Frequency	Plug Type
Afghanistan			
Albania			
Bahrain			
Barbados			

Country	Voltage	Frequency	Plug Type
Belize			
Bermuda			
Brazil			
Burkina Faso			
Canada			
Chile			
Hong Kong			
India			
Kazakhstan			
Libya			
Myanmar			
Oman			
Paraguay			
Poland			
Russia			
Singapore			
United Arab Emirates			

CHAPTER 4

Data Network Signals

Learning Objectives

After reading this chapter you will be able to:

- Comprehend how encoding and modulation generate voice and data signals, and how analog and digital signals are generated and converted from one to the other

- Detail the differences between baseband and broadband communications

- Define broadband networking categories such as CATV and DOCSIS

- Explain POTS and other technologies using the PSTN

- Understand high-speed networking technologies such as ATM, FDDI, ISDN, and xDSL

- Understand the place of T-carrier and E-carrier systems within the public copper-based digital hierarchy

This chapter introduces you to the various types of signals used to transmit data across networks, the properties of those signals, how the signals differ from one another, and the specific situations in which the signals are used. It is imperative for cabling technicians or architects to understand the basics of signal transmission and how to apply that knowledge to projects that range from small to enterprise-scale installations.

4.1   Understanding Data Network Signals

Data communication is the transmission of data signals over a transmission medium, which requires interfaced devices to exchange data. Data terminal equipment (DTE) includes terminals and computer equipment as well as intermediary devices such as those that handle routing and switching. Data communications equipment (DCE, also data-circuit-terminating equipment), which includes hubs and routers, facilitates the actual transmission of the data or signals. Figure 4.1 illustrates a transmission medium with DTEs and DCEs.

Signal transmission in data or telecommunications is the process of communicating voice and data over frequencies by means of analog and/or digital signals. Communications between a sender and a receiver can be categorized by direction of the signal and state of the transmissions. The following general communications categories are simplex and duplex. **Simplex** transmissions travel in only one direction, allowing the sender to use the whole communication path for the transmission. Examples of simplex transmissions include broadcast television and stock ticker-tape machines. **Duplex** transmissions (also called bidirectional communication) move in

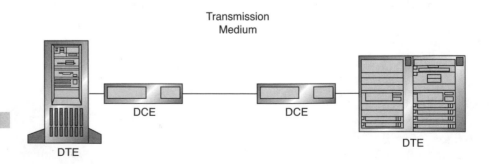

Figure 4.1

DTEs and DCEs on a network

both directions along the same communication path. There are two types of duplex transmissions: **half duplex** and **full duplex**.

- **Half duplex:** Data can move in both directions; however, it can move in only one direction at a time. The senders on each end are able to use the entire communication path. A special signal must be given and acknowledged to change the direction of the communication. The time it takes to hand control over to the other side is called the line turnaround time. Turnaround time can become crucial in certain transmissions. Citizens band (CB) radio transmissions are a prime example of half-duplex communication.

- **Full duplex:** Data can travel in both directions simultaneously, with the sender and receiver each using half of the communication path. This is analogous to a one-way street that is turned into a two-way street, in which each lane goes in opposite directions. Each of the lanes is half the size of the original one-way street. Full-duplex communication includes modem connections and telephone conversations.

Echoplex, often referred to as local echo, is a specialty error-checking mode in which characters typed for transmission are sent back by the receiver to the sender's screen, allowing comparison with what was originally typed. You find this method as an option in older modem-based terminal communication programs.

4.1.1   Transmission Modes

A **transmission mode** refers to the specific order in which data are transferred across whatever physical media is being used, whether the data communication is simplex or duplex. The information being passed across the medium must be transferred in a particular order, or mode. Transmission modes are designed for digital communications, but can be transferred to and from analog signals at any point, depending upon the underlying architecture.

Parallel Transmission

In parallel transmission mode, the bits making up a byte are transmitted at the same time, each bit on a different wire. Although parallel transmissions have specific limitations (for example, 25 feet for standard printing cables and longer for IEEE-based cabling), they are the preferred means for the communication between computers and peripherals. This mode is commonly used for communicating with printers and external network adapters as well as for a computer's internal bus communications. Parallel transmissions are often subject to crosstalk interference.

Serial Transmission

In serial transmission mode, bytes are broken down into individual bits, which are transmitted one at a time, in a predetermined sequence (from least to most significant bits, or vice versa), and reassembled into a byte at the receiving end. Serial transmissions are used in communication with modems (telecommunications), some printers, and some mouse devices. Serial transmissions have distance limitations as do parallel transmissions.

4.1.2   Signal Formats

A signal is an electric current or electromagnetic field that conveys electronic information from one place to another. The simplest signal is a direct current (DC) that can be switched on and off, such as the signals used in telegraphy. More complex signals consist of an alternating current (AC) or electromagnetic carrier carrying multiple data streams.

To convert data into signals that can be transmitted, the data have to be superimposed on a carrier current or wave using modulation, which entails creating either an analog or digital format. Digital modulation has been more popular recently because of the decreasing cost of implementing data networks. Analog modulation methods have been reserved primarily for carrying data over the telephone network.

An electrical signal is a change in voltage or current over time. The signal is described by the levels (amplitudes) the voltage or current reaches as well as by the pattern of level changes. The following characteristics of amplitudes are distinguished in describing electrical signals:

- **Peak:** The highest level reached by a signal.
- **Peak-to-peak:** The difference between the highest and lowest levels reached by a signal.
- **Average:** An arithmetic average of the absolute magnitude of signal levels, independent of positive or negative charge.

- **Root mean square (RMS):** A weighted measure of amplitude, RMS is the value used in describing a power supply. Mathematically, the RMS is the square root of the mean of the square of the signal taken during one full cycle. In the United States, for example, a wall outlet passes about 117 volts RMS, alternating at 60 times a second, or 60 Hz. The peak amplitude for the U.S. power supply is 165 volts.

Peak values are single values, whereas average values summarize amplitudes over time. A signal pattern is a waveform representing levels over time. The two types of waveforms most commonly used in networking contexts are:

- **Sine:** The waveform of a *clean* AC signal direct from a reliable power company.

- **Square:** The waveform of a perfectly encoded digital bit. A square waveform is produced with instantaneous voltage or current changes. Ideally, your network interface card or a transceiver sends such a signal along the network.

Wavelength is the distance between two points in contiguous cycles of a waveform signal propagated in space or along a wire. In wireless systems, wavelength is usually measured in meters, centimeters, or millimeters. In infrared, visible light, ultraviolet, and gamma radiation, a wavelength is more often described in nanometers (units of 10^{-9} meters) or Angstrom units (units of 10^{-10} meters). These waveforms, when graphed, most often take the form of sine waves.

For a digital signal, the shape of the signal is established by the *rise time*, which is the time a signal takes to go from 10 to 90 percent of peak strength. A square wave has a rise time of 0 seconds. In the real world, the waveform will be more trapezoidal. (The counterpart to rise time is *fall time*.)

In real-world situations, signals are accompanied by noise that distorts and weakens the signal, which might cause loss of information, transmission errors, and electrical disruption. Noise also complicates the task of signal amplification, or strengthening. This problem cannot be resolved by just amplifying a weakened signal; this amplifies the noise along with the signal.

Figure 4.2 shows examples of a sine (analog) wave, square (digital or pulse) wave, and noise.

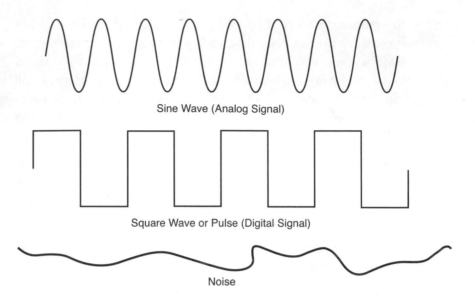

Sine Wave (Analog Signal)

Square Wave or Pulse (Digital Signal)

Figure 4.2

Comparison of a sine wave, digital signal, and noise

Noise

Analog Signals

An analog signal has values that are continuous over time, representing a level on some variable (i.e., voltage or intensity) and ranging between a minimum and a maximum value. In contrast, a digital signal takes only a limited number (usually two) of discrete values.

An analog signal can be periodic, that is, repeating in a regular pattern, or aperiodic. A periodic signal's repetition behavior is measured in cycles per second, or hertz (Hz). A 50-Hz periodic signal repeats its pattern 50 times a second. Each repetition is a cycle consisting of a continuous process in which the value of the signal changes continuously from peak to trough, and back to the peak.

The amplitude (volume), frequency (pitch), and phase (starting time) of an analog signal can each vary; therefore, the data are not discrete. For an electrical signal, amplitude is expressed in volts (voltage) or amperes (current). In computer contexts, current is usually expressed in milliamperes.

Digital Signals

The possible levels of a digital signal are represented by discrete values within a limited range that are created using sequences of 0s and 1s. The number of possible values that can be represented depends on the number of bits that are allocated to represent a single value; 256 possible values can be represented using 8 bits.

A digital signal must recognize the difference between the values 0 and 1. At the electrical level, different voltage levels generally represent the values. For example, 1 might be represented by +5 volts and 0 by 0 volts; or 1 might be represented as either +5 or −5 volts, and 0 represented as 0 volts.

Digital networks carry data, which are information of any size that is transferred in digital form. Circuits designed solely for transmitting digital information are called digital circuits.

4.1.3   Encoding Methods

In review of the chapter thus far, data can take two forms: digital or analog. Data can be transmitted using analog communication or data communication. Thus, encoding data into signals falls into one of four categories:

- Digital to digital
- Analog to digital
- Digital to analog
- Analog to analog

We describe each encoding method in the sections that follow.

Digital-to-Digital Encoding

Digital-to-digital encoding can be unipolar, polar, or bipolar. In the context of digital encoding, polarity means positive or negative. Digital-to-digital encoding methods are described as follows:

- **Unipolar:** This type of encoding uses polarized voltage to encode the value 1 and an idle line to encode the value 0. In certain implementations, this could be the other way around. One of the chief disadvantages of this method is the potential synchronization problem when data have long strings of 1s or 0s. Today, unipolar encoding is regarded as an overly simplistic method and rarely used.

- **Polar:** This type of encoding uses two voltage levels, one positive and one negative. This increases integrity and helps reduce the synchronization problems found in unipolar encoding.

- **Non-Return to Zero (NRZ-L):** This type of encoding represents the value 1 with either a positive or negative voltage, and the value 0 takes the converse voltage. The common computer serial port, RS-232 uses the NRZ-L encoding scheme.

- **Non-Return to Zero (NRZ-I):** This type of encoding is less frequently used than NRZ-L. It uses voltage transition to represent an inverted 1 value and no voltage transition to represent the value 0. The NRZ-I encoding method has good synchronization on long strings of 1s and poor synchronization on long strings of 0s.

- **Return to Zero (RZ):** This type of encoding uses three levels, with positive voltage encoding the value 1 and negative voltage encoding the value 0. RZ also uses the zero voltage action, by which the signal goes to 0 volts midway through the bit interval. This ensures synchronization, but the downside is a decrease in bandwidth efficiency.

- **Manchester:** This type of encoding, used in the Ethernet standard, represents the value 1 by a negative-to-positive transition in the middle of the bit duration and the value 0 by a positive-to-negative transition in the middle of the bit duration. The Manchester encoding scheme has a transition at the beginning of the bit duration to get back to its proper state.

- **Differential Manchester:** This type of encoding, used on Token Ring networks, is similar to Manchester encoding with a few exceptions. The value 1 is represented by the absence of a transition at the beginning of the bit encoding and the value 0 is represented by a transition at the beginning of the bit. In both Manchester and Differential Manchester schemes, there is a bit transition in the middle of the bit for the purposes of synchronization.

Figure 4.3 illustrates several of the digital-to-digital encoding methods.

Analog-to-Digital Encoding

Analog-to-digital encoding converts an analog wave into digital pulses. Two common analog-to-digital encoding methods include quantization and pulse code modulation, which are described as follows:

- **Quantization:** In this most common method of analog-to-digital encoding (ADC), a signal is sampled (recorded) and an integer value assigned to the sample. The integer value is changed into a

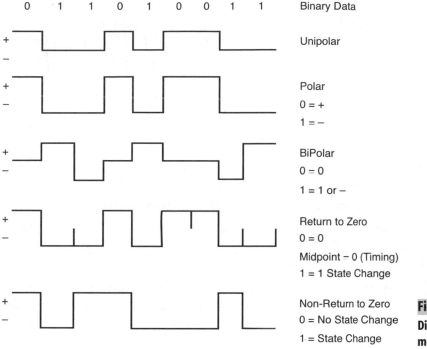

Figure 4.3
Digital-to-digital encoding methods

binary string encoded to a digitally transmitted block of information. Modern networks frequently use this method to transfer voice and other analog signals across modern digital lines, such as Integrated Services Digital Network (ISDN) or T-1 carriers. (ISDN and T-1 technologies are covered later in this chapter.)

- **Pulse code modulation (PCM):** PCM samples an analog signal at fixed intervals and quantizes the value into an integer ranging from −127 to +127 (a 7-bit number), using pulse amplitude modulation (PAM), the signal value is quantized. This number represents an absolute value, with the 8th bit determining whether the sign is positive or negative. The resulting 8-bit number is digitally transmitted. For bandwidth purposes, the smallest number of bits possible, consequently the smallest number of samples, is preferable, but this limits the quality and precision of the decoded signal. The Nyquist-Shannon sampling theorem states that the minimum sampling frequency must be greater than twice the maximum analog (input) frequency.

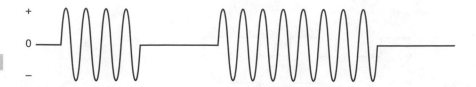

Figure 4.4
Amplitude shift keying

The most common analog-to-digital converter (ADC) is a codec, which is a contraction of coder/decoder. Codecs are used in digital telephone systems, such as ISDN, to transmit voice signals over digital lines. In making the conversion, a codec must use some kind of signal-sampling technique. It converts the samples into discrete signals for transmission.

Digital-to-Analog Encoding

Digital-to-analog encoding modulates a fixed carrier signal that has a specific amplitude, frequency, and phase value. The receiving end compares the signal to the carrier signal, a process called shift keying, which is accomplished in several ways, as follows:

- **Amplitude shift keying (ASK):** Creates two specific amplitude levels for the value 1 and the value 0 (Figure 4.4). The amount of data that can be transmitted is one bit per baud. (We discuss baud in Chapter 5.) ASK is highly susceptible to error and somewhat inefficient. It is typically on voice-grade lines for speeds up to 1,200 bits per second (bps) and generally used to transmit digital data over optical fiber.

- **Frequency shift keying (FSK):** Encodes digital information by shifting between two carrier frequencies and is not as susceptible to interference as ASK. FSK is generated by applying unipolar digital signals to one carrier, inverse digital signals to another, and creating a summation of the results. FSK's downside is that it requires much higher bandwidth requirements. It is generally used up to 1,200 bits per second on voice-grade lines, as shown in Figure 4.5, and for high-frequency (4 to 30 megahertz [MHz]) radio transmission or full-duplex operation over voice-grade lines.

- **Phase shift keying (PSK):** Encodes data by modulating the phase—the location of a position on an alternating waveform—of the carrier. A phase shift indicates a change in the data stream. A 180-degree phase shift, for example, indicates a 180-degree shift in the data stream from a state designating a value of 1 to a phase designating a value of 0 (Figure 4.6).

PSK has advantages over ASK in not being susceptible to interference and over FSK in requiring lower bandwidth. It is often displayed as dots on a

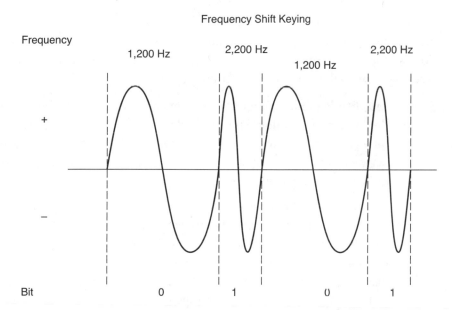

Figure 4.5
Frequency shift keying

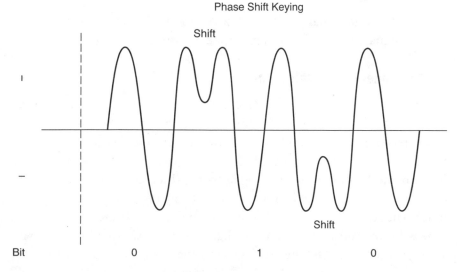

Figure 4.6
Phase shift keying

Cartesian coordinate system in which the phase equals the angle of line from its origin to the specific coordinate, with the line length measuring the amplitude. On voice-grade lines, data rates up to 9,600 bits per second are normal; however, data rates as high as 19.2 kilobits per second have been attained.

Analog-to-Analog Encoding

Analog-to-analog communication starts with a fixed carrier signal that is then modulated using its frequency, amplitude, or phase. Radio transmission is the most common form of analog-to-analog communication. The three types of analog-to-analog communications are described as follows:

- **Amplitude modulation (AM):** AM encodes signals by modulating the carrier's amplitude. This method is susceptible to noise; such as the static and whistling you might hear on AM radio. The space between the carrier signals is based on the frequency range. AM radio has a frequency range of 5 KHz and 10 KHz between carriers.

- **Frequency modulation (FM):** In FM, carrier frequency changes with the amplitude of the information signal. Noise is not as much a problem with FM as with AM radio. The space between the carrier signals is based on the frequency range plus a guard band. We discuss guard bands in the section titled "Broadband Services" later in the chapter. The frequency range of FM radio is 15 KHz, with 200 KHz between carriers.

- **Phase modulation (PM):** Phase modulation changes the carrier signal's phase with the amplitude of information signal.

4.1.4   Modem Communications

A modem was for some time the most common means of communicating data over an analog network. "Modem" is an acronym taken from the first two letters of modulator and the first three letters of demodulator. *Modulation* is a means of transmitting data from one place to another. In a modem, modulation converts a digital signal (high and low, or logic 1 and 0 states) originating from computer media to analog audio-frequency (AF) tones. The AF tones are then sent over media. Digital 1s are converted to a tone with one constant pitch, while digital 0s are converted to a tone with another constant pitch. This occurs so fast that it produces the hiss or roar

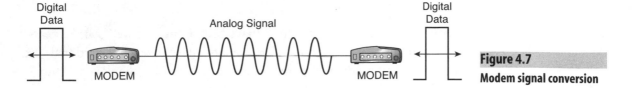

Figure 4.7
Modem signal conversion

familiar to dial-up Internet users. The demodulation process converts the tones back to digital signals that a computer can directly receive. Modems have to use sound waves to transmit data over voice-grade lines; these lines cannot convey computer signal highs and lows directly. Figure 4.7 illustrates the process of modem signal conversion.

To transmit data at speeds greater than 600 bits per second (bps), modems must collect bits of information and transmit them via a more complicated *sound*, which allows many bits of data to be transmitted at the same time. Computers can transmit data to modems much faster than modems can transmit the same data over a phone line, which gives modems time to group bits together and apply compression algorithms. Two common data compression protocols used by modems are MNP-5, which has a compression ratio of 2:1, and v.42*bis*, with a 4:1 compression ratio. MNP-5 produces a lot of overhead, so it actually increases the time it takes to transmit precompressed (e.g., ZIP) files. V.42*bis* can tell when compression is unnecessary, so it can speed up the transfer of precompressed files. If v.42*bis* is unavailable, it is best to disable MNP-5 data compression when transmitting precompressed files.

The process by which modems verify whether the information sent to them has been corrupted during the transmission is error correction. Error-correcting modems break up information into small units, called frames, to which the sending modem applies a cyclical redundancy check (CRC) or checksum. The receiving modem verifies whether the checksum matches the information sent. The entire frame is resent if there is no match. Although error correction might slow data transfer on noisy lines, it does offer greater reliability. Historically, common compression protocols include MNP2-4 and the ITU v.42*bis*, which determine how modems verify data. As with data compression protocols, an error-correction protocol must be supported by modems at both ends of a connection.

4.1.5   Fax Machines

Fax (short for facsimile) machines have been in use for well over 30 years, but modems with fax capability are an alternative to traditional fax machines. Fax modems are smaller and less expensive than fax machines. Fax devices are categorized by group and by class, as follows:

- **Group 1:** The original fax machines, which were slow and often took more than five minutes to send one page.

- **Group 2:** These machines appeared in the late 1970s and converted an image to digital signals, which were transmitted by modem. Most widespread fax use began with Group 2 faxes.

- **Group 3:** These machines use the V.17 protocol, which makes them capable of sending faxes at 14,000 bps. As of this writing, this is the most popular method.

- **Group 4:** These faxes are designed for digital ISDN lines.

The classifications for fax modems are as follows:

- **Class 1:** Modems whose command sets have special extensions that allow them to act as Group 3 fax machines. Almost all the processing is left to software in this class. The official standard for Class 1 fax modems is EIA/TIA-578.

- **Class 2:** Host or bus-mastering devices that handle much of the processing previously left to the central processing unit (CPU). Many fax modems were produced according to unofficial standards before there was an official standard. Thus, Class 2.0 is used to refer to fax modems that strictly meet the standard, whereas Class 2 applies to those that were produced before an official standard was established.

- **Class 3:** This is a proposal designed to handle the conversion of data streams into images.

4.2   Digital Circuit Communication

Early experiments in electricity used glass tubes to carry electronic signals. In 1855, Heinrich Geissler, a German glassblower, was the first to conduct such experiments. Thomas Edison discovered that inserting a small, positively charged metal plate in one of his light bulbs caused current to flow

TABLE 4.1	Important Dates in the Development of Circuitry
Date	**Event**
1904	Sir John Fleming develops the vacuum electron tube (diode). The British called it a "valve."
1907	Lee DeForest develops the audion, designed to weaken, boost, and amplify signals.
1947	Bell Labs inaugurates the era of solid-state electronics with the development of the transistor by Walter Brattain, John Bardeen, and William Shockley.
1958	Jack St. Clair Kilby invents the integrated circuit at Texas Instruments.

from the filament to the plate. This was the first thermion diode, which would become a forerunner to the modern semiconductor used in digital circuitry. Table 4.1 shows the significant dates in the evolution of digital circuitry.

4.2.1 Circuits

Initially, the term circuit referred to a closed path through which electricity can flow. Later on, the term later was broadened to refer to components such as computer chips that could create such a path. In digital communication, the term also refers to a logical data stream between hosts on a network.

Circuit terminology revolves around two concepts. Circuits can be either switched or dedicated in terms of access. They also can be either point-to-point or multipoint, referring in terms to the number of end devices that connect through them.

Switched Circuit

Dial-up telephone connections are prime examples of switched circuits. Someone calls a particular number, and equipment at the telephone company establishes the connection between the caller and the person at that number. Certain telephone equipment is assigned to that circuit for the duration of the call. When the caller hangs up, the call is disconnected and the equipment is available for other calls.

Dedicated Circuit

A telecommunications provider can establish a permanent connection between two parties. This type of connection is called a dedicated or leased line and is frequently used to connect networks at different branches of a business. The parties pay a fixed monthly rate that is usually less than the total cost of a large number of individual calls.

Point-to-Point Circuit

A point-to-point circuit connects only two devices and can be either switched or dedicated. In digital networks, modems commonly connect the two devices, but they can be connected using null modem technology, in which a standard RS-232 serial cable is used for the connection, without requiring any additional devices.

Multipoint Circuit

A multipoint circuit connects several devices, with a telecommunications provider establishing the connections. The lines making up the connections have to be dedicated because they use special installed devices that must use a protocol to prevent interference arising from simultaneous transmissions.

4.2.2   Baseband Services

Baseband uses only one carrier frequency, and sends digital signals without complicated frequency changing. It is also referred to as narrowband. Ethernet is an example of a baseband network. In general, the entire cable capacity is used for a single signal, as shown in Figure 4.8.

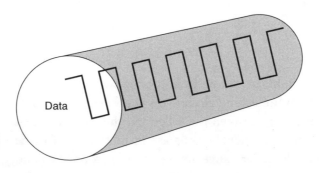

Figure 4.8

Baseband transmission

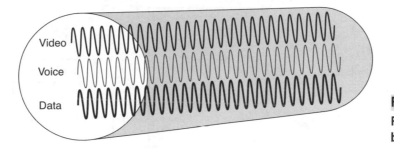

Figure 4.9
Frequency bands in a broadband transmission

4.2.3 Broadband Services

Transmission service categories exist that govern the strategy of transmission channels in addition to signal and encoding categories. A **broadband** transmission is an analog communication strategy that uses multiple channels simultaneously. The data transmissions are modulated into frequency bands called channels. Figure 4.9 illustrates how information is transmitted through the channels.

Small bands of unused frequencies, called guard bands, are allocated between data channels to provide a buffer against interference arising from signals from one channel drifting or leaking over into another. Cable TV uses broadband transmission. Each channel gets a 6 MHz bandwidth. Broadband transmissions can include voice, data, and video, and use coaxial or fiber-optic cable as the physical medium.

In the transmission of digital data, a modem or other device demodulates signals back to digital form when the signals are received. Modems used in broadband transmissions need two bands—one band for sending and the other for receiving—each with at least 18 MHz bandwidth.

Multiplexing

Time division multiplexing is a technique for maximizing the number of voice channels that can be transmitted over a single line, as shown in Figure 4.10. It grew from the discovery that during a communication connection there was a lot of time in which no information was received (that is, empty space, or silence). Using this empty space, devices and methods were used to maximize the amount of information sent over these lines where the

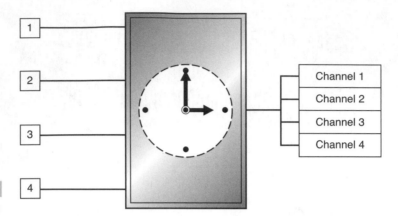

Figure 4.10

Time division multiplexing

diverse signals could be combined without losing information or interfering with other information traveling over the same line. A time slice is a subdivision of an aggregated circuit.

Frequency division multiplexing is another way of dividing up information. It was discovered in 1925 following the invention of the vacuum tube as a means to multiplex. Frequency division multiplexing divides the entire range of the frequency that a line uses into subparts, each of which has a frequency range that can contain one channel's worth of bandwidth.

Dense wavelength division multiplexing (DWDM) puts data from different sources together on an optical fiber, and each signal is carried on its separate light wavelength. On systems using DWDM, up to 80 (theoretically more than 80) separate wavelengths (channels) of data can be multiplexed into a lightstream that is transmitted on a single optical fiber. In a system in which each channel carries 2.5 Gbps (a billion bits per second), up to 200 billion bits per second can be delivered by a single optical fiber. DWDM is also sometimes called wave division multiplexing (WDM).

Because at the end of the transmission each channel is demultiplexed into the original source, these systems can transmit different data formats at different data rates at the same time. Internet Protocol (IP) data, Synchronous Optical Network (SONET) data, and Asynchronous Transfer Mode (ATM) data can all travel simultaneously along the optical fiber.

4.2.4   Community Antenna Television (CATV)

Cable TV was developed to answer the problem of getting broadcast television to viewers located at long distances from television stations. CATV positions a powerful antenna in a high place that can receive distant TV signals and distribute the signals to users over a cable.

CATV operators originally offered services such as movies and televised sporting events. They now also offer high-speed Internet access, alarm systems, and telephone services.

Some of the newer technologies require upgrading the older cables to higher bandwidth systems, such as those provided by fiber optics. Cable upgrading is one field of opportunity for cable installers.

Data Over Cable Service Interface Specification (DOCSIS)

To avoid having too many competing data technologies over cable, the larger cable companies and a number of manufacturers agreed on standards in advance. The standards were meant to create interface specifications for interoperable, data-over-cable TV network products. These standards documents are the **Data Over Cable Service Interface Specification (DOCSIS)**.

TV channels typically have a bandwidth of about 6 MHz wide. A channel with a lower limit of 54 MHz, for example, will have an upper limit of about 60 MHz. DOCSIS 1.0 technology provides data throughput capacity upstream of about 5 Mbps per 6 MHz cable channel. DOCSIS 1.1 increases downstream technology to 10 Mbps.

DOCSIS 2.0 provides up to 30 Mbps of upstream throughput per 6 MHz cable channel. These designations describe very advanced modulation schemes that can pack more data into smaller bandwidths.

DOCSIS-compliant modems are known as Cable Labs Certified Cable Modems. Although DOCSIS 1.1 continues to be in effect, the DOCSIS 2.0 standard is now used to certify cable modem products.

4.3   Telephony and Broadband Networking

The copper wire of the plain old telephone system (POTS) can provide several times its current capacity with advanced modulation systems, which means telephone circuits can provide broadband circuitry for network interfaces.

Cable installers need to know something of these new techniques as they are used more frequently due to the cost savings and increased opportunities that come from using new technology to get more use out of existing facilities. Most upgrades can take place without much rebuilding; however, any service that affects every line in a facility can be complicated.

4.3.1   Digital Subscriber Line (DSL) and xDSL

Traditional phone service connects homes or small businesses to a telephone company Central Office (CO) over copper twisted-pair cables. This service network was designed to accommodate voice transmissions and is optimized to pass signals from approximately 300 Hz to 3,000 Hz. Telephone network equipment customarily discards frequencies outside of this range as "undesirable noise."

The copper itself, however, can handle a much broader band of frequencies; analog transmission uses only a small part of the accessible information capability of the cable. **Digital subscriber line (DSL)** technology takes advantage of standard copper telephone lines to provide high-speed data access, offering improvements in speed to other methods of network access methods by means of the same physical media available to legacy voice services.

There are several varieties of DSL technology, the chief difference among them being symmetry of communication and speed. The type found most commonly is Asymmetric DSL (ADSL), which transmits from 1.5 to 6 megabits per second (Mbps) downstream and 16 to 640 kilobits per second (Kbps) upstream (hence the term asymmetric).

DSL is offered by Incumbent Local Exchange Carriers (ILECs), or through Competitive Local Exchange Carriers (CLECs), which are smaller companies taking advantage of the Telecommunications Act of 1996. This act permitted other companies to lease lines owned by large phone companies. Basically, CLECs pay to piggyback on ILEC equipment and sell unique services.

One DSL system, called G.Lite or DSL-Lite, provides traditional POTS analog services using the same set of wires and equipment. DSL-Lite works on the premise that the POTS and DSL signals can remain mixed in the lines right up to the computer or phone on the line. At that point, an appropriate filter (high-pass for computers and low-pass for phones) determines what is on the line, and only the correct set of frequencies will enter the devices. The G.Lite family of DSL services is often called *splitterless DSL* because this system requires no external splitter to break the signal.

TABLE 4.2   ISDN Equipment Categories

Component	Description
Terminal Equipment Type 1 (TE1)	A device compatible with an ISDN network (a specialized device created for ISDN). A TE1 connects to a network termination of either Type 1 or Type 2 (NT1 or NT2).
Terminal Equipment Type 2 (TE2)	A device Incompatible with an ISDN network that requires a terminal adapter (TA).
Terminal Adapter (TA)	A device that converts standard electrical signals into a form used by ISDN, allowing non-ISDN devices to connect to the ISDN network.
Network Termination Type 1 (NT1)	A device that connects four-wire ISDN subscriber wiring to the conventional two-wire local loop facility.
Network Termination Type 2 (NT2)	A device that connects four-wire ISDN subscriber wiring to conventional two-wire local loop facility. NT2 is typically found in digital Private Branch eXchanges (PBXs), and performs Layer 2 and Layer 3 protocol services.

4.3.2   Integrated Services Digital Network (ISDN)

ISDN allows the transmission of digital signals over existing telephone wiring, which became feasible when the telephone companies upgraded their switches to handle digital signals. ISDN is usually seen as an alternative to leased lines, and can be used for telecommuting and networking small and remote offices into LANs. There are a number of benefits to using ISDN.

ISDN can carry a variety of user traffic signals, while also providing access to digital video, packet-switched data, and telephone network services. ISDN is much faster at call setup than modem connections because it uses out-of-band (D, or delta, channel) signaling. Some ISDN calls can be set up in less than a second.

ISDN uses the bearer channel (B channel of 64 Kbps) and thus has a faster data transfer rate than modems. Using multiple B channels, ISDN can offer more bandwidth on wide area networks (WANs) than some leased lines. For example, two B channels provide a bandwidth capability of 128 Kbps; each B channel handles 64 Kbps.

Table 4.2 describes the various categories of ISDN equipment; Table 4.3 covers the protocol series that pertain to ISDN.

TABLE 4.3   ISDN Standards

Protocol Series	Description
E	Protocols that recommend telephone standards for ISDN. The E.164 protocol describes international addressing for ISDN.
I	Protocols that cover concepts, terminology, and general methods. The I.100 series includes general ISDN concepts and the structure of other I series recommendations; the I.200 series deals with ISDN service aspects; the I.300 series describes network aspects; the I.400 series describes how the user network interface is provided.
Q	Protocols that deal with how switching and signaling should operate. Signaling in this context means call setup.

ISDN Rates

The two ISDN services are BRI (Basic Rate Interface) and PRI (Primary Rate Interface). BRI service has two 8-bit B channels and one 2-bit D channel, and delivers a total bandwidth of a 144-Kbps line in three separate channels. The BRI B channel service operates at 64 Kbps and is designed to carry user data and voice traffic.

The D channel is a 16-Kbps signaling channel used for carrying instructions that tell the telephone network how to handle the B channels.

Broadband Integrated Services Digital Network (B-ISDN)

Broadband Integrated Services Digital Network (B-ISDN) is the new broadband variant for ISDN developed in anticipation of the increased bandwidth required for on-demand video and voice. With the birth of the Internet, many B-ISDN services could be performed on a personal computer. However, B-ISDN has taken a back seat to CATV and xDSL provisions.

4.3.3   T-Carrier and E-Carrier Systems

In 1861, Johann Phillip Reiss said, "The Sun is made from Copper," which started something that has completely changed the world. Reiss spoke this message into his new invention and was clearly heard. The telephone was born, although the first usable telephone was considered nothing more than a toy. Alexander Graham Bell patented the telephone in 1876; his patent is listed in the U.S. Office of Patents as one for the electrical and magnetic transmission of sound.

Right from the beginning, network providers have been faced with a steady increase in the user numbers and telephone/telecommunications traffic. This increasing demand for services has led to the development of various methods and technologies designed to meet market demands while remaining as economical as possible.

The analog telephone signal has a bandwidth of 3.1 KHz on a voice-grade line. On a data-grade line, the same information is sampled, quantized, and encoded in a format capable of being transmitted at 64 Kbps. When 24 such coded channels are collected together, you realize a transmission rate of 1.544 Mbps. This technology is incorporated into a **T-1** circuit or line.

The 64-Kbps channel was called DS0 and became the basic unit of digital telephony. Voice analog signals are sampled 8,000 times per second and converted into 8 bits of data to produce the 64-Kbps data rate.

Twenty-four DS0s are aggregated into a T-1 circuit. T-1 lines are generally used to connect company branches and to connect Internet service providers (ISPs) to the Internet. The T-1 transmission rate is 1.544 Mbps, the equivalent of 24 T-0 circuits. One of the circuits can be used for D channel signaling and an extra bit is added for framing. Multiple T-1s can be joined to form **T-3** lines, with a data rate of 44.736 Mbps, the equivalent of 672 T-0s or 28 T-1s.

E-1 is a European digital transmission system devised by the International Telecommunication Union-Telecommunication Standardization Sector (ITU-T) that is equivalent to the North American T-carrier system format, with some exceptions. E-1 carries 32 T-0s, two of which can be used for signaling and framing. An E-1's transmission rate is 2.048 Mbps. The Japanese hierarchy is similar, based on J-carriers.

Occasionally, a full T-1/E-1 provides more bandwidth than is necessary and a fractional T-1/E-1 line is a less expensive option. A fractional T-1/E-1 line is a digital phone line leased at a fraction of its data-carrying capacity. Fractional T-1s are a series of 64-Kbps channels that can be joined in aggregates up to the full complement of 23 B channels (and one 64-Kbps D channel), which is called a Primary Rate Interface (PRI) ISDN.

Table 4.4 lists the digital speed hierarchy for DS0, T-carriers, E-carriers, and J-carriers.

TABLE 4.4   Digital Speed Hierarchy

Speeds and Channel Counts—Common Copper Hierarchies' Speed	DSOs	North America	Europe	Japan
64 Kbps	1	—	—	—
1.544 Mbps	24	T-1	—	J-1
2.048 Mbps	32	—	E-1	—
6.312 Mbps	96	T-2	—	J-2
7.786 Mbps	120	—	—	J-2 (alt)
8.448 Mbps	128	—	E-2	—
32.064 Mbps	480	—	—	J-3
34.368 Mbps	512	—	E-3	—
44.736 Mbps	672	T-3	—	—
97.728 Mbps	1,440	—	—	J-4
139.268 Mbps	2,048	—	E-4	—
274.176 Mbps	4,032	T-4	—	—
400.352 Mbps	5,760	T-5	—	—
565.148 Mbps	8,192	—	E-5	J-5

4.3.4   Mid-Range Plesiochronous Digital Hierarchy (PDH)

T-carrier networks had a problem with the ever-increasing and variable demand for bandwidth that required more stages of multiplexing. The result was an asynchronous hierarchy in which small timing differences meant justification or stuffing was required to form the multiplexed signals.

The advantage of this situation was that bit justification, support for separate clocking signals, and switching between the two permitted easier implementation of Layer 1 gateways between J-, T-, and E-carriers. There are many problems with PDH, among which are lack of flexibility, propagation delays, and an absence of consistent management standards.

4.3.5 Synchronous Optical Network (SONET)/SDH

After the introduction of PCM technology in the 1960s, communications networks were gradually converted to digital technology and a complex hierarchy evolved to contend with ever-higher bit rates. Because there was not enough capacity given the physical media characteristics, the hierarchy was asymmetrical and had various issues of reliability.

The initial field trials for **Synchronous Optical Network (SONET)** took place in the late 1980s. SONET represents the Synchronous Optical network, or the North American Public Optical network. SDH is the European synchronous digital hierarchy. Both SONET and SDH take advantage of advances in semiconductor and fiber-optic technology to provide a much better public network than the earlier public switched telephone network (PSTN). SONET/SDH uses a base channel called an OC-1 (Optical Carrier-1), one channel of which can carry 51.84 Mbps. In situations in which networks use time division multiplexing, SONET/SDH uses wavelength division multiplexing. Tables 4.5 and 4.6 list the multiplexing levels of SONET/SDH.

TABLE 4.5   Optical Signal Hierarchy

Optical Signal Hierarchy	Data Rate	SONET	SDH	OC Level
Level 0	155.52	STS-3	STM-1	OC-3
Level 1	622.08	STS-12	STM-4	OC-12
Level 2	2488.32 Mbps	STS-48	STM-16	OC-48
Level 3	9953.28 Mbps	STS-192	STM-64	OC-192

TABLE 4.6   Optical Carrier Rates

Optical Carrier	Data Rate	Payload
OC-1	51.84 Mbps	50.112 Mbps
OC-3	155.52 Mbps	150.334 Mbps
OC-12	622.08 Mbps	601.344 Mbps
OC-48	2488.32 Mbps	2.4 Gbps
OC-192	9953.28 Mbps	9.6 Gbps

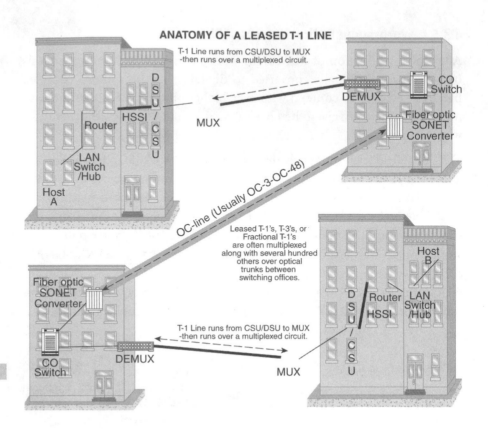

Figure 4.11

Anatomy of a leased
T-1 line

Telecommunications and network service providers use a combination of trunk lines and fiber-optic connections to provide digital circuit leasing options. A connection to a remote location might in fact travel over many different grades of circuits, through a number of routing and multiplexing schemes, as shown in Figure 4.11.

4.3.6   Fiber Distributed Data Interface (FDDI)

The **American National Standards Institution (ANSI)** developed the **Fiber Distributed Data Interface (FDDI)** in the mid-1980s, when a new LAN media was needed to support high-speed engineering workstations. To understand the importance of this technology, you need to know how FDDI works, how the Copper Distributed Data Interface (CDDI) works, and the differences between FDDI and CDDI.

Using fiber-optic cable, FDDI specifies a 100-Mbps token-passing dual-ring LAN. Because it supports high bandwidth, FDDI is commonly used as

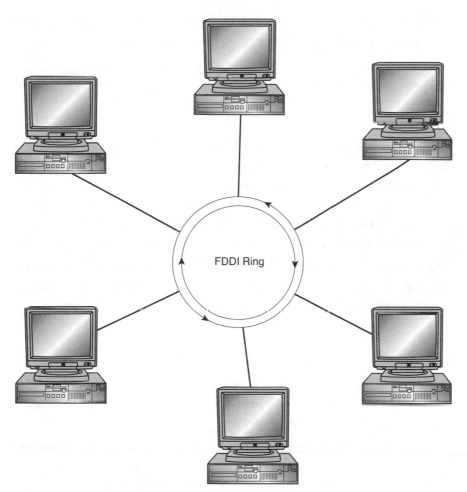

Figure 4.12
FDDI technology

high-speed backbone technology. It has a greater distance than copper, and uses a counter-rotating, dual-ring architecture for redundancy, with traffic on each ring that flows in opposite directions, as shown in Figure 4.12.

CDDI is also called Twisted-Pair Distributed Data Interface, in part because CDDI is the implementation of FDDI protocols over twisted-pair copper wire. Like FDDI, CDDI uses dual-ring architecture and provides data rates of 100 Mbps over distances of about 100 meters.

4.3.7   Asynchronous Transfer Mode (ATM)

Asynchronous Transfer Mode (ATM) is a communications networking technology that carries voice, video, and data in 53-byte protocol data units called cells. Cells are of a fixed length even when the data payload does not meet the fixed length, in which case cell padding is added.

A network can carry any type of information within fixed-length cells, which also provide rigorous service qualities that can differ with the application.

Standardization

ATM includes a set of rules for handling the transmission of information directly over private LANs via optical links or through LAN emulation (ATM LANE). In public and private WANs, ATM is a type of service designed to run over SONET and a communications protocol for sharing transmission resources by means of asynchronous multiplexing, interconnecting those resources via switching.

The ATM Forum, as discussed in Chapter 2, "Cabling Standards and Specifications," is responsible for all the various documentation and standards associated with ATM. FORE Systems (now Marconi) was instrumental in the commercial development of ATM switches and other products.

4.4   Chapter Summary

- Practical applications of signals are often categorized as both voice and data signals. Voice and data signals are generated through encoding and modulation.

- Converters allow analog and digital signals to be converted between one another. The most common analog-to-digital converter is a codec, which is a contraction of coder/decoder.

- Signals can be transmitted using the full range of the media (baseband) or multiplexed (combined) with other signals over the same, shared media.

- Baseband technologies include local area network architectures such as Ethernet. Broadband technologies include xDSL and DOCSIS (which is used with CATV networks.)

- Voice communications predominately take place over POTS and through the PSTN. For higher speed bandwidth, companies use T-carriers, E-carriers, and SONET.

- Data networks used to share the same public network, but that is rapidly being replaced by technologies such as xDSL, ISDN, ATM, and FDDI.

4.5   Key Terms

Asynchronous Transfer Mode (ATM): A type of service used for high-speed network switching using 53-byte cells.

baseband: A single, unmultiplexed channel dedicated to sending a single signal.

broadband: A multiplexed channel that can send more than one signal at a time over the same media.

community antenna TV (CATV): A technology used for pay TV services and high-speed data services using a hybrid fiber-coaxial technology.

data communications: The transmission of data signals over a transmission medium, which requires interfaced devices to exchange data.

Data Over Cable Service Interface Specification (DOCSIS): The standard for CATV data networks. DOCSIS versions include 1.1 and 2.0.

digital subscriber line (DSL): A technology that offers speed improvements to other network access methods using the same physical media available to legacy voice services.

duplex: In terms of communications (also called bidirectional communication), transmissions that can move in both directions along the same communication path. Two types of duplex transmissions are half duplex and full duplex.

Fiber Distributed Data Interface (FDDI): A dual-ring fiber physical topology using token-passing logical topology.

full-duplex: Communications in which data can travel in both directions simultaneously, with the sender and receiver each using half of the communication path.

half-duplex: Communications in which data can move in both directions; however, it can move in only one direction at a time. The senders on each end are able to use the entire communication path.

Integrated Services Digital Network (ISDN): ISDN is a type of service that provides circuit-switched video, voice, and data access.

simplex: Transmissions that travel in only one direction, allowing the sender to use the whole communication path for the transmission. Examples of simplex transmissions include broadcast television and stock ticker-tape machines.

Synchronous Optical Network (SONET): A subscriber WAN service that aggregates multiple signaling types into a single large pipe. Developed in the late 1980s, SONET represents the North American Public Optical network.

T-1: A high-speed digital circuit equal to 24 DS-0 (64-Kbps) channels, or 1.544 Mbps.

T-3: A high-speed digital circuit equal to 672 DS-0 channels (44.736 Mbps), or 28 T-1s.

transmission mode: The specific order in which data are transferred across whatever physical media is being used, whether the data communication is simplex or duplex. Transmission modes are designed for digital communications, but can be transferred to and from analog signals at any point, depending upon the underlying architecture.

turnaround time: In half-duplex communications, the time it takes to hand control over to the other side. Turnaround time can be crucial in certain transmissions such as CB radio transmissions.

wireless signals: Electromagnetic signals that travel over free space and do not require a cable.

4.6   Challenge Questions

4.1   What is an optical signal?

4.2   Define and describe ISDN.

4.3   What is the approximate bandwidth of a T-1 line?

4.4   What is the standard fiber-optic networking for optical networking using a ring topology and is used for extremely high-bandwidth applications?

4.5   What is the networking protocol that defines a uniform cell size of 53 bytes and allows the inter-mixing of various services such as voice, video, and data?

4.6   Transmissions that can travel in both directions across the same communication path are referred to as _____ communications.

a. simplex

b. duplex

c. echoplex

d. None of the above

4.7 The highest level reached by a signal is its _____.

4.8 Which of the following of an analog signal can be varied?

a. Amplitude

b. Frequency

c. Phase

d. All of the above

4.9 _____ communication can travel in both directions, but not at the same time.

a. Half-duplex

b. Full-duplex

c. Simplex

d. Echoplex

4.10 Which of the following technologies can be deployed over the existing modern PSTN network cabling infrastructure? (Choose all that apply.)

a. xDSL

b. POTS

c. FDDI

d. ISDN

4.11 An optical carrier channel represents _____ of bandwidth for transmission.

a. 64 Kbps

b. 56 Kbps

c. 1.544 Mbps

d. 51.84 Mbps

e. 155 Mbps

4.12 A T-3 line operates at 44.736 Mbps or _____ DS-0 channels.

a. 24

b. 28

c. 1.544

d. 672

4.13 The analog telephone signal has a bandwidth of
_____ on a voice-grade line.

a. 56 KHz

b. 64 KHz

c. 3.1 KHz

d. 1.544 MHz

4.14 In which method of analog-to-digital encoding is a signal sampled (recorded)?

a. Modulation

b. Demodulation

c. Amplitude

d. Quantization

4.15 A common packet-switched network over digital circuits that offers variable bandwidth is _____.

a. ATM

b. Frame Relay

c. ISDN

d. PPP

4.16 One DSL system, called _____, provides traditional POTS analog services using the same set of wires and equipment.

a. SDSL

b. ADSL

c. G.Lite

d. None of the above

4.17 _____ transmission is an analog communication strategy in which multiple channels are used simultaneously.

a. Baseband

b. Broadband

   c. Duplexing

   d. Deplexing

4.18 What technique was developed to maximize the number of voice channels that could be transmitted over a single line?

   a. Frequency division multiplexing

   b. Time division multiplexing

   c. Wavelength division multiplexing

   d. None of the above

4.19 A _____ is the most common device used for converting analog signals to digital form.

   a. modem

   b. codec

   c. multiplexer

   d. fax

4.7   Challenge Exercises

Challenge Exercise 4.1

In this exercise, you identify various methods of shift keying and encoding, as described in detail in this chapter. You need a pen and paper to record your answers.

   4.1 Describe the method of shift keying illustrated in Figure 4.13.

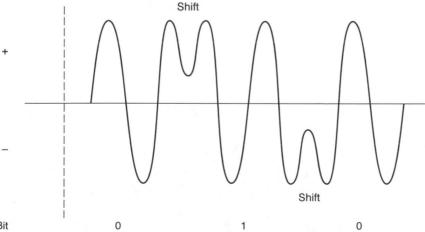

Figure 4.13

4.2 What is taking place as illustrated in Figure 4.14?

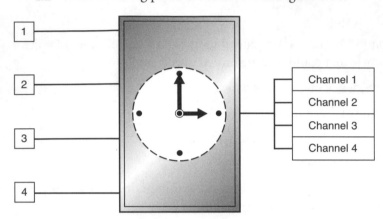

Figure 4.14

4.3 What method of shift keying is shown in Figure 4.15?

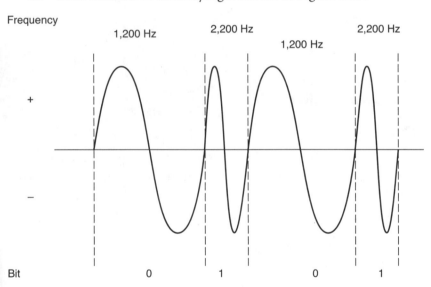

Figure 4.15

4.4 What method of shift keying is occurring in Figure 4.16?

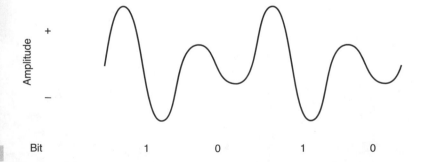

Figure 4.16

4.5 What method of encoding is shown in Figure 4.17?

Figure 4.17

4.6 What method of encoding is shown in Figure 4.18?

Figure 4.18

4.7 What method of encoding is shown in Figure 4.19?

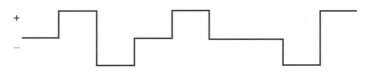

Figure 4.19

4.8 What method of encoding is shown in Figure 4.20?

Figure 4.20

4.8 Challenge Scenarios

Challenge Scenario 4.1

You are the chief information officer of a medium-size business with several offices located throughout a 250-mile radius. Many of your employees travel and require both fixed remote access and portable remote access. You plan to use both wired and wireless solutions. The service providers for your coverage areas are offering a variety of services, many of them analog-based and some digital-based. You are trying to decide which type of technology to use: analog, digital, or both. Research the many technologies that have gone from analog to digital. What was the reasoning for each conversion?

Challenge Scenario 4.2

Continuing from Challenge Scenario 4.1, how would you respond to the question of whether to use analog or digital for wireless services? What factors will influence your decision-making? Use this information to help gauge your decision.

CHAPTER 5

Copper Media: Twisted-Pair Cabling

Learning Objectives

After reading this chapter you will be able to:

- Understand the basic terminology of copper cabling
- Understand how copper cabling is manufactured
- Identify and understand cable labeling and certification
- Distinguish different cable plants
- Discuss cabling categories
- Discuss cabling performance

This chapter concentrates on copper cabling, of which there are two main types: coaxial and twisted-pair. **Coaxial cable** is the type generally used in audio, video, and broadband data networks. It is most often made up of one conductor of either stranded or solid wire (some are braided). **Twisted-pair cable** is almost always found in voice and baseband data networks, and is made up of pairs of small copper wires. All copper cabling has the following physical characteristics:

- **Cable jacket or sheath:** For protection
- **Insulation:** For preventing shorts and other electrical problems
- **Spacers:** For protecting cabling

5.1 Copper Cable Basics

Copper is the most common medium for electrical wiring, acting as the conductor carrying a signal from one point to another. Copper is an appropriate material for electrical transmission for several reasons:

- **Conductivity:** Copper, such as that found in cooking utensils, can heat up quickly because of its strong conductivity. Copper cabling uses solid or stranded copper.
- **Strength:** The physical properties of copper are not affected by very hot or cold climates. Copper has high tensile strength and can retain its properties up to 400°F.
- **Malleability:** Copper is very malleable and can be spun, forged, and hammered into shape whether it is worked hot or cold.
- **Corrosion level:** Copper does not rust and resists most corrosion. Although extended exposure to open elements causes oxidation and a patina to form, it does not affect copper's properties.
- **Ductility:** A ductile material is one that can be finely spun without breaking. Copper can be spun into wires whose diameters are measured in microns.

5.1.1 Solid versus Stranded Cores

Copper cabling has either a **solid core**, or a **stranded core** that consists of a bundle or braid of strands. Large solid cores are usually found in heavy-

duty electrical cabling; small cores are in baseband cabling carrying iso-
lated signals. Stranded cores are commonly found in coaxial cabling,
audio/video cabling, and broadband cabling.

5.1.2 Cable Insulation

Every kind of cabling is enclosed in material called insulation (cable dielectric) that is highly resistant to electrical charge. Insulation protects the
cable's **core** by preventing an outside current from flowing into the inner
core or from external elements. If a cable contains more than one conductor, an inner insulation layer is added to protect the conductors from cross-
interference. As we mentioned in Chapter 3, among the different materials
that can serve as insulators, only certain durable materials are used for
cabling. The insulation material selected depends upon how the cable is to
be used. Cabling insulation materials are separated into one of the following categories:

- **Elastomers:** Very flexible plastic and rubber-like materials that
  retain their shape.

- **Thermoplastics:** The most common insulation materials, and the
  most common thermoplastic is **polyvinyl chloride (PVC)**. Polyethylene is also used in cable insulation. Thermoplastics can resist
  sunlight, artificial light, oil, dust, and other potentially damaging
  elements. PVC insulation is less elastic than elastomers, and is
  therefore easier to cut and strip. PVC insulation comes in a variety
  of colors and is a very practical and cost-effective choice for cable
  insulation.

- **Fluoropolymers:** More frequently categorized as plenum-rated. (A
  plenum-rated cable has a wire coating that burns at a much higher
  temperature and emits fewer fumes than non-plenum–rated
  cabling. The National Electrical Code (NEC) requires that
  plenum-rated cabling be used in plenums, which is the space above
  the ceiling tiles and below the next story into which two or more
  air ducts run. The most common fluoropolymers are Halar and
  Teflon, both of which are fire-rated and therefore can be used in
  high-temperature areas. Fluoropolymers are less elastic than thermoplastics and elastomers.

5.2   Creating Cable

The first step in creating a copper-based cable is **extrusion**, in which wire is stretched and spun into a thin strand. The thin wire is then passed through a crosshead extrusion machine that adds insulation.

5.2.1   Extrusion Process

Extrusion differs from the drawing spin process (discussed in the next section) in that the metal is worked by pushing under tension, rather than being pulled. The steps in the extrusion process are:

1. Raw metal ore is melted and refined in a furnace.

2. The purified metal is smelted into a molten state and then cast in molds to yield slabs, billets, or ingots.

3. The solid metal is worked to acquire the desired conditions or properties.

4. The metal is forced through a shape-forming die, and then melted and compressed in the die cavity.

The extrusion process can be used with either hot or cold metal; however, it is generally carried out with hot metal in a process called ram extrusion.

5.2.2   Wire Drawing (Pulling)

Drawing wire is a method similar to extrusion. Copper wire often starts as coils of thick wires produced by hot rolling, with successive coils welded together to maintain continuous production. Copper wire is then fed through an insulate line. Each insulate line performs several different functions, such as additional wire drawing, annealing (softening), and applying insulation.

The first step in the pulling process is to reduce the size of the copper wire by drawing it again, which uses diamond dies and reduces the wire to another size based on **American Wire Gauge (AWG)** codes (discussed in Section 5.3, "Labeling and Certifying Cable").

After being drawn, wire is very brittle and can be easily fractured if flexed. Finished copper wire has to be flexible to be useable, so the wire is annealed by passing a large electrical current through the wire for a fraction of a second. This raises the wire's temperature briefly to 1000°F. Wire is annealed in water to prevent oxidation, and also to cool and clean the wire before applying insulation. Wire that is not properly annealed tends to be brittle and break easily.

The wire is then passed through an extruder, where a thin coating of plastic containing high-density pellets of the insulating material is applied. As the wires are pushed through the extruder, the insulation pellets heat until they melt onto the wire. The coated wire exits the extruder at approximately 60 mph, passes through a cooling trough, and is coiled on take-up reels.

The wire and insulation diameter is measured, and the wire is tested for electrical properties such as capacitance and resistance before the reels move to the next manufacturing step.

5.3 Labeling and Certifying Cable

The tested cabling is rated and certified according to its diameter and resistance. The United States and other countries use the Universal Service Order Code (USOC) standard (as shown in Table 5.1). Table 5.2 lists the AWG metric gauges. Basically, the AWG principle is "the higher the number (or gauge), the thinner the wire." Thicker wire can carry larger amounts of current over greater distances with less signal loss (attenuation). Cable signal loss is often rated in ohms per 1,000 feet, and for a particular type of wire, lower gauges (larger wires) are resistant to current flow. Table 5.3 lists the NEC AWG maximum amperage.

Figure 5.1 illustrates a cable with the AWG grade labeled on the cable jacket.

5.4 Cable Plant

Cable is bought and implemented in a number of ways by commercial and residential consumers. In commercial installations, cable that is collected and bundled together is called the **cable plant**, as shown in Figure 5.2. *Cable*

TABLE 5.1   **American Wire Gauge Ratings**

AWG Code	Diameter (mm)	Diameter (inches)	Square (mm²)	Resistance (ohm/km)	Resistance (ohm/1,000 ft)
46	0.04	—	0.0013	13,700	—
44	0.05	—	0.0020	8750	—
42	0.06	—	0.0028	6070	—
41	0.07	—	0.0039	4460	—
40	0.08	—	0.0050	3420	—
39	0.09	—	0.0064	2700	—
38	0.10	0.0040	0.0078	2190	—
37	0.11	0.0045	0.0095	1810	—
36	0.13	0.005	0.013	1300	445
35	0.14	0.0056	0.015	1120	—
34	0.16	0.0063	0.020	844	280
33	0.18	0.0071	0.026	676	—
32	0.20	0.008	0.031	547	174
30	0.25	0.01	0.049	351	113
28	0.33	0.013	0.08	232.0	70.8
27	0.36	0.018	0.096	178	54.4
26	0.41	0.016	0.13	137	43.6
25	0.45	0.0179	0.16	108	35.1
24	0.51	0.02	0.20	87.5	27.3
22	0.64	0.025	0.33	51.7	16.8
20	0.81	0.032	0.50	34.1	10.5
18	1.02	0.04	0.78	21.9	6.6
16	1.29	0.051	1.3	13.0	4.2
14	1.63	0.064	2.0	8.54	2.6
13	1.80	0.0720	2.6	6.76	2.2

TABLE 5.1   (continued)

AWG Code	Diameter (mm)	Diameter (inches)	Square (mm^2)	Resistance (ohm/km)	Resistance (ohm/1,000 ft)
12	2.05	0.081	3.3	5.4	1.7
10	2.59	0.102	5.26	3.4	1.0
8	3.73	0.147	8.00	2.2	0.67
6	4.67	0.184	13.6	1.5	0.47
4	5.90	0.232	21.73	0.8	0.24
2	7.42	0.292	34.65	0.5	0.15
1	8.33	0.328	43.42	0.4	0.12
0	9.35	0.368	55.10	0.31	0.096
00	10.52	0.414	69.46	0.25	0.077
000	11.79	0.464	83.23	0.2	0.062
0000	13.26	0.522	107.30	0.16	0.049

TABLE 5.2   AWG Metric Gauge

Metric Gauge	Diameter (mm)	Square (mm^2)	Resistance (ohm/m)
5	0.5	0.20	0.0838
6	0.6	0.28	0.0582
8	0.8	0.5	0.0328
10	1.0	0.8	0.0210
14	1.4	1.54	0.0107
16	1.6	2.0	0.00819
20	2.0	3.14	0.00524
25	2.5	4.91	0.00335

TABLE 5.3   NEC AWG Maximum Amperage

AWG Wire Size	Two Current-Carrying Conductors (ampere)	Three Current-Carrying Conductors (ampere)
18	7	10
16	10	13
14	15	18
12	20	25
10	25	30
8	35	40
6	45	55
4	60	70
2	80	95

Figure 5.1

Twisted-pair cable showing AWG grade label

Figure 5.2
A cabling plant

plant usually refers to cable bundles, whereas *structured cabling system* refers to the cable plant as well as all connectors, the devices that it connects, and the materials and components used to facilitate the cable's connectivity.

Buildings are frequently connected to external services and utilities as well as to other buildings, so the cabling system might include both inside and outside cabling.

Cabling that is run inside a building is called the inside plant; likewise, cabling run outside is called the outside plant. Inside plant cabling material is thinner and reasonably cost-effective. Outside cabling is thicker, has a more durable insulation jacket, and often has extra protection to prevent corrosion and other elemental interference, so it is more expensive.

5.5 Overview of Twisted-Pair Cabling

As we noted in Chapter 1, twisted-pair cabling is the most commonly used kind of cabling type for voice and data networks. The different types of twisted-pair cabling vary in number of pairs, wire resistance, amount of twisting, color codes, and type of insulation used. Twisted-pair cabling is terminated with modular **registered jack (RJ)** connectors (a USOC identifier).

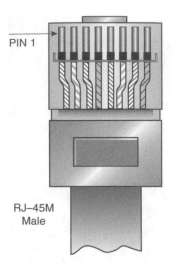

PIN 1

RJ–45M
Male

Figure 5.3

**RJ-45 connector wires are
color-coded**

The RJ-11 is the most common RJ connector found in voice networks, such as telephones. The most common RJ connector used in data networks, the RJ-45 connector, is shown in Figures 5.3 and 5.4.

Connectors make cabling modular and portable. Fixed twisted-pair structured cabling uses termination jacks (Figure 5.5), which furnish permanent termination, and patch panels (Figure 5.6). The patch panel is the point at

Figure 5.4

**Outside view of RJ-45
connector**

Figure 5.5

Twisted-pair cable termination jack

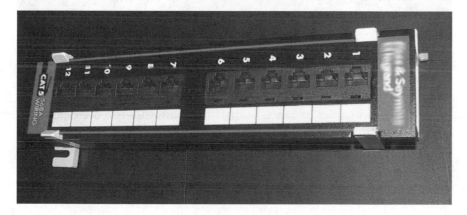

Figure 5.6

Patch panel

which individual wires are untwisted and punched down into a termination block. We discuss cable connectors and termination in Chapter 8, "Cabling System Connections and Termination."

5.5.1 Unshielded Twisted-Pair (UTP)

Unshielded twisted-pair (UTP), as noted in Chapter 1, is the most common type of twisted-pair cabling, and is used mostly in two-pair and four-pair combinations; often even when only two pairs are necessary, four pairs

Figure 5.7

Unshielded twisted-pair cable

will be used to allow for expandability (Figure 5.7). UTP has no protective shielding—only the outer sheath and the inner insulating wires—so it relies on the principles of cancellation, or the **cancellation effect**, to limit **electromagnetic interference (EMI)** and radio frequency interference (RFI). To facilitate the cancellation effect, the number and degree of twists per foot must follow specifications outlined by categories. The greater the twists, the more copper the cable uses and the thicker the overall wire. The thicker the finished wire, the more expensive the cable becomes.

The twists can increase the actual distance that data can travel, making each individual wire length greater than the bundle as a whole in raw measurement. *Skew* can occur if the cable is twisted improperly because signals traveling across different pairs in the same cable don't arrive at endpoints at the same time. This can affect the clocking mechanisms of the sending and receiving devices on data networks, and cause communication problems. Unshielded twisted-pair cabling uses thin wires (AWG 22–24 grades). Cable technicians must be careful not to over-bend or degrade cabling physically during installations, as it can result in the pairs separating and

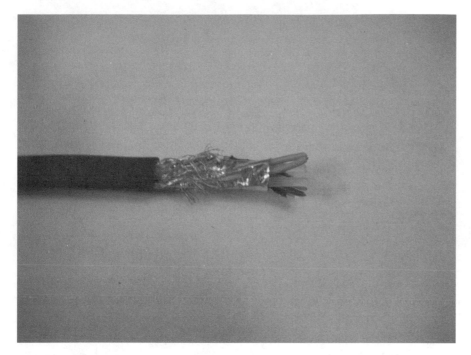

Figure 5.8
Shielded twisted-pair
cable

lead to problems. There are several categories of UTP cables, which we will discuss later in this chapter.

Often, cable installers will encase UTP cabling inside a metal conduit or piping to reinforce the cable and add protection from interference. When this is done, it is important be sure that the metal conduit or piping is grounded properly.

5.5.2   Shielded Twisted-Pair (STP)

Many installations use shielded twisted-pair (STP) cabling (Figure 5.8), rather than running the cable through a metal conduit. At the time of this writing, UTP is the most common type of twisted-pair used in North America; however, Europe and other parts of the world use STP cabling. Over time, more installations will move to STP because of the extra shielding it provides against EMI and RFI. STP is often used as a lower-cost alternative to fiber-optic cabling in situations where a UTP installation will not be adequate.

When an installation requires extra cabling, it is best to use STP cabling. In STP cabling, the individual cables are wrapped with shielding and then the whole set of pairs is wrapped with more shielding. Although this arrangement has many advantages, STP also has disadvantages:

- All layers of shielding have to be properly grounded to reduce signal degradation, and the grounding itself is a possible source of interference if differences in ground are detected.

- STP is more easily damaged during installation, so it requires great care in pulling the cable. STP's shielding makes it less flexible than UTP. Flexing the cable can bunch or cut the shield, and a damaged shield might expose the cable pairs to increased interference.

- STP is more expensive than UTP.

There are three kinds of shielded twisted-pair cabling: screened twisted-pair, 100-ohm twisted-pair, and 150-ohm twisted-pair:

- **Screened twisted-pair (ScTP):** ScTP cabling has only a screen of foil covering all the pairs; inner wires do not carry their own shields. As a result, this cabling is less expensive and has a smaller diameter than STP.

- **100- and 150-ohm shielded twisted-pair:** The 100-ohm STP is similar to UTP. Because the shield is not part of the data circuit, it is grounded at only one end, usually at the telecommunications room or hub. The shield acts as a protective sleeve that intercepts EMI and RFI and takes them to ground before either can affect the cable's data signals. The 100-ohm STP is found in Ethernet networks and 150-ohm STP is most commonly found in IBM Token Ring networks. The shielding on 150-ohm STP is grounded at both ends and is not part of the signal path. It is very bulky and has smaller distance limits than its counterparts, so it is difficult to install and use.

5.6   Categorizing Cable

The Electronic Industries Alliance and the Telecommunications Industry Association (or EIA/TIA, collectively) categorize twisted-pair cabling into what is often called *cat levels*. Each cat level is defined by the number of wires in the cable, the number of twists in the wires, and the wire's bandwidth.

Cable in lower categories can be more susceptible to noise and carry less bandwidth; therefore, these are reserved for voice-grade communications. Higher categories have higher speeds and bandwidth, and are used for data transmission. The EIA/TIA cable categories were developed in response to problems that arose with earlier attempts to standardize cabling. Anixter devised the categories that would later be adopted by the American National Standards Institute (ANSI), EIA, and TIA. The original standards were:

- **Level 1:** For plain old telephone service (POTS)
- **Level 2:** For low-speed computer terminals and networks
- **Level 3:** For Ethernet operating at 10 Mbps and Token Ring

These standards were useful for awhile; however, EIA and TIA established standards for building wiring in the early 1990s. One of their first actions was renaming the Anixter levels Category 1, Category 2, and Category 3 cables. Later, additional cable categories were devised, as outlined in the following sections.

5.6.1   Category 1

Category 1 wiring was once used exclusively for voice-based communications and is rarely seen today. You will still find it in older homes and businesses as well as in home appliance systems and audio systems. Category 1 wiring is not recognized as part of ANSI/TIA/EIA-568-B.1 and ANSI/TIA/EIA-568-B.2, and is therefore not part of modern structured cabling systems.

Category 1 wire is usually 22 AWG or 24 AWG untwisted wire, with a wide range of impedance and attenuation values. Wire in this category is not recommended for data communication, particularly for any signaling speed over 1 Mbps.

5.6.2   Category 2

Older IBM Token Ring networks used **Category 2** wiring, which has a maximum data rate up to 4 Mbps (16 Mbps in passive applications). As with Category 1, it is not recognized as part of ANSI/TIA/EIA-568-B.1 and ANSI/TIA/EIA-568-B.2, and transmission characteristics are not specified.

Category 2 uses 22 AWG or 24 AWG solid wire in twisted-pairs, with a maximum bandwidth of 1 MHz. Apple LocalTalk (256 Kbps) networks have used Category 2 cable.

5.6.3   Category 3

Category 3 cables are 100-ohm UTP cables, consisting of four pairs of 24 AWG solid copper wire in twisted pairs that support bandwidth through 16 MHz.

Although this cable category was once considered the minimal acceptable cable for Ethernet 10BaseT installations, its use is no longer recommended. Category 3 was once extremely popular for telephone wiring, and is still being installed for that purpose.

5.6.4   Category 4

Category 4 cables were designed for 16 Mbps Token Ring networks, but were universally used during their generation. Category 4 is not recognized as part of ANSI/TIA/EIA-568-B.1 and ANSI/TIA/EIA-568-B.2, and no transmission characteristics are specified.

Category 4 cable has four pairs of 22 AWG or 24 AWG wire, with a typical impedance of 100 ohms and a 20 MHz bandwidth.

5.6.5   Category 5

Category 5 wiring is the most commonly installed medium and most cable installed between 1995 and 2004 uses Category 5 UTP. Category 5 consists of 100-ohm twisted-pair cables with four pairs of AWG 24 copper wire, and has a bandwidth up to 100 MHz. Although Category 5 cable can operate a 1000BaseT installation (Gigabit Ethernet specification) under certain conditions, such use is not recommended. Category 5 has now been replaced by Category 5e, but Category 5 is still popular because it can be used for voice networks, security systems, and home automation and control systems as well as for data networks.

5.6.6   Category 5e (Enhanced Category 5)

Category 5e has more (hence, tighter) twists than Category 5 wiring, which improve performance by giving the cable better resistance to external interference and from the other wires within the cable. This kind of twisting also makes Category 5e cables resistant to separation and bunching during installation. Category 5e is used for Ethernet and other data networks that are likely to expand into higher-speed Asynchronous Transfer Mode (ATM) and Gigabit Ethernet networks

5.6.7   Category 6

The **Category 6** specification has not yet been released or ratified, although Category 6 wiring is already available for installation and use. Wire manufacturers have access to the draft standards and have developed physical standards that meet or exceed the draft performance standards. However, manufacture in advance of ratified standards means that some characteristics will vary among cable manufacturers.

Category 6 cable has four pairs of 24 AWG copper wires, is bulkier than Category 5e cable, and is more expensive than either Category 5 or Category 5e cable. However, the extra twists reduce crosstalk, making Category 6 cable a more reliable medium for 1000BaseTX (Gigabit Ethernet), the fastest copper-based Ethernet standard.

5.6.8   Future Categories

Cabling and cabling equipment manufacturers want their products to last as long as possible, so they try to remain not only current, but also to stay ahead of the latest ratified standards. As of this writing, many products carry "Cat 6 and Cat 7 ready" on their packaging; however, using these products involves risk that increases in specifications will occur by the time the standard is ratified. Exercise care when choosing cable created to meet future standards. You don't want to install an infrastructure expected to last 10 to 15 years that could wind up in noncompliance with industry standards.

5.7   Improving Cable Performance

The EIA/TIA and ANSI work constantly to improve manufacturing quality because that leads to improvements in cabling quality. Furthermore, the organizations work with researchers to find ways to increase the quality of cable performance in several areas:

- **Signal reconstruction:** Redundancy introduced in signals sent over noisy channels allows signal reconstruction. Higher and higher frequencies used for signal reconstruction can cause signal

degradation. Noisy data lines will achieve the same effect by causing more and more signal reconstruction.

- **Ingress and egress:** Ingress and egress in this context refers to
undesirable signals entering and exiting the cable. These signals are
the most frequent cause of near-end crosstalk (NEXT). Materials
called flooding compounds and techniques such as bonding are
used to prevent this problem.

- **Manufacturing tolerances:** Cables have better resistance to external
and internal noise when wires of the same gauge are grouped close
to each other, a principle that is fundamental to twisted-pair
cabling. Some manufacturers fuse the insulation of each pair to help
maintain the number of twists and keep the wires from shifting.

- **Reducing twist-induced skew:** This is the process of mathematically constructing cable with the optimum twist ratio to reduce the
skew delay.

- **Separating pairs physically:** Uniform spacing is being used to
reduce crosstalk, and some manufacturers have experimented with
plastic spacers and fillers to separate the wires.

5.7.1   Safety Ratings

In addition to using the categories for selecting cable, you should also
take cable safety ratings into consideration. Underwriters Laboratory
(UL) includes twisted-pair cabling in its certification program, and
cables are evaluated for performance with reference to IBM and TIA/EIA
performance specifications and NEC safety specifications.

Twisted-pair cabling falls into one of two UL categories:

- **UL 444:** The standard for safety for communications cable
- **UL 13:** The standard for safety for power-limited circuit cable

The UL LAN certification program addresses performance as well as safety.

5.7.2   Cancellation Effect

Cancellation is one of the fundamental principles on which twisted-pair
cabling operates. As long as the signals are traveling across the same twisted
pair, the process works. Untwisted or split pairs can cause problems; the
ANSI/TIA/EIA bodies use color codes to prevent this from happening.

When wires are twisted together over a length of cable, the fields of one wire are rapidly alternated with the fields of another to limit the exposure of any one cable to the electromagnetic fields of another.

Twisting also protects the cable from electromagnetic interference produced by other devices or electrical currents. As long as the cable remains twisted, both wires in a pair will likely be affected equally by any interference. The signals in each wire in a pair are moving in opposite directions, so the interference adds to the signal in one wire as it resists the signal in the other, which tends to cancel out interference such as EMI and RFI.

5.7.3   Color Codes for Twisted-Pair

The old-fashioned patch panel switchboards used in telephone offices have given rise to some modern wiring terminology. Operators inserted probes, or plugs, into receptacles, or jacks. One conductor fastened to the tip of the plug and the other conductor attached to a ring around the plug. In time, these wires came to be known as tip and ring, a designation that continues to this day for twisted pairs.

In a four-pair cable, such as Category 5e, the "primary color" is the tip, and is generally white with a tracer, or stripe, in the same color as the pair's solid color wire, which is the ring. Although on some cables the tip wires are opaque and the rings translucent, most twisted-pair cables have this color-coding scheme.

Some manufacturers have taken liberties with the marking system and do not put a white stripe on the tip wire. Other manufacturers use translucent shading that tints the wire that shows through it, rather than use a solid color. In still some cases, manufacturers might distinguish wires by using periodic splotches of the companion color, while in others they use all solid colors.

5.7.4   ANSI/TIA/EIA T568A and T568B

Bell Telephone established the technique for terminating twisted-pair cabling that is called the Bell Telephone Universal Service Order Code (USOC), which logically organizes the wires into a modular plug. The first wire pair goes into the center two pins of the plug and splitting each pair down the middle, the rest follow from left to right.

Unfortunately, this technique of separating data wire pairs can lead to crosstalk. The wiring scheme has since been modified to keep pairs close together in order to continue using the standard RJ-45 connectors and plugs. The result is two wiring patterns called **T568A** and **T568B**, designating the order in which pairs should be mounted in modular plugs and jacks. Do not confuse these wiring schemes (T568A and T568B) with the TIA/EIA standards that specify them (TIA/EIA-568-B).

In most cases, either wiring scheme is appropriate for new cabling jobs. Use the wiring scheme already in use when you are working on an existing network. In either case, make sure the same wiring scheme is used for every termination in the project.

5.7.5   Multi-Pair Cables

Telecommunications cable comes in many sizes, from a single pair of wires to 4,200 pairs. In most cases, running cables with more than 900 pairs can cause difficulties—especially if you are running it between buildings—because most countries' wiring codes require using surge-protecting fuses at the wires' point of entrance. This precaution is necessary to reduce the hazard of external electrical sources, such as lightning strikes or downed power lines that can endanger a building, its occupants, and its equipment. Nine hundred or more protection devices occupy a lot of space and take more time to install, which increases installation costs. A single fiber-optic cable can carry as much traffic and is not affected by lightning or other induced voltages.

5.7.6   Color Codes for 25-Pair Cables

The standard four-pair cable colors are a subset of a larger scheme of colors. Pairs 1 through 4 of a four-pair cable use the same color system that is used in a 25-pair cable. As discussed earlier, one wire from each pair is the tip, and the other is the ring. The colors alternate for each. The tip wire has a stripe of the ring color in it and vice versa. (Sometimes the stripes are actually rings or color bands; other times they are smudges.)

When indicating colors for a pair, the tip color comes first, because that is the order in which the cables are punched on a punch block. Ring colors are just the opposite. That is, if pair 22 has tip colors violet-orange, the ring colors for that pair would be orange-violet.

5.7.7   Color-Coded Binders for High Pair-Count Cables

Fiber-optic cable is becoming more and more popular, especially as a replacement for large pair-count cables such as those used in outside plant cables. However, the large pair cables still in use must sometimes be repaired. One of the biggest problems you will encounter in such a repair is trying to distinguish one pair from another (simply because of the large number of pairs). However, you can use the 25-pair color-coding scheme described in the previous section as a guide.

Cables with more than 25 pairs group the wires in 25-pair units, each wrapped in colored tape to form binder groups, which follow the same color code as the wires of each pair; the first binder is blue/white, the second binder is orange/white, and so on.

In cables with 200 pairs and more, binder groups are bound into units of 50, 100, or more pairs, which are wrapped with colored tape following the same scheme as binder groups.

5.8   Chapter Summary

- Copper cabling is the standard for transmitting electrical energy because of its conductivity, strength, malleability, and corrosion level.

- Copper cabling is implemented in voice and data networks as either coaxial cabling (solid or stranded core) or twisted-pair (bundled and separate). Both types of cabling are made up of a conductor along with cable jackets, insulation, and spacers. Some cabling may have additional shielding to protect from interference.

- Cable insulation varies in accordance with safety regulations. The most common material used for insulation is PVC, although plenum (fluoropolymer) cabling is used for fire protection and where required by the fire codes.

- Cabling is created by extrusion of a metal (usually copper) either by pulling or pushing. After the extrusion, the wire is coated with insulation, and then tested and certified. The wire's diameter and resistance initially determine the cabling rating, which is documented in the United States and other countries according to the AWG Standard.

- Cables are installed inside building structures using a model called a structured cabling system. The cabling system consists of both outside and inside cabling, referred to as cable plants. The outside cable plant is often provided with additional protection from the elements and corrosion.

- Twisted-pair cabling is the most commonly known and implemented cabling in both voice and data networks. Twisted-pair is terminated for permanent cable runs using patch panels and jacks. RJ connectors are used for modular and patch twisted-pair cabling. The most commonly used RJ connectors are the RJ-11 for voice and the RJ-45 for data.

- Twisted-pair can be either unshielded (UTP) or shielded (STP). Twisted pair contains wire of AWG 22–24 core gauge. UTP cabling is more widespread than STP in that it is inexpensive, easy to install, and very flexible. STP costs more and is more difficult to install, but provides for extra protection against interference.

- The ANSI/TIA/EIA standards bodies developed cable categories for determining wire grades for use in twisted-pair cabling. Category 1 is for voice-based communication using POTS. Category 2 is used for low speed data networks and terminals. Category 3 is used for Ethernet operating at 10 Mbps and Token Ring networks. Category 4 was designed to support the 16 Mbps Token Ring network. Category 5, the current dominant medium, supports most voice and data network architectures. Category 5e is replacing Category 5 and provides more reliable performance for high-speed networks.

- Emerging cable categories include draft specifications for Categories 6 and 7. Manufacturers are working ahead of the ratification of these categories to make cabling that will either meet or exceed the draft standards. The purpose of devising standards to maximize UTP and STP is to ensure continued improvement in the manufacturing process.

- In order to assist in cable management and termination, color-coding is used to separate signal travel paths. Color-coding usually associates a color with a pair, where one strand in the pair has a solid color and the other has the same color striped with white. Wiring schemes for color-coding twisted-pair are standardized by TIA/EIA standards 568-A and 568-B.

5.9   Key Terms

American Wire Gauge (AWG): The standard for cabling dealing with cable diameter and electrical resistance.

cable jacket: The surrounding insulator of a cable. It is the first line of a cable's defense from the outer elements.

cable plant: The collection of bundled cabling inside a commercial building.

cancellation effect: The process by which the magnetic fields of one wire are crossed with the fields of another, limiting the exposure of any one wire to the other.

Category 1: Twisted-pair cabling used exclusively for voice-based communications. It can support transmission speeds up to 1 Mbps.

Category 2: A type of twisted-pair cabling used in older IBM networks. It can support transmission speeds up to 4 Mbps.

Category 3: Formerly the minimum standard for voice and data networks. A type of twisted-pair cabling that supports transmission speeds up to 10 Mbps.

Category 4: A type of twisted-pair cabling used to support the 16-Mbps Token Ring networks.

Category 5: The most commonly installed medium. A type of twisted-pair cabling designed to support Fast Ethernet and other networks.

Category 5e: Enhanced Category 5 cabling. A recent standard designed to support all high-speed voice and data networks.

Category 6: A forthcoming standard of cabling designed to support Gigabit Ethernet and other technologies over greater distances.

coaxial cable: A type of electrical cabling consisting of a single solid or stranded core surrounded by shielding and inner and outer insulation.

core: The inner conductive element of a cable.

drawing: A process of wire creation in which the metal is pulled under tension to form a wire.

elastomers: Insulation materials that are plastic and very flexible.

electromagnetic interference (EMI): Any electromagnetic disturbance that disrupts, impedes, or degrades the electromagnetic field of another device by being in close proximity with it.

extrusion: A process of wire spinning in which the wire is stretched to form a thin strand through pushing.

fluoropolymers: Plenum-rated materials that are required for fire-rated insulators.

insulation: The outer material that makes up the cable jacket.

plenum-rated cable: Contains a wire coating that burns at a much higher temperature and emits fewer fumes than non-plenum–rated cabling. The NEC requires that plenum-rated cabling be used in plenums, which is the space above the ceiling tiles and below the next story into which two or more air ducts run.

polyvinyl chloride (PVC): A common thermoplastic insulator.

registered jack (RJ): A connector type used with twisted-pair cabling.

screened twisted-pair (ScTP): A type of twisted-pair cabling that contains an outer screen or shield.

shielded twisted-pair (STP): A type of twisted-pair cabling that uses inner and outer shielding for protection against electromagnetic interference.

solid core: A core consisting of a dense single wire.

stranded core: Tiny fibers of copper packed together to form a single conductive core.

T568A: The first of two ANSI/TIA/EIA standards for twisted-pair color-coding.

T568B: The second of two ANSI/TIA/EIA standards for twisted-pair color-coding. T568-B is most commonly used today.

thermoplastics: The most common type of insulation material.

twisted-pair cable: The predominating cable type in voice and baseband data networks. It is composed of pairs of twisted cabling.

UL 13: The safety standard for power-limited circuit cable.

UL 444: The safety standard for communications cable.

unshielded twisted-pair (UTP): The most common type of twisted-pair used for voice and data networks. UTP relies solely on the cancellation effect.

5.10   Challenge Questions

5.1   Which of the following are important elements of copper that make it a good metal for cabling? (Choose all that apply.)

a. Strength

b. Ductility

c. Malleability

d. Corrosion level

5.2   What are the characteristics associated with the following types of cables:

a. Category 1

b. Category 2

c. Category 3

d. Category 4

e. Category 5

f. Category 5e

g. Category 6

5.3   _____ wires are made up of very fine copper threads twisted together to create the wire. It makes the wire very flexible.

a. Stranded

b. Solid

c. Fiber-optic

d. Clear

5.4   _____ wires are made of a single piece of copper. They are more prone to breakage because they're not as flexible as stranded. They can hold a connector better than stranded.

a. Striped

b. Solid

c. Fiber-optic

d. Clear

5.5   What is the term used to describe all the wiring in a building?

5.6 What is the type of cable that should be used for high-temperature applications?

5.7 Which categories of cabling are no longer in the standards?

a. 1

b. 2

c. 4

d. All of the above

e. None of the above

5.8 What does AWG stand for and why do we care about it when we talk about cabling?

5.9 In a UTP pair of wires, which wire is the tip wire and which is the ring wire?

5.10 Which of the following are the colors used for the TIA/EIA-568 standard?

a. Blue, red, yellow, green

b. Black, blue, white, orange

c. Red, blue, orange, green

d. Orange, blue, green, brown

5.11 What is grounding? How do you typically ground a cable/wire?

5.12 Twisted-pair cabling is terminated using modular connectors known as _____ connectors.

a. RG

b. RJ

c. BNC

d. AUI

5.13 The most common type of twisted-pair cabling is

_____

a. STP

b. ScTP

c. STU

d. UTP

5.14 UTP relies solely on the principles of _____ to prevent interference.

   a. surge suppression

   b. surge protection

   c. EMI

   d. cancellation

5.15 _STP is the most common type of cable found in IBM Token
Ring networks.

   a. 200-ohm

   b. 100-ohm

   c. 150-ohm

   d. 50-ohm

5.11   Challenge Exercises

Challenge Exercise 5.1

In this exercise, you inspect the interior of a length of UTP cabling. The
instructor or lab assistant will provide several 1-foot samples of various
types of cable. For each sample, you will use a wire stripper to remove the
outer sheathing at one end of the cable so that the construction of the
cables can be examined. To complete this exercise, you need a length of
twisted-pair cabling and a cable-stripping tool or a pair of scissors.

Notice that there is a minimum and maximum cutting edge on the cable-
stripping tool. Use the minimum cutting edge to insure that the conductors
are not nicked. Make sure a maximum of two 360° turns are used with the
cable-stripping tool to prevent nicking the conductors. Upon doing so,
clearly inspect each cable and answer the following questions:

5.1.1   What is the marking on the outer sheathing of this cable?

5.1.2   How does this cable differ from other categories of UTP cabling?

5.1.3   How many layers of shielding does it have?

5.1.4   How many strands of copper are within each wire?

Challenge Exercise 5.2

In this exercise, you test your knowledge of cabling categories, AWG diameters, and uses. Complete the following chart:

Category	AWG	Uses
1		Telephones, no data
2		4-MB IBM Token Ring networks
3	24	
4		16-MB Token Ring networks
5	24	
5e		10/100-MB Ethernet networks
6	24	

Challenge Exercise 5.3

In this exercise, you learn to create and test a network patch cable. Your instructor will guide you through the process of cutting the cable, aligning the pins, and finally crimping the cable to the connector. You will then test your cable to verify that it can be used as a network patch cable. To complete this exercise, you need a length of twisted-pair cabling, a connector (e.g., RJ-45), a pair of scissors, a crimping tool, and a cable tester.

TIA/EIA-568-A/568-B and AT&T 258A define the wiring standards and allow for two different wiring color codes, as listed in Table 5.4.

5.3.1 For the male cable connector, hold the cable in one hand with the connector pointing up and the cable dangling down.

5.3.2 Turn the connector around so that the flat side is toward you and the plastic tang is away from you. Pin 1 is to your left.

NOTE There is no specified color-coding for cables under 10BaseT. Color-coding for wiring under the TIA/EIA-568-B standard is given for use with color-banded twisted pairs. When using solid color wires, follow the appropriate wiring diagrams.

5.3.3 It is important that wiring be done using twisted pairs. Pins 1 and 2 should be wired from one twisted pair, and pins 3 and 6 should be wired from a second twisted pair. Insert the wires into the connector.

TABLE 5.4   Wiring Color Codes

Pin #	TIA/EIA 568-A and AT&T 258-A	TIA/EIA 568-B
1	White/Green	White/Orange
2	Green/White	Orange/White
3	White/Orange	White/Green
4	Blue/White	Blue/White
5	White/Blue	White/Blue
6	Orange/White	Green/White
7	White/Brown	White/Brown
8	Brown/White	Brown/White

5.3.4   Crimp the connector onto the wires using a crimping tool.

5.3.5   Test the cable with a cable tester. Your instructor will assist you with this step.

5.12   Challenge Scenarios

Challenge Scenario 5.1

You are surveying a new commercial building purchased by your company. Cable plants exist within the building. You are trying to determine paths for cable runs and locations for wiring closets and cable plants. You want to minimize potential risks for heavy EMI and RFI as much as possible. What types of equipment and external factors will contribute to EMI and RFI?

Challenge Scenario 5.2

You are charged with finding a cost-effective solution for high-speed data bandwidth within a large multistory commercial building. The actual cable installation is at least one year away. You would like to use media that is durable, easy to install, and inexpensive. Using the Internet, research emerging technologies to determine if you have a cost-effective solution that will support high-speed bandwidth through the next 10 years.

CHAPTER 6

Copper Media: Coaxial Cabling

Learning Objectives

After reading this chapter you will be able to:

- Understand coaxial cabling construction and how it is different from twisted-pair cabling
- Identify the types of coaxial cabling
- Identify coaxial cabling categorization
- Describe coaxial cabling uses
- Describe Thinnet cabling and its uses in networks
- Describe Thicknet cabling and its uses in networks
- Understand the historical importance of coaxial cabling and its future

Coaxial cabling is the most frequently used multipurpose cable in residential audiovisual and entertainment systems, and will continue to be used for this purpose for some time in the future. Twisted-pair cabling, however, is commonly used for all new commercial and residential computer installations because it can support much higher bandwidths than coaxial.

This chapter examines the construction and properties of coaxial cabling, the different types and categories of coaxial cabling, the various connectors used with coaxial cables, and finally, the uses of coax.

6.1   Coaxial Cable History

Coaxial cable was invented in 1929. AT&T laid its first transcontinental coaxial transmission system in 1940, and by 1941, coaxial cable was being used commercially. For some time, coaxial cabling was the dominant choice for cable installations for networks and it is still pervasive in broadband networks, such as community antenna television (CATV), and Data Over Cable Service Interface Specifications (DOCSIS) data networks, and surveillance networks.

6.2   Coaxial Cable Construction

Coaxial is the name of both a general category of solid-shielded core cabling and a specific type of cabling. The name describes the cable's appearance. When a piece of coaxial cabling is cut, you see a copper or metal conductor core, surrounded by a layer of insulation, then a layer or braid or shielding, and finally an outer insulation jacket. All the layers are built up around the copper wire, which forms the central axis. Figure 6.1 shows a piece of coaxial cable stripped.

As noted in Chapter 5, "Copper Media: Twisted-Pair Cabling," the wires in coaxial cabling in gauge (AWG 22–24) are larger in diameter than twisted-pair cabling. As a result, coaxial cables are more durable than twisted-pair cabling. This greater durability, along with its design and common termination methods, make coaxial cabling the rational choice for controlled radio frequency (or land-line) technologies. Coaxial cabling is also commonly found in outdoor installations and for device links to antennas for wireless technologies. It can carry a wide range of frequencies, from the very low to the ultra-high (UHF). Wide-area data networks use coaxial

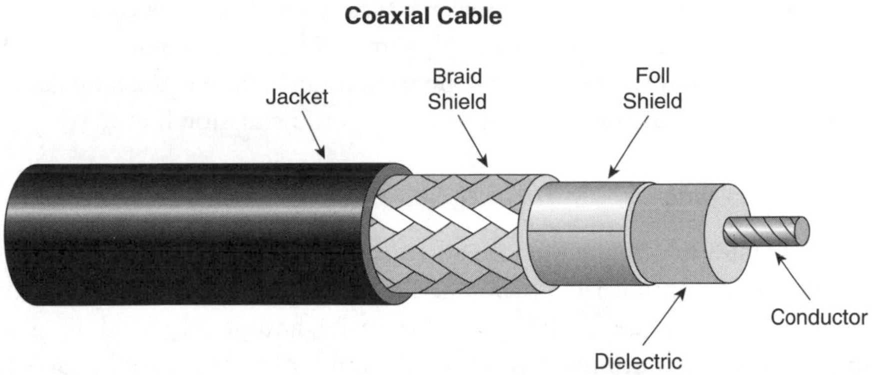

Coaxial Cable

Figure 6.1
Stripped coaxial cable

cabling, as do VHF and UHF broadcast and CATV networks. Closed-circuit and other analog and digital video are likely to use coaxial cabling. Some modern digital voice and data trunk communications systems—such as T-3—use coaxial.

Coaxial cabling has many design variations; however, all coaxial cabling includes a core conductor, dielectric material, and shielding.

6.2.1 Conductor

The core of coaxial cable is made up of solid or stranded wires, depending on the cable's intended use. Cabling that is to be used over a long distance usually has a solid core, because that reduces attenuation. However, a solid-core cable is less flexible than stranded-core cables, so the latter are more suitable for voice and data network installations, although they cannot be used over the same distances as solid-core cabling.

A cable's core is often referred to as copper, but only premium cabling actually has copper cores. To keep costs down, cable manufacturers are using less expensive, more plentiful metals in coaxial cabling, such as steel and aluminum, and covering them with copper. This less expensive cabling is often perceived as cheap as to both cost and quality.

6.2.2 Dielectric Material

The material that surrounds a cable's core is called the **dielectric material**, which must have a stable dielectric constant and strong resistance. The insulating material is usually a plastic or foam-like material used as an insulator, most commonly polyethylene. Most coaxial network cabling uses foam or solid polyethylene, although some use Teflon or polystyrene. A dielectric material is an excellent support substance for electrostatic fields, and is used in capacitors and radio-frequency transmission lines.

6.2.3   Shielding

Coaxial cables vary in the ways that their cores are shielded. Braided wire is the common approach to shielding, although foil is also used either in place of or alongside of the braided wire for additional shielding. The material used for shielding depends on the range of frequencies the cable is to carry. Braiding is a better shield for lower frequency transmissions; higher frequencies are better shielded by foil. More expensive coaxial cabling might use metal tubing, because the tubing provides the best possible shielding while dramatically reducing attenuation.

6.2.4   Insulation Jacket

A black insulation jacket covers most coaxial cable, although jackets also can be beige, gray, or yellow. Most coaxial cabling jackets are made from polyvinyl chloride (PVC), but a plenum-rated jacket is used when fire safety codes require it.

6.3   Coaxial Cable Properties

As is the case with twisted-pair, coaxial's ability to transmit information is a function more of the cable's physics than its physical arrangement. Coaxial cable can carry signals using parallel conductors that develop fields between themselves using either two specific conductors or the space between shield and conductor. This controlled relationship, involving specific engineering of every property involved within the cable, allows the transmission of information signals as radio waves following the path of the wires.

The physical shapes of the cable alone can control electrical effects within the cable. The variables for control are:

- **Wire spacing:** The diameter of the wire diameter as well as the spacing between conductors and shields must be carefully arranged and controlled.

- **Insulation quality:** Before the use of internal insulation, air was the only spacing between conductors and shields. This kind of coaxial cabling (often referred to as twin-lead) was found in less modern radio frequency cable; TV antenna cable used this type of coaxial cabling, too. Insulation quality has advanced over the years, and the type of insulation varies to comply with safety codes and regulations.

- **Climate:** Electricity is notably affected by the external elements such as temperature and humidity, which has lead to the use of insulation to space between shields and conductors.

6.4 Coaxial Cabling Types

According to official standards, any cable that has two conductive components sharing a common axis is a coaxial cable. The design of coaxial cables varies and the variants are all categorized as *coaxial cabling*. These designs usually have different names to easily tell them apart.

Coaxial cables comply completely with the basic design described in the previous section, and contain an inner conductor surrounded by a dielectric material. A conductor that also acts as a shield covers the dielectric. A colored protective jacket covers the outer conductor and acts as insulation. Figure 6.2 illustrates a coaxial cable.

6.4.1 Dual-Shielded

Dual-shielded coaxial cable has two shields covering the dielectric insulator. The additional layer provides extra shielding and more resistance to physical damage. It also safeguards against outside interference sources and attenuation. Figure 6.3 shows the design of dual-shielded coaxial cable.

Figure 6.2
Coaxial cable

Figure 6.3

Dual-shielded coaxial cable

Figure 6.4

Twinaxial cable

6.4.2 Twinaxial

Twinaxial cable contains two insulated conductors that either run in parallel or are twisted together, inside a common shield and insulator jacket. The insulation is different from that used in twisted pair, and resembles the "filler" used in standard coaxial cable. Figure 6.4 illustrates a twinaxial cable. IBM uses this type of cable with its AS/400 midrange computers.

6.4.3 Triaxial

Triaxial cabling has a single core, like standard coaxial cabling, and two shields like dual-shielded coaxial. The major differences are in its application and function. Triaxial cable uses both the inner conductor and the inner shield in transmitting data; the outer shield provides ground potential. Triaxial cable is used primarily with video cameras and transmission equipment.

6.4.4 Multi-Cable

Multi-cables are custom-made coaxial cable bundles that encompass a range of coaxial cable types and style variances based on a customer's requirements. The cables are shielded and bundled with a common insulation jacket. A drawback to multi-cables is that they can be difficult to install and run because of the thickness and loss of flexibility in the bundle.

6.5   Coaxial Cable Categorization

Coaxial cable varies as to its impedance, diameter, shielding, temperature rating, and application. Modern coaxial cable chiefly comes in varieties with 50-ohm, 75-ohm, and 95-ohm impedances, which are the most common impedances used by radio, networking, and video. Fifty-ohm is used mostly in networking and 75-ohm cable is used with CATV networks. ARCnet, an outdated data network technology, used 93-ohm coaxial cable. Twinaxial cabling comes with impedance variances of 78 ohms, 98 ohms, 100 ohms, and 124 ohms.

Coaxial cabling is categorized by an RG grade, the different designations of which apply to all types of coaxial cable. RG, or Radio Guide, designations originated in older military standards for secure radio frequency communication. Most cables today are referred to as RG-# "type," which means they don't strictly conform to the original Radio Guide specifications. RG designations today are so nonspecific that the cable's RG number gives no assurance as to how the cable will perform. The acronym RG now stands for **Radio-frequency Government**.

The most common RG designations are RG-6, RG-8, RG-11, RG-58, and RG-59. RG-58 and RG-8 are 50-ohm coaxes, used for radio transmission and in computer networks. RG-6, RG-59 and RG-11 are 75-ohm cable types used to transmit video. RG-6 and RG-59 are commonly used for home audiovisual systems, because their sizes are compatible with a variety of connectors. Both RG-6 and RG-59 are available in a number of forms, with different shields, jackets, dielectrics, and center conductor materials. Table 6.1 includes a more comprehensive listing of the RG types.

All the cabling types in Table 6.1 use a single braided shield except for RG-151 and RG-223, which have a double braid. The specification /U often follows the RG standard; /U means the cable corresponds to the universally adopted RG standard. When a cable grade standard has been revised, the /U will be preceded by an A, B, etc. (for example, RG-6/AU, RG-6/BU).

There is a specific connector and termination method for coaxial cable, depending on how the coax will be used. Among the more common are T-connectors, BNC connectors, and terminators found in data networks. Type F connectors are prevalent in RF/video/CATV networks. The connectors might look the same across technologies; however, like the cable, they

TABLE 6.1   RG Classifications, Characteristics, and Applications

Grade	Twinaxial or Coaxial	Dielectric Material	Impedance	Temperature Rating	Diameter (inches)	Uses
RG-6	Coaxial	Solid polyethylene	75	−40 to +85	0.332	Video, CATV (recommended for VHF/UHF/800 MHz commercial, CATV, and scanner use)
RG-8	Coaxial	Solid polyethylene	50	−40 to +85	0.289	Thicknet (trunk), CATV (better quality)
RG-11	Coaxial	Solid polyethylene	75	−40 to +85	0.405	Video, network (recommended for UHF/VHF antenna and RF use)
RG-22	Twinaxial	Solid polyethylene	95	−40 to +85	0.420	
RG-34	Coaxial	Solid polyethylene	75	−40 to +85	0.630	
RG-58	Coaxial	Solid polyethylene	50	−40 to +85	0.195	Network, Thinnet
RG-59	Coaxial	Solid polyethylene	75	−40 to +85	0.242	Video, network, CATV
RG-62	Coaxial	Air space Teflon	93	−40 to +80	0.242	Network, ARCnet
RG-71	Coaxial	Air space Teflon	93	−55 to +85	0.245	
RG-108	Twinaxial	Solid polyethylene	78	−40 to +85	0.235	Military
RG-122	Coaxial	Solid polyethylene	50	−40 to +85	0.160	
RG-133	Coaxial	Solid polyethylene	95	−40 to +85	0.405	
RG-164	Coaxial	Solid polyethylene	75	−40 to +85	0.870	
RG-165	Coaxial	Solid Teflon	50	−55 to +250	0.410	
RG-174	Coaxial	Solid Teflon	50	−40 to +85	0.110	Recommended for HF short runs, internal connections, portable RF use, and test cables
RG-178	Coaxial	Solid Teflon	50	−55 to +200	0.072	
RG-179	Coaxial	Solid Teflon	75	−55 to +200	0.100	

TABLE 6.1 (continued)

Grade	Twinaxial or Coaxial	Dielectric Material	Impedance	Temperature Rating	Diameter (inches)	Uses
RG-180	Coaxial	Solid Teflon	95	−55 to +200	0.141	
RG-187	Coaxial	Solid Teflon	75	−55 to +250	0.100	
RG-188	Coaxial	Solid Teflon	50	−55 to +200	0.098	
RG-195	Coaxial	Solid Teflon	95	−55 to +200	0.141	
RG-196	Coaxial	Solid Teflon	50	−55 to +230	0.072	
RG-212	Coaxial	Solid polyethylene	50	−40 to +85	0.332	
RG-213	Coaxial	Solid polyethylene	50	−40 to +85	0.405	Recommended for HF/VHF commercial, ham radio, and CB radio use
RG-214	Coaxial	Solid polyethylene	50	−40 to +85	0.425	
RG-216	Coaxial	Solid polyethylene	75	−40 to +85	0.425	
RG-217	Coaxial	Solid polyethylene	50	−40 to +85	0.545	
RG-218	Coaxial	Solid polyethylene	50	−40 to +85	0.870	
RG-223	Coaxial	Solid polyethylene	50	−40 to +85	0.212	TV studio use
RG-225	Coaxial	Solid polyethylene	50	−55 to +250	0.430	
RG-302	Coaxial	Solid Teflon	75	−55 to +200	0.202	
RG-303	Coaxial	Solid Teflon	50	−55 to +200	0.170	
RG-316	Coaxial	Solid Teflon	50	−55 to +200	0.098	
RG-365	Coaxial	Solid Teflon	50	−55 to +85	0.425	
RG-393	Coaxial	Solid Teflon	50	−55 to +200	0.390	
RG-400	Coaxial	Solid Teflon	50	−55 to +200	0.195	

differ in impedance and resistance. Do not make the mistake of assuming that because a cable and connector are capable of making a physical connection they are compatible. You cannot use a 50-ohm terminator to terminate 100-ohm cabling. They are not electrically compatible.

6.6   Coaxial Cable Connectors

The most commonly used types of connectors for coaxial cabling are those used in Ethernet networks and video systems. These connectors include the **BNC** and **attachment unit interface (AUI)**, commonly used in data networks and video, and the **Type F** connector, used for video, radio frequency, CATV, and audiovisual entertainment systems. Other coaxial connectors covered in the following sections include the DIN, Type N, SMA, TNC, and UHF connectors.

6.6.1   Attachment Unit Interface (AUI) Connector

The AUI connector is used with thick Ethernet (10Base5) architectures, most often to connect to a computer's network interface card (NIC). Figure 6.5 shows an AUI connector.

6.6.2   BNC

BNC can be spelled out in several different ways; however, we will use the most common name: British Naval Connector. BNC connectors are used with thin Ethernet (Thinnet, or 10Base2) and vary from application to application. All types of BNC connectors share the twist-and-lock method. Figure 6.6 shows an example of a BNC T-connector.

Figure 6.5

AUI connector

Figure 6.6

BNC T-connector

Figure 6.7
DIN connector

Thinnet networks also use these barrel BNC connectors for coupling Thinnet cable links. BNC T-connectors are used within a network bus topology to connect network nodes, devices, and computers to BNC cable links.

6.6.3   Deutsche Industrie Norm (DIN)

The **Deutsche Industrie Norm (DIN)** is an interface originally developed in Germany in the 1960s for high-performance military applications and later adopted for commercial applications in analog cellular systems. The 7/16-inch DIN connector is still in wide use in European land mobile radio applications. Figure 6.7 illustrates a DIN connector.

6.6.4   Type F

Type F connectors are threaded, and typically used with consumer audiovisual electronics such as televisions and VCRs. This connector can be used for signals as high as 1 GHz. Type F connectors are inexpensive because the pin of the connector is actually the center conductor. Figure 6.8 illustrates a Type F connector.

6.6.5   Type N

Type N connectors were developed at AT&T soon after World War II ended and are one of the oldest coaxial cable connectors. Type N connectors are

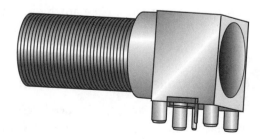

Figure 6.8
Type F connector

Figure 6.9

Type N connector

Figure 6.10

SMA connector

waterproof, and have good signal stability and low attenuation through 11 GHz. They perform reasonably well through a wide frequency range of power (300 watts through 1 GHz), and use a threaded interface connection style. Figure 6.9 illustrates a Type N connector.

6.6.6   Subminiature A (SMA)

SubMiniature A (SMA) connectors were developed in the mid 1960s primarily for semi-rigid small diameter metal-jacketed cable. SMA connectors are quite small and have an extended frequency range. Figure 6.10 illustrates a SMA connector.

6.6.7   Threaded Neill Concelman (TNC)

TNC connectors are an enhanced version of the BNC, with a threaded interface rather than a twist crimp lock. The TNC is smaller than a Type N

Figure 6.11
TNC connector

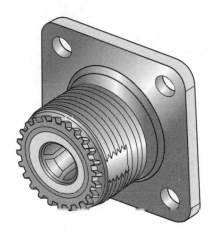

Figure 6.12
UHF connector

connector and has a commensurate reduction in power handling capacity. Figure 6.11 illustrates a TNC connector.

6.6.8　Ultra High Frequency (UHF)

The old industry standby, the UHF connector, was the first connector developed for RF communications above 50 MHz. It was developed during World War II, when 100 MHz was considered UHF.

The UHF connector is a fairly inexpensive, multipurpose screw-on connector that is not exactly 50 ohms. It is used in places in which there is no moisture, and is implemented only with applications that use moderate power, and rarely used for frequencies above 300 MHz. The UHF male connector is also called a PL-259. Figure 6.12 illustrates a UHF connector.

Table 6.2 shows comparisons of basic characteristics among common coaxial cable connectors.

TABLE 6.2   Coaxial Connector Series and their Typical Specifications

Series	Frequency (GHz)	Power (watts*)	Diameters (inches)	Cost
BNC	0–4	80	0.6	Low
F	0–0.9	N/A	0.44	Low
N	0–11	300	0.8	Moderate
SMA	0–18	100	0.4	Moderate
TNC	0–11	100	0.6	Low
UHF	0–0.3	500**	0.85	Low
DIN	0–7.5	2,500	1.25	High

*At 1,000 GHz (see text)

**At 300 MHz

6.7   Coaxial Cable Uses

As mentioned earlier in this chapter, the predominant use of coaxial cable today is for audiovisual and entertainment systems, such as cable television and VCRs. The following sections cover the data, video, and radio frequency uses of coaxial cable.

6.7.1   Data

The two types of coaxial cabling used for data transmission purposes, as mentioned in Chapter 1, are Thicknet and Thinnet, which are described as follows:

- **Thicknet:** An early form of coaxial cable used in data networking, officially known as 10Base5 Ethernet. Thicknet at one time was the standard for backbone cabling and interoffice links because it spans up to 500 meters per link between repeaters, which was quite a distance in its day. Thicknet has been replaced by fiber-optic as the recommended backbone cabling in Ethernet networks.   Thicknet has the largest diameter of all coaxial cabling, hence the name. It is easily recognizable by its bright yellow insulation jacket, and is sometimes called "frozen yellow garden hose." Compared to other media, Thicknet is stiff and non-pliable, which makes it difficult to install. A sharp bend radius can damage the cable. The cable is rigid and fragile, so it is usually placed in tile ceilings.

To attach network devices to Thicknet, cable technicians have to actually drill through the jacket and into the cable, and then push through the insulation with a pair of clamp-on probes to make contact with the central conductor. This two-toothed, penetrating attachment method has given rise to the name "vampire taps." The vampire taps are connected to the computers via the more flexible AUI, which connects to network devices via drop cables.

- **Thinnet:** Officially known as 10Base2 Ethernet, can span only 185 meters. However, it is more flexible and easier to install than Thicknet. Thinnet was nicknamed "cheapernet," because installation was quicker and the components and cabling were less costly.

Coaxial cabling does not have the modularity, flexibility, and cost-effective installation that twisted-pair cabling offers (see Chapter 5). Thick and thin Ethernet topologies are limited to a linear bus topology, in which the cable bus connects to the network card in each computer on a network via a T-connector. This means that unhooking the BNC connector on any one computer will bring down the entire network. In addition, each end of the cable bus requires a 50-ohm terminator to absorb signals that have reached the end of the line, preventing them from causing data collisions by reflecting back into the network. The terminators often become loose or even fall off, causing the network to crash.

6.7.2 Video

The wide range of frequencies used complicates transmitting video signals over cable. Coaxial cable is a popular choice for closed-circuit analog video, because all of the frequencies carried by analog video have to be treated equally. A wiring scheme that treats any of the bands differently would cause audio problems, distorted pictures, or both.

Older analog video systems used three or four wires. In the four-wire scheme, three of the wires carried the color signals (red, green, and blue) and the fourth wire was used for synchronization. In the three-wire scheme, the green cable is also used for synchronization. This is the highest quality of video signal transportation; it is very delicate to work with, and is used only in tele-production facilities and in some computer graphics environments. The wires must be equal in length and carefully terminated. Two-cable systems are basically similar to three- and four-wire schemes, but carry chrominance (color signals) on one cable and luminance (black and white) on another. Both cables carry synchronization information; if one wire is jammed, there is still a picture.

Composite Video

In **composite video**, chrominance, luminance, and synching information is carried across the same wire. Cable length mismatch is not as important with composite video, but it is still important to terminate and connect the cables properly. Composite video is the most common type of closed-circuit analog video.

Digital Video

Digital video signals are more robust than analog video signals, but the trade-off is an extremely high bandwidth. Installers need to be particularly careful when routing and terminating cables for digital video. There are also software issues involved to reduce the overhead that comes with traditional digital transmissions. Unlike packets on data networks, digital video stream packets cannot be retransmitted, as they use connection-less oriented technology. If packet transmission fails, the packet is dropped, resulting in a gap in video quality.

Other Video

The least demanding video signals seem to be in systems for security cameras and surveillance. However, if there is a possibility that a video signal from one of these systems will ever have to be made into a hard copy print, it is important to start with high-quality video equipment and cabling. High-quality cable is especially important in these applications because security and surveillance systems are usually mounted in hard-to-reach locations, which makes service calls difficult and expensive.

6.7.3   Radio Frequency (RF)

Coaxial cable is uniquely suitable for connecting **radio frequencies (RFs)** to antennas, and is used for almost all such applications. Wireless networks use coaxial for antennas and most cable TV systems use coaxial cable as the wiring system. The main trunk lines that run from a cable provider to neighborhood distribution boxes might be fiber-optic, but coaxial cables are more likely to be used between the distribution boxes and the end user.

The most common coaxial connectors are Type N, BNC, and PL-259. It is becoming less common to install RF connectors in the field because that often requires specialized adjustments to either the radio device or to the antenna, using special equipment and training.

The one case in which RF connectors are used with structured cable systems is in wireless networking. Wireless networking uses radio waves to carry signals, which means that each device must have an antenna, and a connector is necessary between the antenna and the antenna cable. However, most wireless networking equipment comes with pre-packaged antennas and antenna cable, which makes it unlikely that an installer will be called upon to terminate the cable.

Generally speaking, you need to take the same care with RF cables as you do in routing other wires. That is, you need to keep them away from sources of interference, and avoid mechanically forcing or damaging the cable. However, an RF cable such as an antenna wire in itself can be a potential interference source, which means it should be kept separate from other data, video, and communications wiring, wherever possible.

6.8   Practicality and Future of Coaxial Cable

The use of coaxial cabling continues to decline, partially due to the care that must go into its installation. Cable installers have to take extra precautions when installing coaxial cabling because the slightest pinch, compression, bend, kink, or tear alters the physical properties of the cable and causes signal flow problems. The weakest links in a coaxial cabling system are the terminators and connectors, and the cabling can pose additional electrical dangers for the cable installer while terminating and crimping connectors. Crimped connections are apt to be more stable than screw-on connectors, although the latter have become very popular with the rise of satellite and digital cable television services, which often double as high-speed broadband Internet connections.

Thus, the installation of coaxial cabling is considerably more difficult compared to twisted-pair cabling installation. Coaxial cabling, within a structured cabling system, is also more costly in terms of both labor and materials to install. For these reasons, there is an ongoing movement within the standards community to remove coaxial cabling from most recommended copper-based standards.

6.9   Chapter Summary

- Coaxial is the name of both a general category of solid shielded core cabling and a specific type of cabling. The name describes the cable's appearance. Coaxial cabling has many design variations; however, all coaxial cabling contains a copper or metal conductor core, surrounded by a layer of insulation, then a layer or braid or shielding, and finally an outer insulation jacket.

- Coaxial cable properties vary and include: wire spacing (the diameter of the wire diameter as well as the spacing between conductors and shields), insulation quality, and the use of shields and conductors to control climate.

- Coaxial cable types include dual-shielded, twinaxial, triaxial, and multi-cable.

- Coaxial cable varies as to its impedance, diameter, shielding, temperature rating, and application. Modern coaxial cable chiefly comes in varieties with 50-ohm, 75-ohm, and 95-ohm impedances, which are the most common impedances used by radio, networking, and video. Coaxial cabling is currently categorized by an RG grade. The most common RG designations are RG-6, RG-8, RG-11, RG-58, and RG-59.

- The most commonly used types of connectors for coaxial cabling are those used in Ethernet networks and video systems. These connectors include the BNC and AUI, which are commonly used in data networks and video, and the Type F connector used for video, RF, CATV, and audiovisual entertainment systems. Other coaxial connectors include the DIN, Type N, SMA, TNC, and UHF connectors.

- The predominant use of coaxial cable today is for audiovisual and entertainment systems, such as cable television and VCRs.

6.10   Key Terms

attachment unit interface (AUI): A type of coaxial cable connector used in Thicknet Ethernet (10Base5) networks.

British Naval Connector (BNC): A type of coaxial cable connector used in video and Thinnet Ethernet (10Base2) networks.

composite video: A type of video where the chrominance, luminance, and synching information is all carried across the same wire.

Deutsche Industrie Norm (DIN): A type of coaxial connector used for European land mobile radio applications and analog cellular systems.

dielectric material: Foam or plastic insulation that spaces between the central core and the shielding inside of coaxial cabling.

dual-shielded coaxial cable: A type of coaxial cable that differs from standard coaxial cable in that it has two shields covering the dielectric insulator.

polyethylene: The most common insulation material used in coaxial cable insulation jackets.

radio frequency (RF): 1. Regarding radio wave propogation, refers to any frequency within the electromagnetic spectrum. Applying an RF current to an antenna creates an electromagnetic field through which communications can take place. 2. Another term for coaxial cabling. RF cable designations usually apply to coaxial cabling used in RF applications.

Radio-frequency Government (RG): The designations used for coaxial cabling that were derived from older military standards; formerly referred to as Radio Guide.

Subminiature A (SMA): A type of coaxial connector designed for semirigid, small diameter metal jacketed cable.

Thicknet: The type of coaxial cabling used in 10Base5 Ethernet networks. This was once the recommended cabling for network backbones. Thicknet is easily known by its bright yellow insulation jacket.

Thinnet: The type of coaxial cabling used in 10Base2 Ethernet networks. Thinnet cabling is more flexible than Thicknet, but does not travel as great a distance.

Threaded Neill Concelman (TNC): A type of coaxial cable connector that is simply an enhanced version of the BNC connector. Instead of a twist crimp-lock, the interface is threaded.

triaxial: A type of coaxial cabling that has a single core like standard coaxial cabling and has two shields like dual-shielded coaxial. It uses both the inner conductor and the inner shield to transmit information while the outer shield provides ground potential.

twinaxial: A type of coaxial cabling that contains two insulated conductors either running in parallel or twisted together having a common shield and insulator jacket.

Type F: A common coaxial connector used in video, radio-frequency, and audio-visual entertainment systems.

Type N: One of the oldest coaxial cable connectors in use. Type N connectors are waterproof and have good signal stability.

Ultra High Frequency (UHF): A very common type of coaxial connector developed for RF communications above 50 MHz.

wire spacing: The carefully controlled spacing between conductors and shields.

6.11   Challenge Questions

6.1   For years, _____ dominated the data network industry as the choice for cable installations.

6.2   Coaxial cabling was first developed in _____.

  a. 1876

  b. 1890

  c. 1929

  d. 1948

6.3   The wires contained in twisted pair cabling are _____ in diameter than coaxial cabling.

  a. smaller

  b. larger

  c. the same size

6.4   Coaxial cabling's design and common termination methods make it the logical choice for controlled _____ technologies.

  a. cellular

  b. fiber-optic

  c. RF

  d. infrared

6.5   The data bandwidth supported by coaxial cable supports only _____ Mbps in Ethernet data networks.

  a. 1

b. 10

c. 100

d. 1,000

6.6   The core material in coaxial cabling surrounded by the
_____ material made up of a plastic or foam-like
material.

a. conductive

b. insulator

c. dielectric

d. jacket

6.7   The most common material used for the dielectric material is
_____.

a. Teflon

b. Polyethylene

c. Kevlar

d. None of the above

6.8   More expensive coaxial cabling may use metal tubing as its
_____.

a. insulator

b. core

c. dielectric

d. shield

6.9   The most common material used with coaxial cabling is
_____, although plenum-rated insulated cabling is
used when fire safety codes require it.

a. Kevlar

b. fiberglass

c. Teflon

d. PVC

6.10 _____ coaxial cable differs from standard coaxial cable in that it has two shields covering the dielectric insulator.

a. Twinaxial

b. Dual-shielded

c. Multi-cable

d. Twisted pair

6.11 _____ cable differs from standard coaxial cabling in that it contains two insulated conductors either running in parallel or twisted together, and have a common shield and insulator jacket.

a. Twinaxial

b. Dual-shielded

c. Multi-cable

d. Twisted-pair

6.12 _____ cable uses both the inner conductor and the inner shield to transmit information while the outer shield provides ground potential.

a. Twinaxial

b. Dual-shielded

c. Multi-cable

d. Triaxial

6.13 Coaxial cable is able to carry signals through the use of _____ conductors.

a. twisted

b. parallel

c. polarized

d. None of the above

6.14 Coaxial cable is categorized according to its _____.

a. impedance

b. diameter

c. shielding

d. temperature rating

e. All of the above

6.15 The various _____ designations apply to all of the various types of coaxial cable.

a. RF

b. RG

c. EIA

d. RJ

6.16 _____ is a grade of coaxial cabling recommended for VHF/UHF/800-MHz commercial, CATV, and scanner use.

a. RG-253

b. RJ-45

c. RJ-11

d. RG-6

6.17 _____ is used with Thinnet cabling.

a. RG-6

b. RG-11

c. RG-58

d. RG-253

6.18 Video and CATV and home satellites use _____ coaxial cabling.

a. RG-34

b. RG-11

c. RG-58

d. RG-59

6.19 Coaxial cabling can pose some additional _____ for the cable installer when terminating and crimping connectors.

a. benefits

b. electrical hazards

c. ease of use

6.20 The _____ connector is used for video, RF, CATV, and audiovisual entertainment systems.

a. DIN

b. Type F

c. AUI

d. BNC

6.21 BNC _____ are also used to connect network nodes, devices, and computers to coaxial cable links within a network bus topology.

a. vampire taps

b. terminators

c. T-connectors

d. jacks

6.22 The _____ connectors are one of the oldest coaxial cable connectors in use.

a. Type F

b. BNC

c. Type N

d. AUI

6.23 The _____ connector is the old industry standby and was really the first connector developed for RF communications above 50 MHz.

a. BNC

b. Type F

c. Type N

d. UHF

6.24 Officially, Thicknet is known as _____ Ethernet.

a. 10BaseT

b. 10Base2

c. 10Base5

d. 10Broad36

6.25 Thicknet is easily known by its brightly colored
_____ insulation jacket.

a. beige

b. black

c. white

d. yellow

6.12 Challenge Exercises

Challenge Exercise 6.1

In this exercise, you install and crimp BNC connectors. Installing BNC connectors onto Thinnet cabling is more challenging than other connector types. This exercise focuses on using crimp connectors rather than soldered connectors, as they are the most popular.

Your instructor will provide you with the Thinnet cabling link, connectors, a crimper for a connector pin, a crimper for a connector, and a BNC tester.

6.1.1 Using a very sharp knife or reliable cable stripper, make a cut into the cable about 5/8 inches or 15 mm from the cable end. You must make sure to cut deep enough to penetrate the outer insulation, but not too deep as to cut or nick the braid. This may take several attempts. Cut completely around the insulation and then slide off the outer ring of insulation.

6.1.2 This next part is tricky. You need to loosen the braid to where you can cut it with cutters. You then need to trim it to where it is 10 mm from the end of the wire. It is important that no loose strands of braid or wire stick out.

6.1.3 Cut the center dielectric material to expose the center core about 5 mm or 3/16 inches from the end of the wire.

6.1.4 Slide the crimp sleeve over the end of the cable.

6.1.5 The center contact pin must go over the center conductor. Make sure that all of the strands of wire stay inside the hole. The wire should bottom out with the pin just above the center insulation. Crimp the center pin with the proper tool and check it for tightness.

6.1.6 Slide the connector body over the center pin, making sure that the braid slides over the mandrel completely. Push until the center pin bottoms out, with the contact pin flush at the connector end.

6.1.7 Slide the crimp sleeve over the braid. Crimp.

6.1.8 Check connections with the BNC coaxial tester. Your instructor will assist you in the proper use of the tester.

Challenge Exercise 6.2

In this exercise, you install Type F connectors. Type F connectors were designed for low cost and easy installation. To save cost and assembly time, there's no center contact pin; the wire of the coaxial cable center conductor serves as the contact.

Your instructor will provide you with RG-59/U cable, F connectors, a crimp tool, continuity tester, and ohmmeter.

6.2.1 Using a very sharp knife or reliable cable stripper, make a cut into the cable about 11 mm from the cable end. You must make sure to cut deep enough to penetrate the outer insulation, but not too deep as to cut or nick the braid. This may take several attempts. Cut completely around the insulation and then slide off the outer ring of insulation.

6.2.2 Fold the braid back over the cable jacket. Cut the braid short about 3 mm or 1/8 inches.

6.2.3 Using a knife, cut through the foil/braid and dielectric insulation to expose a conductor length of about 8 mm. It is important not to cut or nick the copper jacket of the center conductor.

6.2.4 Slide the one-piece connector over the jacket. Press it down or trim it so that corner of the cut is smooth and provides a good lead-in for the connector mandrel. Make sure that no metal pieces are present that might cause an electrical short.

6.2.5 Install the connector mandrel over the foil and underneath the braid. The end of the cut should be approximately even with the inner shoulder surface of connector, and the center wire should project out about 1 to 2 mm past the end of the connector.

6.2.6 Check cable with continuity tester or ohmmeter. Your instructor will assist you in the proper use of the ohmmeter.

6.13   Challenge Scenarios

Challenge Scenario 6.1

We have learned in this chapter that coaxial cable can fail for many reasons. You are about to embark on a cabling project that will require the use of coaxial cabling. You will be using broadband CATV for Internet access and you will be transmitting closed circuit video throughout the entire building. You are devising a plan of action to prevent the various failure modes that can come from coaxial cabling.

For each failure mode, list the root causes and the ways to prevent them:

- Installation failure
- Environmental failure

Challenge Scenario 6.2

What does BNC actually stand for? It may seem like a simple question, but it does not have a simple answer. Many resources document it incorrectly, but there are actually multiple correct explanations. Using the Internet and other research sources, find out what BNC stands for in regard to coaxial cabling standards.

CHAPTER 7

Fiber-Optic Media

Learning Objectives

After reading this chapter you will be able to:

- Understand fiber-optic cabling and how it differs from copper cabling
- Understand the theory behind fiber-optic media
- Understand how fiber-optic cables are manufactured
- Discuss enclosures and patch panels
- Identify fiber-optic cabling connector types
- Understand the increased bandwidth capacity of fiber-optic cabling
- Understand the advantages fiber-optic cabling provides over copper cabling

In recent years, fiber-optic cabling has been replacing copper cable as the media of choice for backbone network connectivity for several reasons. Fiber-optic cable has an expanded bandwidth capacity to reach longer distances, and it resists outside interferences such as electromagnetic noise.

A fiber-optic cable system is a similar concept to the copper-based system. The most notable difference is that the fiber-optic system utilizes light waves instead of electrical pulses to carry data. At one end of the cable is either a light emitting diode (LED) or a laser, which modulates an electronic signal and transmits it down the length of a fiber-optic cable to a receiver at the other end, where it is reconverted to an electrical signal.

Fiber-optic cabling has been in commercial use since the early 1970s. Originally, fiber-optic cable was designed to provide additional bandwidth capacity for the major telephone companies; it has been adopted for use by data communication carriers as well as most major businesses. Cable television companies are also taking advantage of the increased capacity of fiber-optic cable to deliver more channels to their viewers, as well as on-demand programming. Medical research has also led to additional advances in fiber-optic technology.

7.1   Fiber-Optic Cable Basics

Fiber-optic cable is made up of three main components (Figure 7.1). The main portion of the cable is called the **core**. The core is usually made of a very high-quality glass and is concentrically surrounded by another glass tube called the **cladding**, which is responsible for reflecting the light back into the core as it travels the distance of the cable. The core of a fiber-optic cable is approximately the width of a human hair, or even smaller for some types of fiber-optic cable. It is fairly fragile, so it is necessary to provide it with some protection. Surrounding the concentric pair of core and cladding is a protective **buffer coating**, which is responsible for adding strength to the core and cladding.

Pulling long lengths of fiber-optic cable puts a lot of stress on the core and cladding. In some cases, the protective buffer coating is supported by a layer of Kevlar fibers to make the cable stronger while it is being pulled. The main issue with fiber-optic cable is the inability for early glass to be of a sufficiently high quality to minimize distortion. To minimize distortion in the data signal, it is necessary to use glass that is extremely pure. Imagine

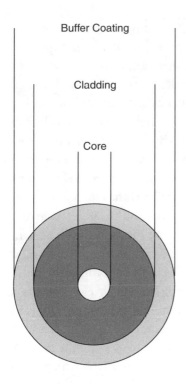

Figure 7.1
Fiber-optic cable, end view

looking through a plate of glass that is cloudy. If the glass is only 1/4-inch thick, it would not present too much of a problem because the length (or in this case, the width) of the glass from one side to the other is relatively short. However, if the glass is 12 inches thick, it would be very difficult to see through. Now, imagine a piece of glass that is several thousand feet thick, and you can begin to understand why it is necessary to use glass that is very pure. Even the smallest impurities can create an obstruction when they are stacked on top of each other. When the feasibility of using fiber-optic cable to transmit data was first discussed, it was decided that in order to be usable, the cable would have to be pure enough to retain 1% of the original signal strength over a distance of 1 kilometer. Put another way, the **attenuation** (or signal loss) of the signal had to be less than 20 decibels per kilometer (20 dB/km).

7.1.1   Refraction and Reflection

The main principles behind how fiber-optic cables function is reflection and refraction. To help you understand the basic concepts of refraction and reflection, picture yourself on a fishing pier over a small pond. If you look straight down, you can see that the poles supporting the dock appear to bend as they enter the water. This is due to the principle of **refraction**. Light changes speed as it moves between dissimilar elements (the air and water), making the poles appear to bend. You can also clearly see fish swimming around the poles. In fact, unless the water is extremely deep or cloudy, you can see all the way to the bottom. However, if you look further out across the pond, you will notice you can no longer see as deeply into the water. In fact, the further out you look, only the images of the trees and sky can be seen on the surface of the water. This is due to **reflection**. The main difference between why you can see objects close to you and yet objects further away are harder to see is the angle at which you are viewing the objects. In other words, reflection is light bouncing at the junction of two dissimilar elements, such as air and water, and refraction is light bending at the junction of two dissimilar elements as it changes speed. Each element has a **refractive index**, or the amount of refraction that it allows to take place. In addition to the refractive index, the other main factor that determines how the light waves are viewed is the angle at which you are viewing the subject. A steeper angle (looking closer in) will result in a more refractive image, and a less steep angle (looking further out) will result in a more reflective image.

Fiber-optic cable uses these principals of reflection and refraction to transmit data. Light waves are guided down the core of the fiber-optic cable by being reflected throughout the length of the cable. The difference in the refractive index of the media between the core and cladding determine the cable's ability to reflect light. Controlling the angle at which the light reflects down the length of the core makes it possible to control how efficiently the light pulses will reach the receiving end.

7.1.2   Transmitters

Fiber-optic transmitters are predominantly LEDs or laser diodes (LDs), although some are vertical cavity surface emitting lasers (VCSELs). The transmitter operates much like a switch—turning on and off to correspond with the binary 1s and 0s that most data devices use to communicate. The transmitter is responsible for converting electrical signals to a light wave, although in some cases the transmitter is acting as a repeater to amplify and forward a light signal that it receives. When you consider the small size of the core of the fiber-optic cable (as small as 8 microns in some cases; a human hair is approximately 50 microns in diameter!), the importance of determining the exact angle the light wave is to travel becomes crucial. This is why the transmitter needs to be lined up exactly with the core of the fiber-optic cable to ensure maximum efficiency. The connector provides this assurance. (Connectors are discussed in the section titled "Fiber-Optic Connectors.") The transmitter must also be powerful enough to transmit the light wave to the receiver over great distances. Lasers transmit at a significantly higher power range than do LEDs, so in long-distance fiber-optic systems, lasers are the transmitter of choice.

! WARNING

Never look directly into a fiber-optic cable that is connected to a laser. Serious eye injury can occur. If you are not sure if it is laser or LED, don't take any chances.

7.1.3   Optical Receivers

Fiber-optic receivers also perform two functions: they must receive the light waves from the transmitter and then convert them into electrical signals. The receiver collects the light signal from the transmitter by using a **photodiode**. A photodiode is an electronic component that is sensitive to light. When light hits the photodiode to indicate the presence of a signal, the circuitry in the receiver converts this light signal into an electrical signal to be processed by other data devices on the network. Usually there is a transmitter and a receiver at each end of a fiber-optic cable so that full-duplex communication can take place. A minimum of two strands of fiber-optic cable are commonly used to communicate at full duplex, although in some cases, a single-fiber cable can be used for transmitting and receiving data. Time division multiplexing is incorporated to allocate the upstream and downstream bandwidth correctly.

7.2   Fiber-Optic Transmission

Fiber-optic transmission works in a similar fashion to copper delivery systems. There is a transmitter at one end of the cable that is either an LED or laser, which converts an electrical signal into a light pulse. There is an optical receiver at the other end of the cable that is used to convert the light pulses back to electrical signals. The light travels down the length of the fiber-optic cable by continuously reflecting back from the junction of the core and cladding using the principals of reflection and refraction. The material in the core has a different index of refraction than the material in

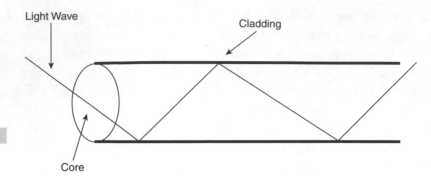

Figure 7.2

Reflection of light waves in fiber-optic cable

the cladding; the difference is what allows the light signal to reflect down the cable. Think of the junction between the core and cladding as a very thin, long mirror. Light reflects off the sides of this mirror at the same angle and at the same strength at which it arrived. In a perfect world this is referred to as **total internal reflection**, but in actuality there is always some light loss or attenuation. Fiber-optic attenuation will be discussed in more detail later in this chapter. As the light waves travel down the length of the cable, they will get reflected off of the junction of the core and cladding at the same angle as they were sent (Figure 7.2), because the refractive index of the cladding is lower than the refractive index of the core. Notice in Figure 7.3 how the signal is able to be reflected off minor curves in the cable; however, it is possible for the cable to be looped so severely that reflection is not possible. That is why there are recommendations regarding the size of a loop in the fiber cable. A loop that is too small or too tight would not per-

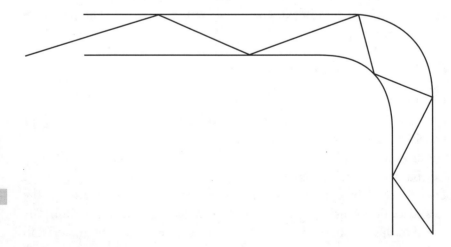

Figure 7.3

Reflection of light waves off of curve in cable

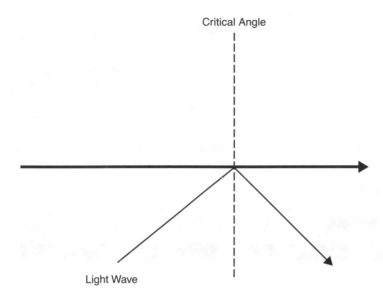

Figure 7.4
Critical angle

mit the signal to be reflected completely and would result in partial or even complete signal loss.

The angle that provides minimum signal to pass is referred to as the **critical angle**, as shown in Figure 7.4. Before the critical angle is reached, some light is reflected and some is refracted. After the critical angle all light is reflected. In addition to the critical angle the light is traveling, the wavelength of the light determines the angle at which the light will strike the cladding and bounce back, so different wavelengths will have different characteristics. Light waves are measured in nanometers (nm) and each different wavelength will travel at a different frequency. Component designers take advantage of these characteristics when they design their products. For example, light traveling at 1,550 nm travels in a straighter line (Figure 7.5) than light traveling at 850 nm, so the light wave on the 1,550 nm wavelength actually travels less distance than the 850 nm light wave travels. Also, because there is less diffusion at the receiving end, there is less light loss at the higher wavelengths. Common light loss levels per wavelength are listed in Table 7.1.

To minimize the impact of diffusion over further distances, it is necessary to periodically amplify and regenerate the signal utilizing repeaters.

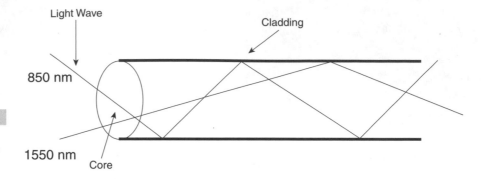

Figure 7.5

Higher-frequency light waves travel in straighter paths

TABLE 7.1   Light Loss Levels per Wavelength

Wavelength	dB Loss	Loss %
850 nm	4 to 6 dB/km	60 to 70
1,300 nm	3 to 4 dB/km	50 to 60
1,550 nm	0.5 dB/km	<10

7.2.1   Attenuation

In fiber optics, attenuation is defined as the loss of light as measured in decibels per kilometer (dB/km). Early adopters of fiber-optic technology determined the largest loss that could be tolerated was 20 dB over a 1-kilometer range. The two most common causes for attenuation are impurities in the glass and loss at fiber-optic connectors, including splices. If the glass were totally pure and there were no need to splice the fiber cable, attenuation would not occur. However, in the real world there is no such thing as perfect glass. To reach thousands of meters, as is usually necessary, fiber-optic cable needs to be spliced or connected (splices are required during repairs). Fiber-optic connectors are rated according to their dB loss and dB loss is cumulative. In other words, if your fiber-optic cable has a loss factor of 5 dB/km with two connectors that have 3 dB loss each, total loss for a 3-kilometer fiber-optic cable run would be 21 dB (5 dB/km × 3 km + 3 dB per end = 21 dB), which is just outside of the specifications for acceptable loss. Another side effect of attenuation is called **back reflection**, which can be caused by impurities in the glass or an improper splice or connector. If the reflective index is changed in the material by either impurities or an air gap caused by an improper splice, a portion of the signal can reflect back to the transmitter causing problems, especially if it is a laser transmitter.

7.2.2   Dispersion

Another factor that impacts total light loss is dispersion. As light waves travel away from their source, they fan out or disperse. Dispersion causes problems with data networks because it can cause data transmission errors. Because of this dispersion effect, it is necessary to amplify and repeat the signal periodically as it travels the length of the fiber-optic cable. However, as mentioned previously, signals over fiber-optic cable require less frequent amplification than signals over copper cable, thus making it a lower overall cost over longer distances.

7.2.3   Bandwidth

As mentioned in the introduction, one of the main advantages fiber-optic cable provides over copper cabling is increased bandwidth capacity. This is due in part to the law of physics stating that light travels faster than electricity. Also, because fiber-optic cable is significantly thinner than twisted pair or coaxial copper cable, more strands can be enclosed in the same size sheathing, thus providing additional carrying capacity. Light travels down the fiber-optic cable in distinct paths, or **modes**. If one ray of light takes path A and another ray takes path B, the ray that followed mode A will arrive first, since it actually had less distance to travel. The more modes of light there are in the cable, the more convoluted the light signal will be at the receiver. That is the basis for the explanation that single-mode cable has a higher bandwidth capability than multimode cable even though single-mode fiber-optic cable only has a single light wave traveling it at any given time. Modes of fiber-optic cable are discussed later in this chapter.

7.2.4   Multiplexing

Before multiplexing was invented, each circuit (copper or fiber-optic) required its own individual strand of cable. One thousand individual circuits would have required 1,000 individual links. Installing this many individual circuits would be cost prohibitive and would have slowed the growth of the communications industry if some better method had not been created. Multiplexing allows a single link to be shared by many circuits. The

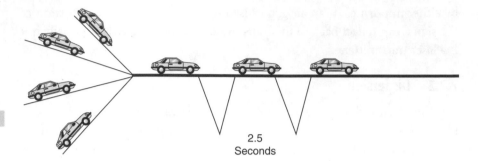

Figure 7.6

Four onramps to expressway

2.5
Seconds

two main types of multiplexing in common use are time division multiplexing (TDM) and wavelength division multiplexing (WDM). (We introduced multiplexing in Chapter 4, "Data Network Signals.")

To more fully understand TDM, think of four parallel onramps that lead to a one-lane expressway, with four parallel exit ramps one mile away from the entrance ramps (Figure 7.6). Let's imagine that, to control traffic, each of the onramps allows a car to enter every 10 seconds, so essentially the expressway is adding a new car every 2.5 seconds (10 seconds divided by 4 ramps). The cars are all traveling at the same speed, and on the one-mile stretch of the expressway the cars are spaced apart with 2.5 seconds between each one. The expressway does not need to support speeds of four times faster than normal to cope with the number of vehicles, since all cars will be moving at the same speed. In this example, we are simply taking advantage of the unused time on the expressway to fit in additional cars (transmit voice or data packets in communications networks). Essentially, TDM is sharing a link and each individual circuit is given its timeslot to transmit data.

In WDM, each individual circuit is assigned a unique wavelength over the link that does not interfere with the other circuits. Each wavelength travels its own unique path. To use our expressway analogy, WDM is using four lanes on the expressway, so each car still has a dedicated lane all to themselves. Again, we do not need to modify the speed (bandwidth) of the cars, we just need to devise a way to share the road between all the cars.

7.3   Types of Optical Fiber

Light waves travel down the core of a fiber-optic cable in a defined path or over multiple paths. These paths are referred to as modes, as shown in Figure 7.7. Light through these modes usually travels at different speeds, arriving at the receiver at different times according to each mode's unique characteristic. Some of the factors that influence the light speed through the modes are the wavelength of the light, the refractive index of the media, and the angle at which the light wave is traveling. If the modes of light all arrive at different times to the receiver, it is easy to sometimes lose the original signal, since the original signal was transmitted from a common LED (Figure 7.8).

7.3.1   Multimode

Multimode fiber has a core that is larger than single-mode fiber. The core of a multimode cable is usually either 50 or 62.5 microns in diameter, although 62.5 microns is in more widespread use. (Fifty-micron fiber is becoming a popular choice with VCSELs.) Multimode fiber was the first type of fiber-optic cable to be used in a commercial environment, and is still common today because of its low cost. Multimode fiber allows hundreds of modes of light to be transmitted simultaneously down the length of the fiber cable. Also, the larger diameter of the multimode core allows for lower cost LED transmitters to send several modes of light simultaneously. Multimode fiber is used to

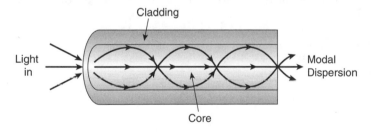

Figure 7.7

Light waves travel in modes through a fiber-optic cable

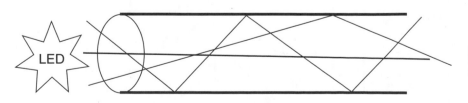

Figure 7.8

Common LED

transmit data over shorter distances than single-mode fiber, and is a lower-cost alternative to the higher-priced single mode fiber solution. In addition to the cable itself being less expensive, it is less costly to manufacture and operate LEDs and LED optical-receiving devices than it is to manufacture lasers.

NOTE Fifty micron and 62.5 micron cables are not compatible. You must either use a 50-micron or 62.5-micron cable when connecting multimode cables. Most fiber-optic cables have the core diameter printed on the cable sheath.

7.3.2   Single Mode

Single-mode fiber has a much thinner core than multimode fiber. The common diameter of a single-mode core is 8 to 9 microns (remember, a human hair is approximately 50 microns in diameter). It is important to note that the size of the core/cladding combination stays the same between single-mode and multimode cable; it is just the size of the core that is different. The common size for the core and cladding combined is 125 microns. Single-mode fiber can carry a light signal over a greater distance than multimode fiber, as discussed in Section 7.2.3, "Bandwidth." This is because single-mode fiber only needs to maintain integrity of a single mode or wave of light and more data can be encoded this way. The problem of multiple modes of light arriving at different times is not present with single-mode fiber; therefore, single-mode fiber is used for longer distances or higher-capacity requirements than multimode fiber.

Figure 7.9 compares single-mode to multimode glass fibers.

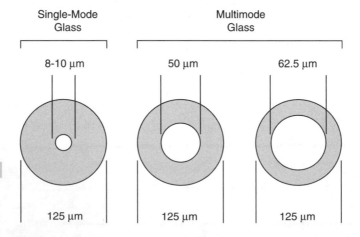

Figure 7.9

Single-mode and multimode glass fiber comparison

7.3.3   Glass Fiber

The main types of fiber core that we have been discussing so far are glass-based fiber optics. Recent innovations in fiber-optic technology have found new materials to use as the core; however, glass remains the material of choice for a majority of fiber-optic networks. Plastics (known as plastic optic fiber [POF]) also have been used as a replacement for glass fiber-optic as a material for the cladding, and in some cases even as a replacement for the core glass. However, POF has not been of much success in the telecommunications industry outside of control cables for industrial robotics, due to POF's limited bandwidth. Some of the other material used for core and cladding are outlined in the following sections.

7.3.4   Silica Fiber

Silica-based fiber-optic cables use silica instead of glass in the core and cladding. The main advantage silica fiber-optic material provides over glass is its ability to withstand higher temperatures, making it more suitable for higher power applications. Due to the higher level of impurities in silica, however, it is not usually recommended for use over long distances. Silica is also more expensive to manufacture and purchase, so it is generally reserved for specific high-temperature applications. At the time of this writing, silica fiber is used primarily in laboratory environments.

7.3.5   Liquid Core Fiber

Liquid core fiber-optic cables use the same principal as a decorative water fountain with illuminated streams of light that appear as if they are colored. We have all seen the type of water display where the colored water appears to dance and change colors as it finds its way back to the pool where it originated. The air around the streams of water is acting as a cladding, and a light that is applied to the base of the stream of water will travel the length of the "tube" of water. Because the refractive index of air is higher than glass, some light will escape the water tube that is formed by the high-pressure jets. This light that escapes presents itself as the different colors the streams of water seem to take on.

The refractive index of water is lower than that of glass, making it a very suitable material to be used as a core. Liquids can be made with virtually no impurities, making it an optimum material to be used as a fiber-optic core. Liquid core fiber-optic cable is more flexible than glass core fiber-optic cables. The drawback to using liquid core fiber is the extra cost involved in the manufacturing process, so most liquid core fiber-optic cables are of shorter lengths. Liquid core fiber-optics is finding use in the medical industry to provide high-quality light that can be directed by a surgeon during an operation.

7.4   Enclosure Systems and Patch Panels

As previously mentioned, fiber-optic cables have a very small core that can be easily damaged if not protected properly. Also, to conform to the minimum size of a fiber-optic loop and not violate the critical angle, we need to have a way to keep excess fiber-optic patch cables, as well as terminated building fiber, neat and protected from damage. Fiber-optic enclosures and patch panels allow the cable installer to protect the delicate fiber cable from damage, while still making it useable for the network administrator. A common device that is used as a fiber-optic cable enclosure is called a **Lightguide Interconnection Unit (LIU)**, as shown in Figure 7.10. The LIU provides a location to terminate individual fiber-optic strands into a patch panel, which will be discussed in the next section. An LIU is generally made of galvanized steel that is then powder-coated to provide durability. Most

Figure 7.10

Fiber-optic cables terminated in a LIU

Figure 7.11
Patch panel for fiber-optic cables

major LIU manufacturers make their devices 19 inches wide so they can be installed in a normal communications rack. If the LIU is to be located in an environment where there is a risk of moisture or corrosives, the LIU can be sealed with gaskets to make it virtually waterproof. Most LIUs have swing-out trays in the front and the back to provide easy access to the patch panel located inside. Also, most LIUs provide a place to route excess cable to ensure that all loops are of a minimum diameter, so the cable will not get damaged and maximum light can traverse the cable.

Patch panels (Figure 7.11) for fiber-optic cables are usually installed into the LIU. Because the core and cladding of two fiber-optic cables that are to be joined together must match perfectly, the patch panel must be manufactured to exact specifications and some standard type connector must be used to ensure a good fit. (Fiber-option connectors are discussed in the next section.) Another patch panel issue deals with attenuation. Remember from the previous discussion that when you splice or join a fiber-optic cable, you can introduce additional light loss or attenuation. The same holds true for the fiber-optic patch panel. The connectors on the patch panel should identify total loss at various wavelengths, and these losses should be added to any other cable loss on that particular cable to ensure compliance with standards and good operation of the fiber-optic cable.

7.5 Fiber-Optic Connectors

Several different types of connectors are available for fiber-optic patch cables, such as LC, FC, and MTRJ, to name a few. However, the two most common types of fiber-optic connectors in use today are the ST-style (Figure 7.12) and the SC-style connector (Figure 7.13). Notice that the ST connector uses a barrel BNC-type adapter and the SC connector uses a block-style adapter. An easy way to reference these is the ST style connector is a stick-and-twist motion to connect (place the female connector over the male connector and twist to lock it in place) and the SC connector is a

Figure 7.12

ST-style fiber-optic connector. Courtesy of Leviton 2005

Figure 7.13

SC-style fiber-optic connector. Courtesy of Leviton 2005

stick-and-click motion to connect (line up the male block with the female block and click in place).

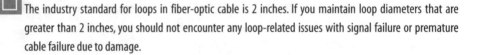

NOTE The industry standard for loops in fiber-optic cable is 2 inches. If you maintain loop diameters that are greater than 2 inches, you should not encounter any loop-related issues with signal failure or premature cable failure due to damage.

7.6   Security Considerations

Security on fiber-optic cables is tighter than that of copper cables. If a fiber-optic cable is tapped into, it would break the core, which would shut down the signal and render it useless. Copper cables, in comparison, can be monitored in band. In addition to the information security aspect of fiber-optic cable, there are physical security benefits that it offers for volatile locations. Fiber-optic cable does not use any electricity, so there is no danger of a spark igniting flammable material or gasses.

7.7   Constructing Fiber-Optic Cabling

How glass fiber-optic cable is manufactured is very interesting. Fiber-optic cable starts out as a cylindrical glass called a blank. The tip of the blank is then heated until a ball of molten glass starts to drip down. The drip of molten glass is what eventually becomes the fiber-optic cable. The largest part of the drip is cut off, leaving a thin strand of glass. The thin strand is

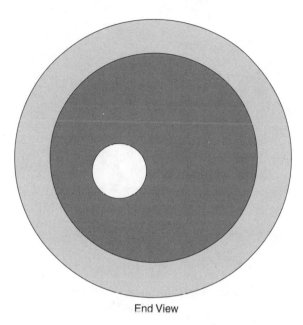

End View

Figure 7.14

Core moves out of center of cladding

then drawn down from the glass blank, and as the tip of the glass is continually heated in the furnace, the heated the drip is taken up on a computer-controlled cupping system where the cladding is added. After the cladding is added, the strand is moved to a spooling system and cooled. During this drawing process, the fiber-optic cable is continually monitored to ensure it maintains an exact diameter for the type of cable being used, and the core and cladding are monitored to ensure concentricity. If the core moves out of the exact center of the cladding (Figure 7.14), it will lead to alignment problems when splicing or connecting the fiber-optic cable. Another problem occurs if the core and/or cladding get out of round (Figure 7.15), which could lead to alignment problems and ultimately to signal loss. Once the fiber-optic cable has been made, it is tested for the following qualities:

- Strength and durability
- Refractive index of the core and cladding
- Uniformity of the core and cladding
- Attenuation, including how temperature effects attenuation
- Bandwidth capacity

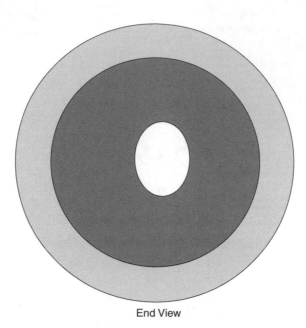

End View

Figure 7.15

Out-of-round core

One common misconception about glass fiber-optic cable is that it is very fragile and weak. In fact, the opposite is true. The high level of purity of the glass that is used to manufacture fiber-optic cable is what gives it its strength. Just as a chain is no stronger than its weakest link, any glass is no stronger than the surface flaws or impurities that would cause it to break. Normal glass has many impurities that cause it to be weak and shatter, whereas fiber-optic glass has far fewer impurities, making it strong.

7.8   Fiber Types

Fiber-optic cable is either single mode or multimode. In addition to this classification, fiber-optic cable is also designated by how it is sheathed. The two main classifications of sheathing are **loose tube** and **tight buffered**.

7.8.1   Loose Tube

Loose-tube cable consists of many individual strands of fiber-optic cable encased in a protective sheath known as a *buffer tube*, which contains a strong material that provides additional strength for the entire sheath. Most loose-tube fiber-optic cables are sheathed in 250-micron weather-protective material that can withstand aerial as well as underground installation. Loose-tube cabling is usually found in outdoor fiber-optic installations.

NOTE

The individual strands of fiber in a loose-tube cable are color coded to make it easier to identify both ends of the cable, because loose-tube fiber-optic cables can contain many individual strands.

7.8.2 Tight Buffered

Tight-buffered cable consists of a single strand of fiber-optic cable enclosed in a 900-micron protective buffer sheathing. Also included are Kevlar strands that provide additional protection to the cable as well as strength. Tight-buffered cable is generally used for indoor applications and is more flexible than loose-tube cabling.

7.9 Common Cable Configurations

Many fiber-optic patch panels use ST style connectors, whereas a significant amount of modern telecommunications equipment utilizes SC style connectors. This does not pose a problem because fiber-optic patch cables are available with ST–ST connectors (ST style on both ends), SC–SC connectors, and SC–ST connectors (also known as hybrids). Fiber-optic patch cables are available in many different lengths. It is a good rule of thumb to use the shortest cable possible to reduce the amount of slack in the cable, thus reducing the number of loops in the cable and reducing the possibility of violating the minimum loop rule.

7.10 Chapter Summary

- Fiber-optic cabling is fairly new in comparison to copper media which has been in use for decades.

- Fiber-optic cable is made up of three main components: core, cladding, and protective coating. Extreme care must be taken during the manufacturing process to ensure that the highest-quality glass with the fewest impurities is created.

- The protective coating can either be loose tube or tight buffered. Loose tube has many individual strands of cable sheathed together and tight buffered has a single strand in the protective sheath.

- Fiber-optic cable has the ability to carry more data over longer distances than copper cable, making it suitable for many data and voice applications.

- Fiber-optic cable is more suitable for locations where a fire hazard exists. There is no chance of stray voltages causing sparks.

- Fiber-optic cabling uses light waves called modes to transmit signals from a transmitter to a receiver.

- Fiber-optic cable is available in two main types: single mode and multimode. Single mode is more suited for higher bandwidth applications over longer distances, whereas multimode fiber-optic cable is more suited for shorter distances.

- Single-mode fiber-optic cable has a smaller diameter core; multimode fiber-optic cable is nearly 10 times as large in diameter.

- Attenuation and dispersion are factors that degrade signal quality in fiber-optic cabling. Attenuation is loss of signal due to impurities in the glass, and dispersion is a law of physics regarding the behavior of light waves.

- There are a variety of fiber-optic connector types available. Different telecommunications equipment manufacturers make their equipment with different styles of connectors, and fiber-optic patch cables are available to meet every requirement.

7.11   Key Terms

attenuation: The reduction of signal strength over the length of a fiber-optic cable. Attenuation is cumulative over the length of a cable run.

buffer coating: A durable outer coating that protects the core and cladding.

cladding: Material surrounding the core of the fiber-optic cable composed of a material with a different refractive index to provide signal reflection.

core: Center of the fiber-optic cable and made of very small diameter, high quality glass. Data are actually transmitted over the core.

critical angle: The angle at which internal reflection will not occur.

Lightguide Interconnection Unit (LIU): A chassis for enclosing fiber-optic patch panels used to maintain cable integrity.

loose tube: A fiber-optic cable type that encloses many individual strands of fiber-optic cable (core and cladding) in a single sheath.

micron: One millionth of a meter (10^{-6}), equivalent to one millimeter.

mode: The path light takes as it travels the length of a fiber-optic cable.

photodiode: An electronic component that is sensitive to light. Light signals hit the photodiode and are converted to electrical pulses.

reflection: Light waves bouncing off the junction of two elements with dissimilar refractive indexes.

refraction: Light waves bending off the junction of two elements with dissimilar refractive indexes.

refractive index: The amount of refraction a given element will allow.

tight buffered: A fiber-optic cable type that encloses a single strand of fiber-optic cable (core and cladding) in a sheath.

total internal reflection: The principal that light will bounce off the junction between core and cladding with no signal loss.

7.12 Challenge Questions

7.1 What is the diameter of the core and cladding combined in a fiber-optic cable?

   a. 8 to 10 microns

   b. 125 microns

   c. 62.5 microns

   d. Depends on whether it is single mode or multimode

7.2 What are the advantages fiber-optic cabling has over copper? (Choose all that apply.)

   a. Fiber-optic cable can carry data over longer distances without the need of repeaters.

   b. Fiber-optic cable has a higher bandwidth than copper cable.

   c. Fiber-optic cable is less susceptible to electronic interference.

   d. Fiber-optic cable has a lower maintenance cost than copper cable.

7.3 How does refraction differ from reflection?

7.4 Which of the following are valid fiber-optic transmitters? (Choose all that apply.)

a. Transceiver

b. LED

c. LD

d. Photodiode

7.5 Explain total internal reflection.

7.6 What two factors most commonly effect attenuation?

a. Cable diameter

b. Glass quality

c. Cable splices

d. Wavelength

7.7 Which fiber-optic mode can carry more bandwidth?

a. Single mode

b. Multimode

c. Fixed mode

d. Duplex mode

7.8 Which type of fiber-optic cable contains many individual strands of fiber-optic cables in a single sheath?

a. Loose clad

b. Loose buffered

c. Tight clad

d. Tight buffered

7.9 Which of the following are fiber-optic transmitter types? (Choose all that apply.)

a. Photodiode

b. LED

c. LD

d. SD

7.10 Which of the following light waves would travel the longest distance in a 100-meter fiber-optic cable?

a. 850 nm

b. 1,200 nm

c. 1,400 nm

d. 1,550 nm

7.11 What is the maximum dB loss that is acceptable on a fiber-optic cable that is 1 kilometer in length?

a. 20 dB

b. 50 dB

c. 0.5 dB

d. 100 dB

7.12 Which of the following methods of multiplexing are utilized in fiber-optic delivery systems? (Choose all that apply.)

a. TDM

b. PCM

c. WDM

d. LDM

7.13 Which of the following characteristics of light describes how light fans out as it gets further away from the source?

a. Attenuation

b. Phase loss

c. Encoding

d. Dispersion

7.14 Which of the following materials is not used in the manufacturing of fiber-optic cables?

a. Glass

b. Plastic

c. Copper

d. Silica

7.15 Why is fiber-optic cable thought to be more secure than copper cable?

a. Fiber-optic cable is difficult to be tapped into midstream.

b. Total internal reflection keeps the signal inside, making it difficult to eavesdrop.

c. Fiber-optic cable does not spark and can be used in flammable areas.

d. All of the above

7.16 After a fiber-optic cable is manufactured, it is tested for which of the following qualities? (Choose all that apply.)

a. Bandwidth capacity

b. Uniformity of the core and cladding

c. Textile strength

d. Attenuation

7.17 Which of the following are issues concern cladding and core in a fiber-optic cable? (Choose all that apply.)

a. Concentricity

b. Critical angle

c. Reflective index

d. Waveguide

7.18 Which of the following cable configurations allow the most amount of bandwidth?

a. 1,550 nm light wave over a single-mode fiber-optic cable

b. 850 nm light wave over a single-mode fiber-optic cable

c. 1,550 nm light wave over a multimode fiber-optic cable

d. 850 nm light wave over a multimode fiber-optic cable

7.19 What is the procedure called that utilizes a common link with many devices sharing the available bandwidth?

a. Time sharing

b. Multiplexing

   c. Duplexing

   d. Circuit switching

7.20 What was the first type of fiber-optic cable to be commonly used
      in a commercial environment?

   a. Phase mode

   b. Liquid core

   c. Multimode

   d. Single mode

7.13   Challenge Exercises

Challenge Exercise 7.1

In this exercise, you investigate the Telecommunications Industry Association (TIA) standards for fiber-optic cables. To complete this exercise, you
need a computer with a Web browser and Internet access.

   7.1.1 Log on to your computer, open the Web browser, and type:
   http://www.tiaonline.org

   7.1.2 On the TIA Web site, click the **Search** hyperlink. Enter search
   terms as appropriate to answer the following questions:

   a. What is the published standard for the TIA/EIA buffered fiber test?

   b. What is the published standard for the TIA/EIA fiber-optic
      knot test?

   Note: It is not necessary to purchase the standards from the
   TIA/EIA for this exercise. You only need to reference the locations
   and published standard locations.

Challenge Exercise 7.2

In this exercise, you polish the end of a fiber-optic cable to reduce the
amount of attenuation on the cable. To complete this exercise, you need a
fiber-optic cable, a fiber-optic cable polishing cloth (or lapping film), a
fiber-optic polishing disk (puck), and a fiber-optic inspection scope

   7.2.1 Insert the end of the fiber-optic cable into the fiber-optic inspection scope and look for dirt or debris on the end of the core. You
   should also be able to clearly identify the core and cladding.

7.2.2   Insert the fiber-optic cable into the puck. Be sure to twist and lock it in place.

7.2.3   Gently rub the lapping film over the puck in a figure-8 pattern or in small circles until the debris is removed.

7.2.4   Remove the fiber-optic cable from the puck and reinsert it into the fiber-optic inspection scope.

7.2.5   Verify that all flaws have been removed from the end of the fiber-optic cable.

7.14   Challenge Scenarios

Challenge Scenario 7.1

You have been asked to recommend a cabling media type for a manufacturing company. This company uses very high voltage generators to produce power for their machines. Would you select fiber-optic cabling or copper? Why? What are the advantages that each provide?

Challenge Scenario 7.2

Design a fiber-optic infrastructure for a company that has three buildings. Building A has three floors with three wiring closets and is 6 miles away from headquarters. Building B has two floors and is 1,000 feet from headquarters. The headquarters building is on one level, with three wiring closets. Each wiring closet is within 100 meters. Where would you use multimode fiber and where would you use single-mode fiber?

CHAPTER 8

Cabling System Connections and Termination

Learning Objectives

After reading this chapter you will be able to:

- Understand the function of twisted-pair connectors and the various types available

- Know what termination devices are and their importance in network cabling

- Understand the function of coaxial connectors

- Discuss the nature of optical fiber connectors

In this chapter, we take a precise look at what you (as a technician or student working with cabling) will need to know about cable connections and termination. In almost every case when working with cabling (and all of its types), you need some form of end point, or termination, to your transmission media and a way for it to physically connect to a device. This end point is called a *connector*, and it terminates the end of the media, creating a way for it to connect to something else such as a patch panel, switch port, and so on. This chapter shows what you need to know about connecting and terminating the most common forms of cable in use today: copper in the form of twisted-pair cable, coaxial cabling, and fiber optics. In this chapter, you learn how to design the termination of twisted-pair cable with an RJ-45 connector as well as termination into an insulation displacement connector, or IDC for short. We cover connecting and terminating coaxial cable as well as optical fiber.

As a cabling guru, you need to know cabling theory and practical application, as most often, you will have to attach the connection on the end of a cable you are making or have ordered for purchase. If you are a network designer who must order a new network switch, patch panel, or router (which are only a few examples), what cable connector type would you plug into it? What connectors would you order with the device? You must establish all of this to determine a design solution before your purchases. Having a deep and thorough understanding of transmission media types, connectors, and terminations is a hallmark of a great network designer or engineer.

In addition, you should understand the differences between new technology and old technology, and know how to use a mix of the two in your lab or production environment. These are covered in this chapter as well. In each example mentioned within this chapter, we bring to your attention why this information is critical for the real-world, production-based environment found in most networked companies today.

8.1   Twisted-Pair Connectors

The most common form of cable used on data or voice networks is twisted-pair (TP) cabling. Common forms of twisted-pair cabling are shielded (STP) and unshielded (UTP); a less common type is screened twisted-pair (STP). You learned about STP and UTP as well as what makes up twisted-pair cabling in Chapter 1, "Data Cabling Basics." The main difference between STP and UTP is that STP offers more protection with additional shielding whereas UTP is more susceptible to interference and other problems encountered on the transmission medium. You may have a more difficult time adding a connector to STP, because the cable is a bit cumbersome to work with and the connector itself is shielded. This section covers what types of connectors you can use and how to terminate twisted-pair as well as how to work with screened ScTP connectors.

We first cover the connector types and then move to the different ways you can place the copper wires (according to the standards) in the connector.

8.1.1   Modular Connectors

When working with twisted-pair cables, you add a connector to the end of the raw (cut) cable, so that you can insert the cable end into a server or workstation network interface card (NIC), a patch panel, switch, or any other compatible infrastructure device. Each device has a connection method that meets the modular connector you select. For example, if you purchase a standard RJ-45 NIC, only a RJ-45 connector can be inserted into it. Therefore, it is imperative that as you learn to make or acquire, and use, cable, you understand the importance of using the proper termination on it. This section covers the various types of connectors you can use and what you need to know to properly terminate the cable with the connector in the trim-out phase.

RJ, which stands for Registered Jack, has different numbers assigned to it to denote what type of cable category can or should be inserted into it, or what type of pair-to-pin layout you should use. This can be confusing to the novice cable technician, but by simply memorizing the most common cabling standards (such as T568A or T568B) and using a book of this kind as a reference, after much practice you can successfully create your own cables.

You can terminate copper media in many ways, so let's look at the one most commonly used for twisted-pair media: the RJ-45 connector.

You will see the terms *connector, plug, jack,* and *pin* used throughout this chapter and in other chapters of this book. It's therefore important that you understand the meanings of these terms. A **connector** (when referred to as RJ-45 connector, for example) is nothing more than a

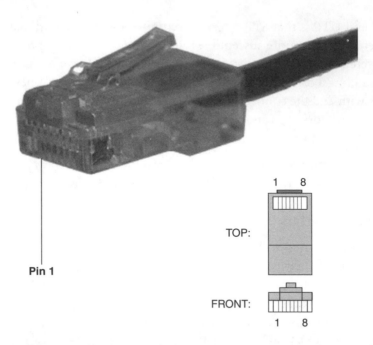

TOP:

FRONT:

Pin 1

1   8

1   8

Figure 8.1

RJ-45 connector

general connection type. The Telecommunications Industry Association/Electronic Industries Alliance (TIA/EIA) specifies an RJ-45 connector for twisted-pair cable. The **plug** is the male component crimped onto the end of the cable, whereas the **jack** is the female component in a wall plate, patch panel, or other device. The **pins** are, counting from number 8, the areas on the connector where the cable is fed into. A commonly asked question is, "Where is pin 1 located?" That is answered in the next section, where you can view a diagram showing the location of pin 1. Once you determine the location of pin 1, all you need to do to wire the jack (instead of the plug) is to reverse the pin layout that the plug follows and apply it to the jack. You reverse the pin placement when looking at the jack end, because it is the female end that you will be inserting the plug into. Remember, the plug is the male end and the jack is the female end, which is commonly seen as the cavity in the wall plate.

RJ-45 Connectors

RJ-45 connectors are currently the most commonly used connectors and will be for some time to come. Figure 8.1 shows what a typical RJ-45 connector looks like.

It is important that you pay close attention to the Top and Front positions shown to you in Figure 8.1. Later in this chapter, you will learn how to place the raw cable into the connector (according to a specific standard and code). Knowing the pin location and how to hold the connector via pin location will help you to avoid inserting the cable upside down or in the wrong mating position, causing you to miss a pair when crimping the cable and connector. An incorrectly crimped cable cannot be used—you will have to recut the cable and attach a new connector. Because this happens when you are new to this process, we suggest you don't cut cable to an exact length; always pull a little extra so that you can have some room to work with when adding connectors.

In Chapter 5, "Copper Media: Twisted-Pair Cabling," you learned about the differences in stranded and solid core copper cable. RJ-45 connectors can be ordered as either solid or stranded core plugs, and you should match up (for best practices) stranded cable with the stranded RJ-45 plug and vice versa.

RJ-11 Connectors

The RJ-11 connector is something that most likely you have seen for years. Every time you pick up your telephone at home, it's a good chance that somewhere on that phone or base unit is an RJ-11 connection. The RJ-11 connector is primarily used in the United States and its surrounding territories; it is not as widely known or used internationally. Figure 8.2 illustrates an RJ-11 connector.

The RJ-11 connector can be used for some types of (mainly older) data networks, but it is highly recommended by the industry that you use a

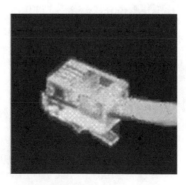

Figure 8.2
The RJ-11 connector

newer technology, which is currently the RJ-45 connector. RJ-11 connectors are most commonly used for the public switched telephone network (PSTN; also called plain old telephone service, POTS), based on copper wires that carry analog voice to the homes of people worldwide, depending on where it's deployed.

Although the RJ-45 can terminate many categories of cabling, most commonly it is used to terminate categories 5, 5e, and 6. Chapter 5 covers the cabling categories in detail.

When creating a cable with an RJ-11 connector, make sure to use RJ-11 stranded plugs with RJ-11 stranded cable and follow the same method for solid core cable. Again, you can inadvertently create a bad connection if you do not match them up properly.

RJ-11 generally uses either one or two wire pairs (Figure 8.3). The two-pair connector is sometimes referred to as RJ-14. Notice that the figure appears to include a three-pair system, but this is actually a 6-pin/6-conductor (6P6C). Category-rated cable is not terminated in any one of these connectors.

When viewing Figure 8.3, notice that the pin position is similar to an RJ-45 connector (as shown in Figure 8.1). However, the RJ-11 connector is smaller and holds only three pairs of wires instead of four. Plus, the RJ-45 has the additional outside pins, labeled 1 and 8, where the RJ-45/6P6C and RJ-11/14 crossover would be on pins 2 through 7 in the RJ-45 scheme. A quick way to remember this is that an RJ-11 connector is missing the brown tip and ring set.

Screened Twisted-Pair (ScTP) Connectors

A ScTP connector is similar to UTP cabling in that ScTP uses the same color codes. However, ScTP is the result of a using shielding over four twisted-pairs, which UTP does not offer. In fact, when you attempt to use UTP cabling and cannot (for whatever reason) overcome a problem with interference (that may be disrupting or corrupting your communications), you should use ScTP or STP instead of UTP.

When working as desktop support technicians years ago, one of your authors was part of a team responsible for cutting, crimping, and installing new cable in a production environment on a network. Because the network cabling already installed in the building was very old, we had to use RJ-11 connectors for our data runs, and used only

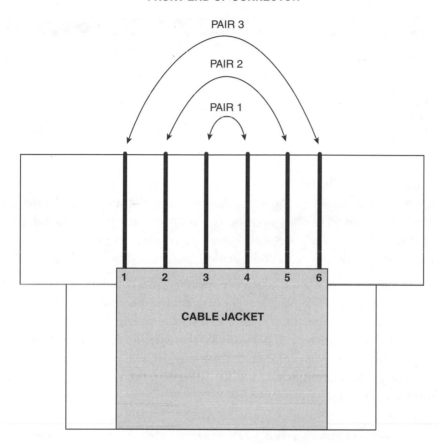

Figure 8.3

Pin and pair layout for an RJ-11 or RJ-14 connector

two of the three available wire pairs. Be aware that, in the field of cabling, you may be confronted with odd connection types. You should be prepared to identify them and provide a solution.

8.1.2   Twisted-Pair Cabling Codes

As you learned in Chapter 2, "Cabling Standards and Specifications," if no standards were in place, every company or organization could have a different cabling scheme in use, causing great confusion. It would be impossible to manage, as all companies would be responsible for maintaining its own standards. With accepted industry standards, everyone who creates or uses a cable does so using the same rules, so no one must recreate the wheel and everyone can share in the same technology.

TABLE 8.1　Twisted-Pair Cabling Tip and Ring Color Scheme

Wire Pair	Tip	Ring
1	White/blue	Blue
2	White/orange	Orange
3	White/green	Green
4	White/brown	Brown

When working with twisted-pair cable in particular, you have to pay close attention to the color of the wires. As you learned in previous chapters, a cable consists of pairs of wires (most commonly, four pairs) surrounded by a cable jacket, and each wire pair has a *tip* and *ring* designation. Each pair uses a specific color or color combination, as specified by cabling standards.

When making a connector it is imperative that you have basic information about the pairs readily available, because before you terminate the end of the cable with a connector, you have to organize your pairs into the correct color scheme for the type of network for that specific cable. Therefore, you need to know the exact color code for each pair, which are listed in Table 8.1. The tip is generally striped with white, and the ring is a solid color: blue, orange, green, or brown.

NOTE

Until you have these pairs memorized, you may want to make a small card to carry with you for quick reference. With experience, you will eventually know this information, as you do it over and over again, but until then, it is helpful to have the card as a quick reference.

Pair 1 (which is the tip and ring for the white/blue and blue colors) is always placed in the middle of the connector, regardless of which type you are using. This is true of RJ-11 and RJ-45 connectors, and of T568A, T568B, and USOC standards, which you'll learn about in the following sections. Let's move on to learn about specific cabling codes, such as those set by ANSI/TIA/EIA and USOC.

American National Standards Institute (ANSI)/TIA/EIA

ANSI/TIA/EIA is responsible for setting standards for the color-coded arrangement of pairs to allow communications to occur (see *http://www.*

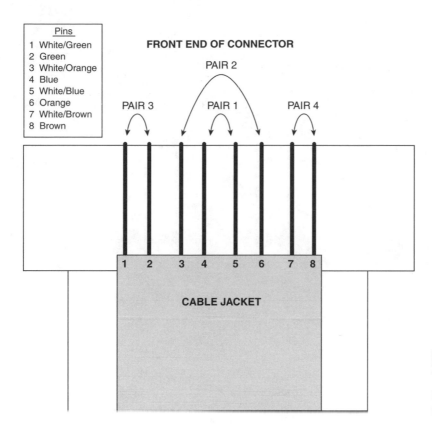

Figure 8.4
T568A standard for four-pair cable within an RJ-45 connector

tiaonline.org/). Two significant ANSI/TIA/EIA wiring standards are T568A and T568B.

T568A   To add a connector to a length of twisted-pair cable using the T568A standard, you need to know how the four pairs are first laid out in the connector. Because you already learned about pairs in Chapter 6, we focus on the termination portion in this section.

Let's look at the pair layout for T568A in Figure 8.4. The numbering in Figure 8.4—where the cable jacket is cut and the copper wires inside exposed—is the reference to see which copper wires make each pair. It is also referenced by pin number, which is important to know because if you do not hold the cable correctly (with pin 1 on the left and pin 8 on the right), then you will place the wires in the connector backwards.

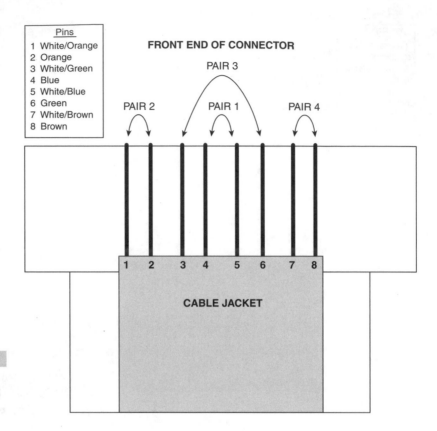

Figure 8.5

T568B standard for four-pair cable within an RJ-45 connector

T568B (also known as AT&T 258A)   Now that you are familiar with the T568A standard and what pair layout is used to create it, let's look at the T568B standard (as illustrated in Figure 8.5). Its main difference is simply the reversal of pairs 2 and 3 in the connector.

You can create a commonly used cable—a crossover cable—using a standard twisted-pair cable with a T568A standard termination on one end and a T568B standard termination on the other. Crossover cables (as their design implies), connect one device to another with similar transmit and receive channels. To visualize this concept, use a phone receiver as an example. You listen on one end and speak into the other end. If you put two phone receivers together, with the similar ends together, the two parties can listen (receive) and talk (send), but not to each other. If you flip one receiver around, each party can hear what the other party is saying, and you have accomplished your desired communication.

In the world of computer networking and data transmission, it's very much the same. When you connect a NIC to a switch port, the design allows for the send channel (TX, for transmit) to connect to the receive channel (RX, for receive). However, if you connected a standard T568A cable from one switch to another, you would have the problem we previously discussed—TX and RX in positions where they could not function with each other. Therefore, to connect a switch port to another switch port (called *uplinking* one switch to another), you would use a crossover cable (which has RX and TX channels crossed internally in the cable) to allow for communications.

A crossover cable uses T568A on one end and T568B on the other to effectively cross the TX and RX channels. **NOTE**

Another option is to buy a switch or hub with an option (sometimes called an MDX) that, when selected, internally crosses the channels. With this option, you do not have to make a new cable; you can use a straight-through patch cable instead.

Universal Service Order Code (USOC)

The USOC was created by Bell Telephone for voice transmissions and is still currently in use although in a relatively limited fashion. A major drawback to USOC is that it is not compatible with Ethernet, and is used on cable that is not category rated and might not have any twists at all. Most data cabling implementations today use the T568 standards. However, just as Token Ring was once widely used but is now found in limited production environments, you may still have the task of supporting it. You should therefore be familiar with its existence and how to work with it.

Figure 8.6 shows the pin and pair layout for the USOC standard. This is by far the easiest standard to memorize and remember.

Now that you have learned quite a bit about twisted-pair termination and connection types, the next step is to learn about the other portion of twisted-pair termination—termination devices such as 66-block or 110-block. The following section focuses on where cabling termination is centralized, what design items you should consider when using them, and important details you will need to know for the trim-out phase of your projects.

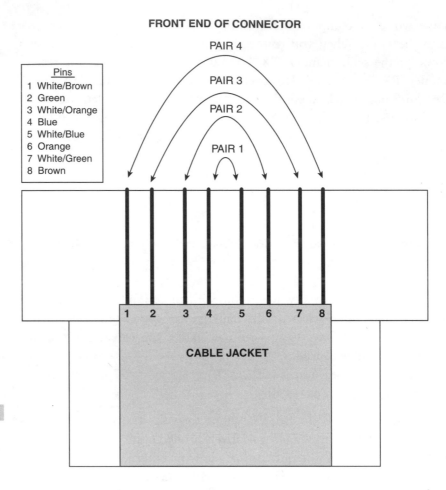

Figure 8.6

USOC standard for four-pair cable within an RJ-45 connector

8.2   Termination Devices

Why is it important to learn about termination devices? Remember, the most important reason to use any of the equipment in this section is because it is human nature to desire change. People change where they sit, they change jobs and leave, and things get relocated, so you need a certain level of flexibility. The cross-connects, blocks, and hardware in this section facilitate that flexibility, so you do not have a single cable run connecting a device on your network to a central device that creates major issues when you want to move things around or incorporate new technology.

Figure 8.7
66-block termination.
Courtesy of Leviton 2005

8.2.1 66-Block Termination

When terminating voice-based connections (for phones that connect to a private branch exchange [PBX], for example), you can use a 66-block. Terminating connections on a 66-block is seen when using 25-pair wire and connecting phone systems to a PBX. In some cases, you can use a 66-block to terminate data connections (we personally have seen this in overseas companies), but this is not recommended and the use of the 110-block (covered in the next section) is preferred. You can see an example of a 66-block termination device in Figure 8.7.

Sixty-six-block cross-connect is a term that is also used in the industry. It describes a block with a 25-pair wire installed on one side, and across from it is where you connect single phones to connect up to the 25-pair cable.

Ordinarily, a 66-block is used with voice applications. In existing systems, you need to avoid untwisting the pairs when terminating Category 5 cabling on a 66-block. Terminating on a block is no different than terminating a jack or plug. You still need to maintain the twists. Punch down the

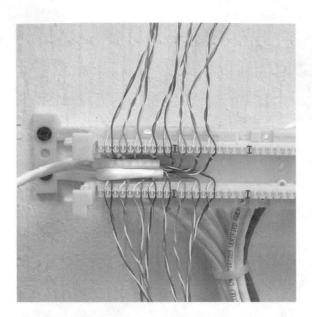

Figure 8.8

110-block termination.
Courtesy of Leviton 2005

cable ends with a punch-down (impact) tool. Chapter 15 describes how to use this tool and how to do this in the trim-out phase. Make sure you use the appropriate blade on the tool. A blade is specific to the 66-block, and most tools have it clearly labeled on the tool itself.

8.2.2   110-Block Termination

Much like a 66-block, your cable can be terminated with a 110-block. In theory, both blocks terminate the connection. Both voice and data can be used on the 110-block; however, many installations use the 110-block for voice systems and patch panels for data systems. When you terminate with a 110-block (Figure 8.8), you are essentially creating the same thing as with a 66-block—a cross connect, or transfer point, from one cable set to another—with the flexibility to easily change your network cabling configuration.

The transfer is accomplished by simply punching down the cable in the 110-block and ensuring it is firmly seated. Once in place, you put a connector block on top of the first set of wires and then punch the second set of wires on top of it. This creates the connection between the two sets of wires (two separate cables).

Important details to remember when working with a 110-block termination device include:

- Make sure you label all of your cables! Although this is something you should do for every kind of cabling job, it is imperative that you label all cables when using a 110-block. Because you are connecting so many cables, it is easy to lose sight of what you are doing because of the unwieldy nature of the block's layout. If you have a dead connection (maybe you didn't punch a wire set down hard enough and as a common example, it did not make contact with the IDC), you will have to trace out your cable and re-punch it down into the block. This being said, if each one was marked or labeled, the job will be quicker and easier to maintain, support, and troubleshoot when necessary. Get in the habit of doing documentation because the extra effort up front helps you in the long run.

- Make sure you maintain the twists on the twisted-pair cabling even up to where you connect the wires to the 110-block. If you do not, there may be problems with the connection. Keep the twists as close to the cut wire as possible—the twists are used to cancel **crosstalk**. Crosstalk is the coupling of one pair's signals to another's, which creates malformed or damaged signaling, impeding the data flow across the cable.

- Remember to use the punch-down (impact) tool with the 110-block cutting blade when terminating your cable. Make sure you face the tool in the correct direction with the cutting edge facing the end of the cable; if you reverse the tool, you will cut into the wire itself and ruin it. If you accidentally cut the wire, pull it out and re-terminate it properly—never use a damaged wire pair. Doing so can create problems that are not easy to trace and fix.

TIP

Take painstaking efforts of excellence when doing a cabling job. This pays off in the end when everything works properly. Less than perfect cabling environments can be problematic for your clients and quite possibly for yourself as the cable technician.

- Although 110-blocks are frequently used, it is recommended that you use patch panels and patch cables when applicable. Figure 8.9 shows the use of a patch panel in a data network. Because it's easier to manage, this provides a more flexible cabling structure. Figure 8.9 shows PC A wanting to communicate with PC B. Each PC is fitted with a NIC and both are located somewhere in the building.

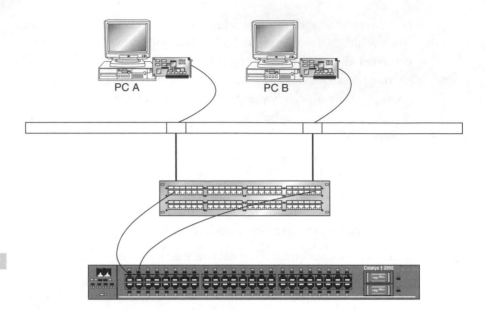

Figure 8.9

A data network with a patch panel in use

Both NICs have patch cables connecting to wall plugs. The main run terminates somewhere in the building at a central location (patch panel) and then a patch cable is used to connect to a concentrator, which in this example is a Cisco Systems switch.

This is the most common layout you will see when working with data networks.

8.2.3　Other Termination Devices

While working in the cabling arena, you may come across other forms of termination devices. Although most frequently you will work with 66-and 110-type punch-down blocks as well as patch panels, you may come across other items like BIX or Krone hardware. BIX hardware (BIX blocks), shown in Figure 8.10, can be used with data and voice, as can the Krone hardware. Make sure that you use the BIX punch-down blade to terminate the cable on the BIX block because it is a unique connection, as are the 66- and the 110-block connections. The BIX and Krone blocks are quite similar to the 66- and 110-blocks; the former blocks are simply proprietary to the companies that make them, for instance, Krone Inc.

Figure 8.10
BIX hardware. Courtesy of HOMACO 2005

Copper Insulation Displacement Connector (IDC) Termination

When terminating connections on most devices (like the 110-block), you should know about insulation displacement. An **IDC** (i.e., **insulation displacement connector or contact**) is a type of connection or connector that allows for the use of a punch-down tool or impact tool to create a near-perfect connection into the terminating block. Using an impact tool, you can firmly seat a wire-making connection to the connector while removing the excess wire and jacketing in one step. This, of course, needs to be an engineered solution, so it is called an IDC termination. If you connect wires to a patch panel or 110-block, you need to ensure that the connectors on the block are IDC connectors as well.

We have now learned the design basics of terminating twisted-pair connections for voice and data with the use of many different cable categories and different types of connectors, plugs, jacks, and blocks, as well as standards. This knowledge is critical before beginning the trim-out phase of your twisted-pair cabling deployment, which is covered in Chapter 15. In the next section, we cover coaxial connectors, which are still commonly in use today.

8.3   Coaxial Connectors

Coaxial cable and connectors are covered in detail in Chapter 6, "Copper Media: Coaxial Cabling." You learned the fundamentals of coaxial cabling and its practical uses, and an overview of the various coaxial cable connectors available today. In this section, we focus on what you need to know about the termination of and connectors used to terminate coaxial cable.

Although many coaxial connector types exist, we cover the F- and N-type connectors as well as the now-infamous BNC connector, in the following sections to limit the scope of this book. However, before beginning our discussion of coaxial connector termination, recall these important details about coaxial cable:

- Coaxial cables consist of an inner conductor (solid as well as stranded wire). This core is separated by a dielectric (core) from its outer conductor (single- as well as double-braided shield). It is covered with polyvinyl chloride (PVC) or a plenum-rated outer sheath.

- Coaxial cable is basically used for audio/visual (AV) purposes, but can support data. The most common coaxial cables are RG-6, RG-11, RG-58, RG-59, and RG-62.

8.3.1   F Connector

F connectors are installed on the end of coaxial cable to form a mating end, so that the cable can be connected to a device such as a security camera or a VCR. The Type F is a threaded connector rather than snap-on and is mainly used for a variety of connection devices. F connectors are used on the following applications:

- Cable television (CATV) boxes
- Cable modems
- Security cameras
- VCRs
- Satellite systems

An F connector is shown in Figure 8.11.

Figure 8.11
An F connector

Figure 8.12
An N connector

8.3.2   N Connector

An N type connector is used in microwave based networks. It was named after Paul Neill of Bell Labs in the 1940s. The N connector was made to be weatherproof, strong, and durable, and became popular in many military-based applications. N connectors can be used to terminate coaxial cable such as RG-8, RG-58, RG-141 as well as RG-225. An example of an N connector is shown in Figure 8.12.

8.3.3   BNC Connector

The most common coaxial cable connector in use today and for some time to come is the BNC. The BNC connector (which stands for many different names, but most often the Bayonet Neill Concelman or British Naval Connector) is small, facilitates a quick connect/disconnect action, and works by being screwed onto a device such as a NIC. BNC connectors were quite popular in the early 1990s, but have since been increasingly replaced by RJ-45 connectors used with twisted-pair cabling. BNC connectors on coaxial cable were common on Thinnet (or 10Base2) networks, but now are more often used on T-3 data network lines. The connector itself is simple to

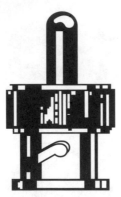

Figure 8.13

A BNC connector

understand; it has two lugs on the female side of the connector that mate with the male side. When the connectors are twisted, they are then locked into place. This connector can be used with RG-58 to RG-179, RG-316, and many others. BNCs run at 50 ohms. An example of a BNC termination connector is shown in Figure 8.13.

In Chapter 7, "Fiber-Optic Media," you learned the fundamentals of optical fiber and its practical uses, and were introduced to fiber-optic connectors. The next section focuses on what you need to know about the termination of fiber-optic cabling, and contains additional details about the connectors used to terminate optical fiber. We cover the most commonly used connector types (SC and ST), when to use each one, and why.

8.4 Optical Fiber Connectors

There are many different fiber-optic connection methods, connector types, and ways to terminate them. Single-mode and multimode connectors create differences in the types and methods used as well. If you plan to work with fiber optics, you should perform in-depth research about fiber before attempting to terminate it. In fact, it is best to take a course in fiber-optic cable termination or learn from an expert.

WARNING ! Be careful when handling optical fiber and its connectors. The cable can easily break, and terminating it can be difficult because proper termination requires masterful levels of skill and experience. Adding connectors is a trade in itself, which requires special tools as well as learned and applied skills. Because fiber is relatively

click

stick

Figure 8.14
An SC-type optical fiber connector. Courtesy of Leviton 2005

costly, and because it's easy to make mistakes when connecting and terminating it, use care and caution when handling fiber cable and connectors.

When terminating fiber and adding connectors, there are two main connector types to choose from: SC and ST. There are many other connection types (fiber has been around since the early 1970s) that have been developed through the years, but to limit the scope of this book to what is most widely used on data networks today, you need to be thoroughly familiar with the SC and ST connectors.

8.4.1 SC Connector

SC is a snap-in connector, meaning that you place it in a receptacle, such as on a network switch, and click it into place; this is also called *stick* and *click*. The SC connector is shown in Figure 8.14.

SC connectors are a relatively new connector-type technology, but are in popular use today. Part of their popularity is that they are cheaper and easier to use than ST connectors, and less prone to damage. You will most likely see these types of connections from large core switches with fiber uplinks to smaller closet switches in a campus network.

8.4.2 ST Connector

ST (which is an old AT&T trademarked technology) is another type of fiber connection type. Like the BNC connector for coaxial cable, it has a bayonet-based mounting end and a long cylindrical ferrule, which is a spring-loaded

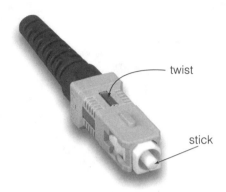

Figure 8.15

An ST-type optical fiber connector. Courtesy of Leviton 2005

sheath used to hold the fiber in place. You insert the connector into a receptacle and twist it to lock it into place. *Stick* and *twist* is considered an older technology, but is still widely used on data networks, and its install base is broad. An ST connector is shown in Figure 8.15.

Real-world production environments (especially when working with Cisco Systems, Nortel Networks, and 3Com) do not include ST in new implementations. In most cases, the only way to use this older technology is with a mediation device, such as a transceiver, which has one end that plugs into an attachment unit interface (AUI) port and an ST-based mounting connection on the other end.

NOTE To remember the difference between SC and ST connectors, think of SC as stick and click, and ST as stick and twist.

8.4.3   Termination

It is not recommended that you try to terminate fiber on your own because it is a difficult process. A simple mistake can ruin the costly fiber, thus rendering the materials useless because there is no way to turn back from most mistakes. However, those of you interested in working with fiber are encouraged to take a fiber cable termination course or learn from a pro. Either method will help you learn this unique and valuable skill with fewer mishaps.

Should you be faced with the necessity to terminate (or even splice) fiber cables, proceed with caution. Many difficulties can arise during the process, such as the following:

- **End gap:** The first problem that you can encounter when terminating or when splicing fiber-optic cable is called **end gap** (see

Figure 8.16, part A). If there is end gap in the fiber-optic cable, the light has to jump across the gap in the space between where the two ends are butted together. Recall from the discussion in Chapter 7 on fiber optics that light has a tendency to disperse, or fan out, as it leaves the source. There will be some light loss in the cable as it fans out over the end of the receiving side. Also, the air gap between the fiber ends cause a difference in refractive index, which can cause some of the signal to actually reflect back to the source.

- **Concentricity:** One of the manufacturing problems with fiber optics is that the core is not always perfectly centered in the cladding (see Figure 8.16, part B). In this case, it is very difficult to line up the cores on both ends of the cable correctly. If they do not perfectly match, signal loss will occur. This is referred to as **concentricity.**

- **Uneven ends:** A third problem that can occur when terminating fiber-optic cable is if one end is not perfectly flat. Refer to Figure 8.16, part C, for an example.

- **Air gaps:** A similar problem arises when the ends of the cable are not flat as with a cable termination or splice that has an air gap, the fact that the refractive index has changed and there will be some light reflecting back to the source. This is also a good time to point out the need for complete cleanliness on the both ends of a fiber-optic cable that is being spliced or terminated. Both ends must be polished completely smooth with no dirt in the way or the light signal will not be fully transmitted from one side to the other (Figure 8.16, part D). Even airborne dirt and dust can cause a serious degradation in signal quality when you consider the small size of the diameter of the core in a fiber-optic cable.

To attach a connector (i.e., terminate) a fiber-optic cable, follow these steps:

1. Polish the end of the cable to a high degree of smoothness.

2. Insert the end of the cable into a microscope to check for any chips, debris, dust, or areas that are not smooth. Remember, with the small diameters of fiber-optic cable even the smallest amount of dirt and dust will negatively impact the performance of the cable.

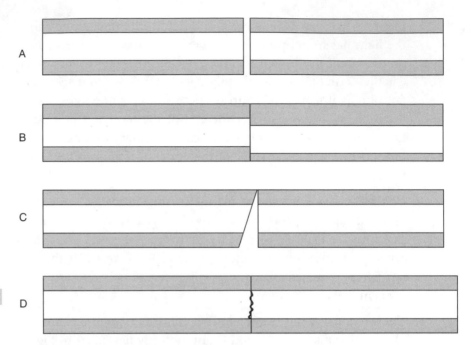

Figure 8.16

Fiber-optic cable termination problems

3. Once you are satisfied with the condition of the cable end, insert the cable connector over the end of the cable, or for some types of connectors, insert the cable end into the connector. Make sure to seat the cable all the way in.

4. Each style of connector has its own unique procedure for termination. For an ST-style connector, for example, slide the ferrule over the end of the cable before attaching the end, and then crimp it using a crimping tool (as discussed in Chapter 13, "Building Your Cabling Toolkit"). A **ferrule** is a metal or fiberglass collar that helps add strength and stability to the cable connector.

5. After the crimp is applied, gently tug on the end to ensure a tight fit, and then heat the end to make a tight connection between the fiber-optic core and cladding with the core and cladding on the terminating end.

NOTE   Follow the manufacturer's instructions when installing connectors on a fiber-optic cable. The procedures outlined in this section are given as an example for a variety of cable connector types and are not all inclusive.

If you need to install a new switch, be aware of the differences in **gigabit interface converter (GBIC)** modules, which are a newer form of a basic transceiver that allow you to install different types of interfaces on a switch, which in turn keeps the switch (or other device) modular by design. When ordering a new switch or other device that you want to connect fiber to, make sure to pay attention to the different types of connectors. You can terminate one end of a fiber patch cable with an ST connector and the other end with an SC-type connector, creating a *hybrid* cable. This makes it easy for you to use fiber-based patch panels with existing wiring runs and newer hardware (such as network switches), and not have to reterminate the older fiber that's already in place.

8.5   Chapter Summary

- Cables by design are made to connect two points to establish a path from one point to another. Because of the different types of cabling in use, all the different things that can be connected, and the many different ways to accomplish that connection, it is imperative that as a cabling technician you understand how connections and terminations are established, and what hardware and tools are necessary to accomplish the job.

- The fundamentals of twisted pair connection methods, connector types, and standards are critical to success during the trim-out phase. Follow the specifications in the standards to help ensure proper and reliable communications. The most important of these standards are T568A and T568B.

- Twisted-pair termination often uses a 66-block, 110-block, or patch panel, but other connection types are used as well.

- Coaxial cable termination includes common connection types such as BNC connectors as well as the F and N type connectors.

- Optical fiber cable termination most often includes ST or SC connectors. As a cable installation technician, if you plan on terminating your own optical fiber connections, you should take a class or learn from an expert before attempting to terminate fiber connections on your own.

8.6   Key Terms

concentricity: Signal loss in a fiber-optic cable due to the core not being perfectly centered in the cladding.

connector: A device placed on a raw (cut) end of a cable so it can physically plug into a network switch or patch panel (among other applications) to transmit voice, data, or media. An RJ-45 connector is a general type of connection.

crosstalk: The coupling of transmission circuits (e.g., wire pairs in a cable) that results in undesired signals, impeding the flow of data across a cable.

end gap: A space between two fiber-optic cables that are spliced or between a fiber-optic cable and a terminating connector. End gap forces light to jump across the gap in the space. In addition, the air gap between the fiber ends causes a difference in refractive index, which can cause some of the signal to actually reflect back to the source.

ferrule: A metal or fiberglass collar that helps add strength and stability to a fiber-optic cable connector.

gigabit interface converter (GBIC): A transceiver that converts one type of connection and signaling to another.

jack: The female component in a wall plate, patch panel, or other device.

pins: The areas on a connector where cable wires are fed into.

plug: The male component crimped onto the end of a cable.

8.7   Challenge Questions

8.1   You are deploying a new cabling architecture. The deployment is brand-new, which means there is currently no cabling in the building whatsoever. You need to use twisted-pair cabling, and there is no immediate need for fiber or coaxial cabling. Considering category type and connection type, which standard would you put into place?

a. TIA/EIA-T568-B

b. USOC

c. TIA/EIA-T768-A

d. TIA/EIA-T768-B

8.2   Using the scenario in Question 8.1, include the need for coaxial cable as well as optical fiber. What connector types would you use?

a. BNC

b. RJ-45

c. RJ-11

d. ST

8.3 You are a technician assigned to install Category 5 UTP cabling in a new building, which is comprised of two core layer Cisco switches running at Fast Ethernet (100BaseT) speed as well as Fast Ethernet to the desktop through 20 Access layer switches in five separate wiring closets. You are responsible for the design. What type of connectors will you use with this installation?

a. RJ-11

b. RJ-22

c. RJ-48

d. RJ-45

8.4 You are installing fiber-optic cable for a client. You need to terminate the fiber on a network switch with two GBICs. What is a GBIC?

a. Gigabyte interface converter

b. Gigabit interface concentrator

c. Gigabit interface converter

d. Gigabit intervening converter

8.5 You need to install a fiber run for a company in your area. It is a metropolitan area, and the fiber will connect two buildings in a campus. You learn that you will be able to use an unused stretch of fiber ran by your local Telco underground, which will save you the cost of having to run new fiber. This is generally referred to as.

a. Hard fiber

b. Hidden optics

c. Dark fiber

d. Light fiber

8.6 Using the scenario in Question 8.1, include the need for coaxial cable as well as optical fiber. What connector types would you use?

a. Two pairs

b. Two wires

c. Three pairs

d. Eight wires

8.7 You are the network engineer assigned to terminate a fiber-optic cable on a network's core switch. This core switch has a click-in type of connector. Which optical fiber connector would you choose to terminate this connection?

a. ST

b. SA

c. SC

d. STP

8.8 When terminating cable on a block, what step should you consider before punching the cable down to keep the process organized?

a. Re-twist the exposed ends to ensure a tighter connection.

b. Label the cables so that you can easily find them while working.

c. Punch down the cable in another block to prepare the cable for termination.

d. Strip the jacket and then reapply electrical tape to the exposed area.

8.9 What is the recommended method for copper termination recognized by ANSI/TIA/EIA for UTP cable termination?

a. Insulation displacement connection

b. Isolation displacement connection

c. Insulation displacement condition

d. Insulation discharge connection

8.10 You are a cable technician assigned to create a UTP cable for your network. You are looking at the freshly cut cable and see eight wires—four pairs—all color coded. Choose the pair that is color coded correctly:

a. Pair 1: Orange and orange with white stripes

b. Pair 2: Blue and blue with white stripes

c. Pair 3: Green and green with white stripes

d. Pair 4: Red and red with white stripes

8.11 To eliminate the possibility of having a crosstalk problem from cable that you made, what should you do?

a. Use only RJ-48 connectors.

b. Make sure to maintain your twists near the termination point.

c. Use connectors with two levels of rubber insulation.

d. Make sure that you untwist the cables at the end to eliminate interference.

8.12 Match the following pin layout with the appropriate standard.

Pin 1: White/green
Pin 2: Green
Pin 3: White/orange
Pin 4: Blue
Pin 5: White/blue
Pin 6: Orange
Pin 7: White/brown
Pin 8: Brown

a. USOC

b. AT&T 258A

c. TIA/EIA-568-B

d. TIA/EIA-568-A

8.13 You are working on a connector with the following layout:

Pin 1: White/orange
Pin 2: Orange
Pin 3: White/green
Pin 4: Blue
Pin 5: White/blue
Pin 6: Green
Pin 7: White/brown
Pin 8: Brown

Which standards apply to this layout? (Choose all that apply.)

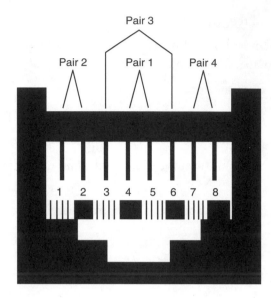

Figure 8.17

a. USOC

b. AT&T 258A

c. TIA/EIA-568-B

d. TIA/TIA-568-A

8.14 You are a cable technician assigned to create a UTP cable for your network. You are looking at the freshly cut cable and see eight wires–four pairs—all color coded. Which of the following pairs is color-coded correctly?

a. Pair 1: Black and blue with white stripes

b. Pair 2: Green and green with white stripes

c. Pair 3: Orange and orange with white stripes

d. Pair 4: Brown and brown with white stripes

8.15 From the pairs illustrated in Figure 8.17, what is the correct standard for the pair-to-pin layout? (Pair 2 is orange/white, Pair 3 is green/white, Pair 1 is blue/white, and Pair 4 is brown/white.)

8.16 You are a cable technician assigned to run and terminate cable in a central location within a small company. The company requires frequent moves of personnel and equipment. You want to plan for

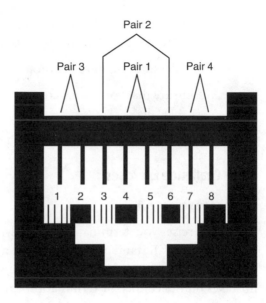

Figure 8.18

a central concentration point for the termination of the cable. What solution provides for a central point of termination?

a. IDC

b. Patch panel

c. CSU/DSU

d. 60-block

8.17 From the pairs illustrated in Figure 8.18, what is the correct standard for the pair-to pin layout? (Pair 3 is green/white, Pair 2 is orange/white, Pair 1 is blue/white, and Pair 4 is brown/white.)

8.18 You are a cable technician assigned to create a UTP cable for your network. You are looking at the freshly cut cable and see eight wires—four pairs—all color coded. Which pair is color coded correctly?

a. Pair 1: Blue and blue with white stripes

b. Pair 2: Black and black with white stripes

c. Pair 3: Green and blue with white stripes

d. Pair 4: Brown with white stripes

8.19 What are some of the most important items to remember when planning a cabling job? (Choose three.)

a. Make sure you plan your cable connections and terminations properly before starting the work.

b. Make sure you have the proper tools to complete the job.

c. Ensure that you have a contact onsite whom you can call if a cable fails.

d. Make sure that all terminations are properly identified and labeled.

8.8   Challenge Exercises

Challenge Exercise 8.1

In this exercise, you terminate a twisted-pair cable using the TIA/EIA-T568-A or T568-B standards. If you make a mistake and have to re-cut the wire and crimp a new connector to make sure it's perfect, do so. Making mistakes on the connector termination can cause phantom-like problems on your systems and network. Make sure you cut, create, and test every cable for perfection. To complete this exercise, you need a 6-foot length of twisted-pair cable, an RJ-45 connector, a crimper, and a cable tester.

To terminate a cable with a TIA/EIA-T568-A or T568-B (or USOC) standard, perform the following steps:

8.1.1 Cut your cable to length, including at least 2 more inches than what you need.

8.1.2 Do not allow the twists in the cable to unravel close to the edge where you cut the dielectric. Keeping the twists within the cable will help to reduce crosstalk and other problems.

8.1.3 Select the standard you need to use: TIA/EIA-T568-A or T568-B. (When you create cabling for a client, ask or analyze what is already in use by the client, and select the standard already in place.)

8.4 Use the chart and figures in this chapter to determine the placement of pairs and the pin layout. Find the appropriate standard, determine which pairs map to which pins, and place the exposed wire in the RJ-45 connector. Make sure that you have cut all wires to the proper length, they are properly seated, and you have kept the twists in the pairs all the way up to where the cable meets the connector. Crimp the connector onto the cable so that the cable is held in position while the wires touch the inside contacts of the RJ-45 connector.

8.5 Test your cable with a cable tester (your instructor can provide you with information about appropriate cable tester settings) against the proper standard to make sure the cable works as advertised before you use it.

8.9 Challenge Scenarios

Challenge Scenario 8.1

You are a cable installer/technician servicing a client in a metropolitan area. Your client owns and operates a six-story building, and has contracted you to help with the cable layout and design portion of the project plan. The client wants to redesign and upgrade the existing network-switching infrastructure from Nortel Networks devices to Cisco Systems devices. Six optical fiber runs are already in place, one on each floor of the building, which lead to a centralized fiber patch panel in the main telecommunications closet on the first floor. You have tested all of the runs and found that they are adequate for the new architecture. You need to replace the Nortel equipment with the newer Cisco equipment.

What do you need to consider and recommend to the client regarding which optical fiber connectors should be used to make this solution successful?

CHAPTER 9

Safety Considerations

Learning Objectives

After reading this chapter, you will be able to:

- Understand safety codes and standards
- Identify personal safety and equipment
- Understand workspace safety
- Use appropriate safety techniques when working with electricity

In this chapter, we cover safety issues that will keep you (and possibly others) safe from many of the dangers that are present in the world of networking and cable installation and repair. As a cable technician, you will eventually find yourself on a company's premises doing work such as installing new cabling. Often your task will be running cable in walls and through conduits, and terminating cable runs in small closets. Danger and unsafe conditions lurk around every corner and it's your responsibility to keep yourself as well as others safe. The way to do this is to always consider safety while working on the job.

In the first section of this chapter, we cover the codes, rules, and regulations that govern safety. We discuss organizations that focus their efforts on worker safety and what you need to know about them while working as a cable technician. The next section breaks down the elements of personal safety and items you need to consider. These include hard hat areas, protecting your back from injury, and how to use a ladder safely. Next, we cover the elements of workspace safety, and how to keep yourself and your coworkers safe in the workplace.

Remember, safety is everyone's responsibility. The burden of safety should not be placed on any one person; it's our collective responsibility to make sure that things go as they should so that nobody gets hurt—including yourself. In the last section of this chapter, we cover safety with electricity. As a cable technician, understanding electricity and how to handle and work with it are very important, because you will be surrounded by it as you install cable. Failing to protect your safety may result in personal injury and possibly death. This chapter's goal is to make you as aware of the many dangers that may exist in your profession, and how you can properly prepare for them to minimize the risk of harm. Always consider safety. Safety comes first.

9.1   Safety Codes and Standards

When in your own workplace or at a client site, you must always abide by safety codes, standards, regulations, and rules that are put in place to keep you and others safe. It's important to abide by these codes, regulations, and so on because failing to do so can involve legal liability. In other words, breaking such rules and laws can result in stiff penalties and fines. Some codes require you to do installations under strict guidelines, so following them is imperative. It's important that you know about them, have an understanding about what they require, and how to implement them. You should also consider that most of these rules are put into place to ensure our safety.

In the following sections, we cover the most common standards and codes, as well as the organizations that create and enforce them. Our coverage includes occupational and workplace safety (OSHA and NIOSH), product safety (UL), and environmental safety (EPA) considerations.

9.1.1 Occupational Safety and Health Administration (OSHA)

OSHA is one of two major U.S. agencies that is responsible for ensuring that rules and regulations exist and are complied with to maintain the safety and health of workers in the United States. The other organization (to be explained shortly) is **NIOSH**. OSHA's standards, rules, and regulations ensure that workers on the job are safe and given the opportunity to know how they are protected as well as when an employer or co-worker is not in compliance.

Cable technicians, while working at a job site, are also protected under OSHA's rules and regulations. As an example, while working and installing cable, it may be necessary for you to use a ladder to perform that task. OSHA has strict guidelines on the use of the ladder that, if followed, will ultimately minimize your risk of injury. Although ladder safety is covered in more detail in the "Ladder Safety" section later in this chapter, we now take a brief look at a small subset of OSHA's rules on ladder use.

The OSHA standard for portable ladders contains specific requirements designed to ensure worker safety. These include maintaining appropriate loads on the ladder, the angle of the ladder, slipping, and so on. For example, when discussing loads, OSHA states that when using self-supporting (foldout) and non-self-supporting (leaning) portable ladders, they must be able to support at least four times the maximum intended load, except extra-heavy-duty metal or plastic ladders, which must be able to sustain 3.3 times the maximum intended load. This is a good example of what OSHA does to keep you, the worker, safe. OSHA provides accepted safety guidelines that keep you safe and hold those who fall under its rules legally liable. If it can be proven that a standard was neglected, stiff penalties can be incurred. In another example, OSHA requires that a ladder you use is clean and free of oil, grease, wet paint, and other slipping hazards to prevent injury. Beyond penalties, it's really your safety that you should consider. Who wants to fall two stories off of a ladder?

You should always consider safety for your own health and well-being and for others around you, and understand that rules and regulations are in place to help ensure your health and safety. OSHA is the agency committed to that purpose. Throughout this chapter, we reference OSHA where applicable when discussing your own personal safety and ways to ensure it. For more information about OSHA, visit the agency's Web site at *http://www.osha.gov/*.

9.1.2   National Institute for Occupational Safety and Health (NIOSH)

NIOSH is the other agency we mentioned in the last section. NIOSH is responsible for attempting to prevent workplace injury through research and prevention. If you visit the Web site for NIOSH (*http://www.cdc. gov/niosh/homepage.html*), you can see that it's actually part of the Center for Disease Control (CDC). NIOSH, similar to OSHA, is committed to making sure that you as a worker are safer than if such rules were not in place.

9.1.3   Underwriters Laboratories Inc. (UL)

UL is a nonprofit organization that focuses on consumer safety. UL safety tests the products that manufacturers release for public consumption. If a representative sampling of a product passes the tests, the manufacturer may include the UL mark (symbol) on its product (Figure 9.1). A product carrying the UL mark has met UL's stringent safety requirements based on UL's own published Standards for Safety. The mark implies that not only has a product been initially tested by UL, but it has been retested as well. To use the logo, the manufacturer needs to comply and stay in compliance. The UL symbol is found on nearly every electronic device in use today.

The symbol in Figure 9.1 is for North America. Because UL works with many organizations and agencies worldwide, UL marks exist for a number of Asian, European, and Latin American countries. You can find a complete listing of all UL marks at the UL marks Web site (*http://www.ul.com/mark/*).

Figure 9.1
The UL mark for North America

UL also certifies local area networking (LAN) cabling to make sure that it meets Telecommunications Industry Association/Electronic Industries Alliance (TIA/EIA) 568-B.2 specifications and is in compliance with their standards and rules. Specifically, UL ensures that the material used to make such cable has met specific specifications to provide for consumer safety. UL 444 (Standard for Communications Cable) is one such standard that shows that cable has been tested and evaluated. After a cable product is granted the UL mark listing, the product is proven to have met the UL 444 standard. You can find listings and reference material for this standard and others at the UL Web site (*http://www.ul.com*).

9.1.4   Environmental Safety Considerations

When considering environmental safety, the EPA should automatically come to mind. The EPA (which stands for Environmental Protection Agency) is a government agency responsible for the detection of violations and enforcement of laws that regulate environmental issues. As a cable installer/technician, you need to understand the EPA's rules and regulations regarding the environment, and strictly adhere to these laws. For example, if you come across a dangerous chemical or substance, such as asbestos in ceiling tiles, and throw it away haphazardly, doing so not only could harm you, others, and the environment, but you also could be heavily fined for the infraction. For more information, visit the EPA's Web site at *http://www.epa.gov/*.

9.2   Safety Fundamentals

Now that you are familiar with the agencies in place to provide safety to you, people around you, and the general environment, it's time to consider "safety" in itself. This includes how you can practice safety and ensure safety for yourself and those around you. In this section, we cover safety consciousness, first aid fundamentals, and emergency rescue planning as well as proper communication in an emergency situation.

9.2.1   Safety Consciousness

Safety begins with you, and is everyone's responsibility. We obviously do not want to harm ourselves or others, so to minimize the risk of harm and danger it's important to analyze what factors could lead to potential harm and minimize those risks if possible. As a cable installer/technician, you will be in situations where you may cut or burn yourself, or may have to apply first aid to yourself or possibly a friend, coworker, or teammate. Being safety conscious simply means that you are aware of the need for safety and are looking for ways to remain safe. You can do this by analyzing your situation while considering risks as well as being properly prepared to handle an emergency if one occurs. This means you need to know how to take care of an issue or problem when it arises. You should also be familiar with a typical emergency response plan (how to systematically deal with an issue) as well as have a plan on how and whom to communicate with in an emergency.

9.2.2   Basic First Aid

First aid is something very simple to perform, but when mixed with panic and unclear thinking may be almost impossible. In extreme cases, you may do more harm than good. The first rule to first aid occurs before you begin applying it to an emergency situation: maintain a clear head and do not panic. Next, you should have some basic knowledge of how to apply first aid. As a cable technician, most likely you will be dealing with simple cuts, burns, and abrasions. However, what if you are running cable, and a fluorescent tube in a lighting fixture breaks and the glass gets in your eye? This is just one scenario of many possible situations you may need to deal with in your career. It's a good idea to be prepared for just about anything that could happen. Having a basic first aid kit is a start.

Figure 9.2
Eye wash station

For first aid, you should be concerned with the basics. Make sure you know where the first aid kit is located, and if you want to maintain a larger level of personal safety, maintain and keep a kit in your bag or in your vehicle while on the job, as you may find that not all locations you visit have one readily available. Next, make sure you know where first aid stations are located, for example, an "eye wash" station (Figure 9.2). Portable eye wash stations are commercially available, and this may be something you want to consider as well when preparing your personal safety plan.

First Aid Kit

If you make your own first aid kit, the following is a list of the most common items found in a typical kit:

- **First aid manual:** Clearly explains how to handle basic problems
- **Basic bandages:** Assorted bandages, athletic tape, and moleskin
- **Basic drugs/lotions:** Aspirin, antiseptic, and antacid tablets
- **Basic first aid tools:** Tweezers, small mirror, razor blade in a sheath
- **CPR shield:** To cover the other person's mouth so no fluid is transferred
- **Additional bandages:** Gauze pads, ace bandages, and butterfly bandages

- **Additional drugs/lotions:** Burn ointment
- **Additional first aid tools:** Sling, basic splint, and instant ice pack

As mentioned previously, having these first aid materials handy will increase your personal safety as well as safety for others. Most vehicles today come equipped with a first aid kit, and most companies have them onsite somewhere. Most homes have first aid supplies.

9.2.3   Emergency Response Plan and Rescue

In an emergency, you have to act quickly and timing can be critical. This means acting in a calm, calculated fashion. The only way to do this effectively is to have a plan and rehearse it. This might remind you of a "fire drill" at your local elementary school, where the school principal would simulate a fire by ringing a bell and have students assemble in places away from the danger. Team leaders (such as teachers) were responsible for an orderly evacuation, providing a head count, and reporting that all were present. In an emergency, you should be able to use the same "fire drill" principles in your place of business today. Prepare an emergency response plan and rehearse it, so that during a time of panic it can be executed with precision. The fire bell rings, the children form a line in single file, they walk out the nearest fire exit, and meet at a designated location for an organized head count. This is the simplest form of emergency response but it only works when everyone knows exactly what to do.

A typical emergency response plan (which can be altered in any way you see fit) generally has the following steps:

1. Take a moment to make sure you have determined the level of emergency. If the injury involves bleeding and is critical (for example, a cut artery), it would make sense to apply direct pressure on the wound and call 9-1-1 or any other emergency contact hotline to get immediate professional help. If the emergency is less urgent, then you may want to deal with it yourself (for example, a minor burn of the skin).

2. Come prepared. Knowing what to do ahead of time will potentially prevent an emergency and possibly save your life. This means you know where first aid mechanisms are located, or have a kit with you, and know how to use the facility and/or first aid kit. Consider taking first aid and cardiopulmonary resuscitation (CPR) training

from your local Red Cross chapter, and taking refresher courses every 3 to 5 years.

3. Remain calm. If you are panicking, you will only make the situation worse for yourself and/or the victim.

Every medical emergency can be handled by remembering four key words:

- Prevent
- Prepare
- Recognize
- Act

Remembering these words may help you maintain your composure and ensure you are ready to act. Remember, your job in any emergency is to remain calm and carry out an emergency response plan. Accidents do happen. It's our job to make sure that we deal with them in the best way to minimize damage to persons and property as well as to minimize the risk of making it worse during an emergency.

Communication

Communicating effectively in an emergency is paramount to dealing with the emergency. During a serious emergency, a trained individual will not be available to help if you have no emergency contact information. Make sure that you keep a list of general contact information for local emergency agencies in case a serious accident does happen. Also, make sure you can communicate the issue properly. If you are panic stricken, you will not be able to describe the problem in a clear manner to the emergency personnel who are trying to help. This only adds time, confusion, and aggravation to the process. In a near fatal situation, you could even cost the victim priceless seconds they may have needed to survive the emergency.

9.3 Personal Safety and Equipment

When working with cabling and in sites where you will be installing cable, you need to consider your own personal safety and what you can do for yourself to ensure that you are safe. This section looks at how to ensure your own personal safety as well as what equipment you can use to make certain that you don't needlessly hurt yourself.

In the following sections, we look at particular items you can use to minimize or prevent your own personal harm, such as: protective clothing, eye and hearing protection, respiratory protection, hand protection, hard hats, and back belts. We wrap up the chapter with a discussion on the proper handling of tools and equipment.

9.3.1   Protective Clothing

While working with anything that could be deemed hazardous, it is essential to protect your body and your skin with protective clothing. As mentioned earlier in the chapter, you will be working within a potentially hazardous environment. Wearing the proper clothing within that environment will provide you with added safety. Some articles of clothing that you should consider are good work shoes or boots. Open toe or soft shoes will not protect you from objects falling on your foot. Make sure you wear clothing that protects your skin (long pants and shirts) as well as clothing that is not too baggy, as it might get caught on equipment, cabling housing (especially in cramped pathways), and other structural framework features.

9.3.2   Eye Protection

Your eyes can be easily damaged by a myriad of possible dangers. Make sure that you protect your eyes while cutting or working with anything that may be dangerous. This includes using power tools or working around machinery, or anything that causes particles to become airborne and possibly get lodged in your eyes. Protective glasses can be worn in these situations to provide adequate protection. As mentioned previously, eye wash stations also should be located prior to working on a job, so that if something becomes lodged in your eye, you know where to go to flush it out quickly.

9.3.3   Hearing Protection

Hearing protection is very important on the job. If you are in a location with loud noise, such as a manufacturing plant, and are not using ear protection, hearing loss may occur (and will occur over time). Noise can not always be reduced due to the nature of the business, so it's your responsibility to consider that damage could occur to your hearing if you do not protect yourself. Hearing protection can be accomplished by using ear plugs. Ear plugs come in many shapes, sizes, and styles. Make sure that you protect your ears in noisy environments to reduce the risk of hearing loss.

9.3.4   Respiratory Protection

You may find yourself working in an environment with poor air quality, such as in a wood-processing factory or in a chemical plant where noxious fumes are in the air. To prevent personal harm, consider using a respirator with good filters. A respirator can help filter damaging particles out of the air so that you can breathe fresh air.

9.3.5   Hand Protection

Hand protection comes in the form of using gloves. Wear appropriate gloves to protect the hands against cuts, burns, abrasions, and so on. A major misconception is that gloves are needed only if a project is very physical in nature, or if there is a lot of known or predetermined abrasion possibilities. Although those are commonsense uses for gloves, the use of a proper type of glove when using any type of chemicals, including fiber-optic epoxies, is widely ignored in today's chemically-sensitive world, even with the all of the associated and known risks to health. Be sure to use gloves whenever possible.

9.3.6   Hard Hat Use

If an object strikes your head with enough force, it can result in head injury and even brain damage. Often you need to wear a hard hat, especially if the work area is designated as a hard hat area.

9.3.7   Back Support Belts

Many times, we lift things quickly and improperly, which puts a strain on the back and may cause injury. Back injuries are common and often can be avoided. Remember to lift properly by bending at the knees, and consider using a back support belt, which can be purchased at most hardware stores.

9.3.8   Handling Tools and Equipment

Tools can be very dangerous as many have sharp edges or can puncture the skin. As a cable installer, you will be in a variety of environments where many different kinds of tools and equipment will be present. Be careful with your surroundings and your own equipment. Never rush, always hold your tools properly, and take care to pay attention to what you are doing while you are working. Always read any safety notes that come with the tools and equipment you are using.

9.4   Workspace Safety Considerations

When working in any environment you need to consider safety at all times. Accidents can happen. To avoid becoming the victim of some disaster, you must be familiar with workplace safety guidelines and requirements. As a cable technician, inevitably you will be faced with climbing a ladder. You may at some time work in an area that is hazardous, especially if you have to deal with a fire or some other type of disaster. In the following sections, we cover the fundamentals of workplace security, general tips on how to stay free from harm, how to work with ladders, and how to use and operate the different classes of fire extinguishers.

9.4.1   General Safety Tips

When first starting a cabling job, assess your surroundings and talk to the building manager to make sure you are complying with local codes. After you have assessed your work area and know you are in compliance with local codes, assess the cabling work you will be doing before starting the actual project. Make sure that if you have to move ceiling tiles (drop ceilings), route cable through walls, or anything else in particular, you have assessed all the possible risks that may occur. Use the appropriate personal protection equipment as mentioned earlier in the chapter.

9.4.2   Ladder Safety

Commonly, most people don't think about a ladder before climbing one. How many times have you checked a ladder for stability before climbing it? This should be something that you do routinely. Ladder safety starts with using an approved ladder that is certified for use and in good shape. You should not use a makeshift ladder. Make sure that the ladder reaches the intended target and provides adequate support for the person climbing it. Inspect the ladder for defects or damage to minimize personal harm. When climbing a ladder, always face the ladder and climb using both hands for support. Do not carry anything up the ladder with you; use a tool belt or the shelf found on most common ladders.

Finally, make sure you use the right kind of ladder. Ladders are made of different materials to make them lighter and/or stronger. Aluminum ladders should be used cautiously if you are working around electricity (which will be covered later in this chapter), because aluminum conducts electricity and potentially may cause shock electrocution.

9.4.3   Hazardous Environments

Unfortunately, you may not always be in a position to work in a nice, clean office building. You may find yourself working outside, on a roof, in a ceiling, in narrow crawl spaces, or in other dangerous areas. If you are in this situation, you need to assess the environment you are going to be working within and take appropriate precautions. For example, if you are working in a very cold area, you may want to wear warm clothing; if you are working in an area with manufacturing equipment nearby, you may want to wear protective glasses and a hard hat. In nearly any work environment, know where the fire extinguishers are located. Simply using common sense will help protect you from potential hazards.

9.4.4   Fire Extinguisher Use

There are basically four different types (or classes) of fire extinguishers and each class extinguishes a specific type of fire. You must understand the different types, because using the wrong type of extinguisher can either not stop or even fuel a particular type of fire. Newer fire extinguishers use a picture and labeling system to designate which types of fires they are to be used on to prevent errors.

Older fire extinguishers are labeled with colored geometrical shapes with letter designations (Figure 9.3) that represent the fire extinguisher class (A, B, C, or D). Because you may need to use an older fire extinguisher at some point, you should memorize the letter designations, symbols, and details about each class. There are also numbers on the extinguisher. These numerical ratings denote something specific to the class of extinguisher.

Fire extinguisher classes are described as follows:

- **Class A:** Used to put out fires that involve basic materials like plastics, wood, or paper. The numerical rating for a Class A fire extinguisher refers to the amount of water the fire extinguisher holds as well as the amount of fire it will extinguish.

Ordinary Combustibles

Flammable Liquids

Figure 9.3
Fire extinguisher class system

Electrical Equipment

Combustible Metals

- **Class B:** Used to put out fires that involve flammable liquids. These materials include but are not limited to grease, gasoline, and oil. The numerical rating for a Class B fire extinguisher refers to the approximate number of square feet of a flammable liquid fire that a person using the extinguisher can expect to extinguish.

- **Class C:** Used to put out fires that are electrical in nature. Class C extinguishers do not have a numerical rating.

- **Class D:** Used to put out fires involving burning metals. Different versions of Class D extinguishers are based on specific combustible metals, such as magnesium, potassium, and titanium. These extinguishers generally have no rating nor are they given a multi-purpose rating for use on other types of fires.

NOTE Class A and Class B fire extinguishers have a numerical rating based on tests conducted by UL. This rating is used to determine the extinguishing potential for each size and type of extinguisher.

9.5   Understanding Electricity

Electricity, if not understood or improperly handled, can be fatal. It is because of this risk that we take a detailed look at what electricity is and how it moves from its source to an end point. Every year, tens of thousands of people are injured or killed in electricity-related accidents. As a cable installer, you are at a greater risk than the general public of exposure to dangerous electrical currents. The more you understand about electricity and how it is a potential risk to you, the safer you will be.

So what is electricity? You can say that electricity is a form of energy. Energy is power and that power is what makes things happen. When you want to send a data file from one PC to another PC over a network cable, electricity is making it happen at the lowest level. Electricity begins with tiny particles called atoms. Atoms are too small to see, but they make up everything around us. The center of an atom has, at a minimum, one proton and one neutron. If you have one electron traveling around the center of the atom (at a very great speed) and use voltage (which is an outside force) to push electrons from atom to atom, you have electricity. We as humans have found a great many ways to harness electricity's power and use it; voice and data networking over cables is only an example.

9.5.1   Voltage

Voltage is the force that creates a flow of current when a closed circuit is connected between any two points. As discussed in Chapter 3, one volt produces one amp of current when acting against a resistance of one ohm. Because this topic is part of a much bigger subject—electrical theory—which is beyond the scope of this book, understand that voltage is what is used to push the electrons from atom to atom, which is what creates electricity.

NOTE

If you are unsure about how much voltage a device may be carrying, use a multimeter to determine the voltage of a device before coming in contact with it. Determining if the device uses a low or high voltage before working on it can save your life.

9.5.2   Conductors

Understanding electricity is only the beginning. As a cable installer, you must also understand what a **conductor** is to help protect yourself. A conductor is a material that electricity can flow through easily. Metals (such as copper and aluminum) are excellent conductors. Cabling is composed mostly of copper wire because it is such an excellent conductor. The phone systems in your home today most likely travel over copper wire. It is also because of this reason that electrical wires are made of metal; they are good for conducting electricity. When we discussed ladder safety previously in this chapter, mention was made to choose the proper ladder. Aluminum ladders can be dangerous because aluminum is an excellent conductor of electricity. If you, the ladder, and electricity all meet at the same time, it can be lethal. Water is another good conductor. Your body is made mostly of water, so electricity can travel through your body because your body acts as a conductor. Therefore, take extra precautions when working around electricity, especially with materials that are good electrical conductors.

9.5.3   Insulators

You can remain safe by not only reducing your risk of becoming or connecting yourself to a conductor of electricity, but also by using **insulators** to shield yourself from electricity. This is why copper cabling is sheathed in plastic coating.

Electricity does not travel easily through certain materials such as special rubbers, plastics, and glass. These types of materials are known as insulators. Insulators are used to keep electricity from leaving the wires it travels on such as voice and data cabling.

9.5.4   Grounding

Grounding is the last piece of the electricity equation that you need to understand to keep yourself safe. Electricity is always trying to get to ground. If you are a conductor and you touch electricity while you are touching ground, electricity will flow right through you. Grounding can be defined as the procedure used to carry an electrical charge to ground (earth ground) through a conductive path, which is typically a grounding rod. Electricity is always looking for a path; the most common path to disperse unwanted or unneeded electricity is ground. Grounding is covered in more detail in Chapter 10, "Electrical Protection Systems."

NOTE   Now that you understand more about electricity, it should be clear why you should use only a class C extinguisher on electrical fires. The material used in other classes of extinguishers may act as a conductor and increase the possibility of harm to you or others.

9.6   Working with Cabling and Electricity

Now that you have a good understanding about the science of electricity and why you can be severely hurt or killed, let us talk about issues that you as a cable installer will deal with directly. Because you can not really predict when a run-in with electricity will be fatal, you should always approach it with caution. Electric shock can cause shallow breathing and rapid pulse, severe burns, and unconsciousness as well as death.

In this section, we cover the nuts and bolts of high voltage, grounding, bonding, wire separation, electrostatic discharge (ESD), electrical code, and working with fiber-optic cable. Many of these topics are covered in depth in Chapter 10, "Electrical Protection Systems."

9.6.1   High Voltage

Now that you understand voltage (the force or pusher of electrons), it should be predictable that you will experience situations where amps are increased, you will be exposed to higher voltage. Since cabling functions on "low" voltage, you will not often come across a situation where you are dealing with high voltage, which could be deadly. This does not imply, however, that you are safe from it. There will be times when you will be exposed to high voltage, and because you will not be expecting it, that is even more of a reason to pay attention to it.

Commonly used Category 5 cable does not carry high voltage. However, while running new cable in attics, ceiling, and walls, you may drill or do something else that puts you in contact with high voltage connections. Those who have done so warn: If you make contact with a high voltage wire, you will definitely feel it. It may even kill you. Use extreme care when working around high voltage sources and pay attention to the fact that, although you may not come in direct contact with it while doing cabling work, you may indirectly come in contact with it and be surprised when you do.

Lightning is also a source for high voltage. Use extreme caution when working outdoors.   NOTE

9.6.2   Bonding

Earlier in this section we discussed grounding and why it is important. To recap, grounding is used to provide a path for electricity to flow safely into the earth. Electricity will always seek a path and a ground is a safe way to disperse it. Understanding grounding is the foundation of understanding bonding.

Bonding is nothing more than the creation of a relationship between two devices that need to seek ground. An approved bonding connector will allow a device to use another device to get to ground. Bonding is the method used to produce a good electrical contact between multiple metallic parts for the purpose of sharing ground. The material used should be a conductor so as to pass the electricity through it. This is why you should use an approved bonding connector.

9.6.3   Grounding and Bonding Standards

Grounding and bonding standards are in place to set minimum safety requirements. NEC article 100 addresses grounding, and NEC Articles 100 and 250-90 address bonding. You can read about these standards online at the NEC Web site at *http://www.nfpa.org* (click the Publications>necdigest link).

9.6.4   Electrical Code

The National Electrical Code (NEC) is a code established to protect persons from the risks of using and working with electricity, which if improperly handled, can result in the damage of property and the loss of life. The NEC falls under the sponsorship of the National Fire Protection Agency (NFPA). You might wonder how the NEC and NFPA are connected. An important connection exists between cabling and fire. Because cable sheathing is often made of a plastic material, when the sheathing burns, deadly gas and smoke can be produced. As you might recall from Chapter 5, you can use plenum-rated cable to prevent this issue.

The position of the communication industry as it relates to grounding is covered in the TIA/EIA-607 document. However, this document does not cover life-threatening topics as do the NEC regulations. The TIA/EIA-607 document only covers the concerns of the communication equipment.

9.6.5   Wire Separations

Make sure that there is a clear separation of the cable you are installing with other cabling in the area. Other cabling can mean any other electrical cable carrying electricity at any voltage.

The problem with not having clear wire separation is that (if the cabling were too close) you could have issues with electromagnetic interference (EMI), which is *noise* from other electrical devices (fans, motors, etc.) that distorts the signal on your cable. EMI can cause problems with data and voice communications. Therefore, ensure that you have clear wire separation when deploying cable.

9.6.6   Electrostatic Discharge (ESD)

Another issue you may come in contact with as a cable installer is ESD, which is the sudden discharge of stored static electricity. ESD can damage electronic equipment and impair electrical circuitry, resulting in complete or intermittent failures of that equipment. If you have ever felt a shock from a doorknob in the winter's dry air, you've experienced ESD. That shock you felt and the spark of electricity it produced is enough to completely damage just about any computer component when exposed to it.

To prevent ESD, you need to ground yourself. Because electricity always seeks ground, this is the best way to discharge the stored electricity you contain. You have two options: either touch something that is grounded, and thus ground yourself, or use an ESD wrist strap and/or a mat or table that is also properly grounded.

9.6.7   Fiber-Optic Safety Considerations

Fiber-optic cabling consists of a glass fiber core. If the glass fibers break, they can become lodged in your skin. In addition, fiber-optic cable transmits light, which is how the data get from one end of the cable to the other. The source of this light is a light emitting diode (LED), laser, or vertical cavity surface emitting laser (VCSEL). Therefore, never look into the end of a fiber-optic cable. Doing so may cause severe eye damage.

9.7   Chapter Summary

- Safety in the workplace, whether in your company's building or at a client's site, involves knowing all applicable safety codes, standards, regulations, and rules, and following them as written. Failing to do so can involve legal liability. Some organizations that govern workplace and environmental safety are Occupational Safety and Health Administration (OSHA), National Institute for

Occupational Safety and Health (NIOSH), Underwriters Laboratories Inc. (UL), and the Environmental Protection Agency (EPA).

- Safety fundamentals include knowing basic first aid and having an emergency response and rescue plan.

- Personal safety includes the proper use of protective clothing: eye, hearing, and respiratory protection; hand protection (gloves); hard hats; and back support belts, among others.

- Before beginning a project, you should assess the workplace environment for potential hazards and determine ways to minimize danger to yourself and those around you.

- Be aware of the different classes of fire extinguishers and the types of fires each should be used for.

- To prevent electricity-related accidents, you must have a firm understanding of what electricity is, how it is conducted, and safe and unsafe voltages. Grounding and bonding work together to direct electricity to a safe place (that is, to the earth, or ground).

9.8   Key Terms

bonding: The creation of a relationship between two devices that need to seek ground. Bonding is the method used to produce a good electrical contact between multiple metallic parts for the purpose of sharing ground.

conductor: A material that electricity can flow through easily. Metals such as copper and aluminum are excellent conductors.

fiber-optics: A technology that uses glass or plastic fibers (also called threads or optical waveguides) instead of metal cables to transmit data. Fiber-optic cables have more bandwidth than metal cables and can transmit data digitally, but they are also much more expensive and fragile. Most Telcos, however, are gradually replacing their regular telephone lines with fiber-optic cables.

grounding: The procedure used to carry an electrical charge to ground (earth ground) through a conductive path, which is typically a grounding rod. Electricity is always looking for a path to travel; the most common path to disperse unwanted or unneeded electricity is ground.

insulator: A material, such as special rubber, plastic, or glass, that does not conduct electricity well but covers a material that does conduct electricity. In cabling, insulators keep electricity from leaving a wire that is in a voice or data cable.

National Electrical Code (NEC): A code established to protect persons from the risks of using and working with electricity, which if improperly handled, could result in the damage of property and the loss of life. The NEC falls under the sponsorship of the NFPA.

National Institute for Occupational Safety and Health (NIOSH): The U.S. agency responsible for attempting to prevent workplace injury through research and prevention. NIOSH is a part of the CDC.

Occupational Safety and Health Administration (OSHA): One of two major U.S. agencies responsible for ensuring that rules and regulations exist to maintain the safety and health of workers.

voltage: The force used to create a flow of current when a closed circuit is connected between any two points.

9.9 Challenge Questions

9.1 What U.S. agency is responsible for ensuring that rules and regulations exist to maintain the safety and health of workers?

   a. NEC

   b. OSHA

   c. AGFA

   d. NOSHA

9.2 What agency is responsible for attempting to prevent workplace injury and enhance worker safety through research and prevention?

   a. IDC

   b. LEC

   c. NIOSH

   d. AFGA

9.3 You are the cable installation technician who has been asked to install 10 new cable runs from a wiring closet to the main computer room. When installing the cable, one of the prerequisites is that the cable is certified against a set of standards. What private institution is responsible for the testing and rating of electrical and electronic products for safety, such as Category 5 cable?

a. UL

b. NIOSH

c. NEC

d. AFGA

9.4 You are installing Category 5 cable in a new location and the cable needs to be certified by Underwriters Laboratories so that you are sure it meets TIA/EIA specifications. What UL listing is the standard for communications cable?

a. UL 440

b. UL 444

c. UL 350

d. UL 390

9.5 You are gong to make a basic first aid kit to keep with you while working as a cable installer. Which of the following items would be included in this basic first aid kit? (Choose all that apply.)

a. First aid manual

b. Scalpel

c. Basic bandages

d. CPR shield

9.6 When an emergency arises, what are the four most common steps to ensure that the emergency is handled properly?

a. Prevent, prepare, recognize, and act

b. Prepare, react, stop the bleeding, and call emergency response

c. Recognize the problem, find a solution, act, and stay calm

d. Act, call emergency response, stay calm, and document the incident

9.7 You are the cable technician assigned to install new cable in a facility. You will be working in a manufacturing plant with a lot of heavy machinery in a hazardous environment. Which of the following items should you use in this environment? (Choose all that apply.)

   a. Hard hat

   b. Hearing protection

   c. Eye protection

   d. Equipment toolbox

9.8 Which of the following materials should a ladder not be made of to reduce the risk of electrocution when an exposed wire touches you or the ladder?

   a. Galvanized metal

   b. Wood

   c. Aluminum

   d. Treated wood

9.9 What element travels around the center of the atom that helps to create electricity?

   a. Electron

   b. Proton

   c. Neutron

   d. Voltage

9.10 What class of fire extinguisher would you use to put out a fire that involves basic materials like plastics and wood?

   a. Class A

   b. Class B

   c. Class C

   d. Class D

9.11 What class of fire extinguisher would you use to put out a fire that involves flammable liquids?

   a. Class A

b. Class B

c. Class C

d. Class D

9.12 What class of fire extinguisher would you use to put out a fire that involves electricity?

a. Class A

b. Class B

c. Class C

d. Class D

9.13 What class of fire extinguisher would you use to put out a fire that involves burning metals?

a. Class A

b. Class B

c. Class C

d. Class D

9.14 When working with electricity, what is defined as the force used to create a flow of current when a closed circuit is connected between any two points?

a. Insulation

b. Impedance

c. Amperage

d. Voltage

9.15 You are working as a cable technician in a facility with exposed electrical wiring. You are worried that the ground you are working on is wet. Why should you be concerned about the water as well as the exposed electrical wiring?

a. Water is a conductor.

b. Water is an insulator.

c. Water magnifies the shock by the power of 10.

d. Water will cause you to slip and fall, which is your only concern.

9.16 When working with a Category 5 cable, you notice a plastic sheath covering the copper wire. The copper wire is the conductor. What is the plastic sheath considered?

    a. Conductor

    b. Isolator

    c. Insulator

    d. Magnifier

9.17 Electricity requires a path. What is the most common path to disperse unwanted or unneeded electricity?

    a. Backup link

    b. Spare circuit

    c. Overflow line

    d. Earth ground

9.18 What is the process of using a device to create a connection between two devices that both seek a ground?

    a. Multi-grounding

    b. Bonding

    c. Overflowing

    d. Sheathing

9.19 What is the code established to protect persons from the risks of using and working with electricity, which if improperly handled, could result in the damage of property and the loss of life?

    a. AGFA

    b. NIOSH

    c. OSHA

    d. NEC

9.20 You are a cable technician installing new cable in a facility that builds and repairs computers. You open the door to the main computer room and create an electrical spark by touching the metal door knob. What is this called?

    a. ASP

    b. ESD

c. ETS

d. NET

9.10   Challenge Exercises

Challenge Exercise 9.1

In this exercise, you practice safe ESD protection. ESD can ruin computer components and circuit boards. To prevent ESD, you need to ground yourself. In this exercise, we look at how to prevent or lessen the chance of accidentally damaging computer components because of a buildup of static electricity. To complete this lesson, your instructor should provide you with an ESD wrist strap and floor mat.

When working on computer equipment, make sure you are properly grounded. This is because when you see or feel a spark/shock and disperse your static electricity, this ESD is powerful enough to damage nearly any component or circuit board.

9.1.1 Ground yourself with an ESD wrist. To do so, fix the strap to your wrist and then connect the cable that is attached to the strap to a grounded source, such the metal part of your desk. The wrist strap allows you to handle sensitive equipment without damaging it because you are grounded via the strap. Electricity naturally follows a path to ground, so instead of it discharging from your finger to the computer (and potentially destroying it), the charge can travel from you to ground via the strap. Remove the wrist strap.

9.1.2 Stand on the ESD floor mat, which can help you remain grounded. As you stand on the mat, it disperses the charge for you without you having to wear a strap. Touch anything that is grounded, thus grounding yourself.

9.11   Challenge Scenarios

Challenge Scenario 9.1

You are the cable installer working at a customer's site, which is in a manufacturing plant. The site is a hazardous area. You see a small fire coming out of an electric socket. Which class of fire extinguisher do you use to stop the fire? What are the consequences if you choose the wrong fire extinguisher?

CHAPTER 10

Electrical Protection Systems

Learning Objectives

After reading this chapter, you will be able to:

- Understand the importance of electrical protection in relation to telecommunications

- Identify different types of protection and their uses

- Recognize the different standards and codes covering electrical protection

- Be aware of when and how to implement these systems

As with any electronic device in your home, telecommunications systems require protection from electrical surges, lightning, built-up static, and other sources of electrical charges. These voltages can damage the installed cabling structure as well as the equipment connected to it. With a properly implemented electrical protection system, you can mitigate and sometimes eliminate the damage to your telecommunications system should it fall prey to one of these dangers.

The most important item to note during our discussion of electrical protection systems is that, although we will cover several styles and methods of electrical protection including written standards, the system you install must conform to your local codes. The standards that we discuss here are guidelines only. Your local codes are requirements even if they are in opposition to the guidelines presented here. The one possible exception to this is our discussion on the National Electrical Code (NEC), as most municipalities have adopted it as their local electrical code; however, some regions may have additional requirements above what is in the NEC. Whenever you are installing any type of telecommunications system or its supporting structures, it is advisable to know and abide by your local electrical codes.

10.1   Grounding and Bonding Connections

As discussed in Chapter 9 "Safety Considerations," a **ground**, according to the NEC, is "a conducting connection, whether intentional or accidental, between an electrical circuit or equipment and the earth or to some conducting body that serves in place of the earth." Given that definition, then grounding, or **earthing** as it's known in the international community, is creating a conducting connection to the earth or a body that serves as the earth. **Bonding** is the creation of a relationship between two devices that need to seek ground, and the method used to produce a good electrical contact between multiple metallic parts for the purpose of sharing ground. With these definitions, you can see how grounding and bonding go together. Bonding creates a permanent pathway for the current to travel, and grounding connects that path to the earth or something acting as the earth.

In the next section, we cover the most important standards and codes that apply to electrical protection systems.

10.2   Standards and Codes

As with any other aspect of network cabling, electrical protection for telecommunication systems has generic requirements outlined in written standards and codes. You may remember from Chapter 2, "Cabling Standards and Specifications," that standards are recommendations, requirements, and codes typically written into local law. Standards are performance-oriented whereas codes are safety-based.

10.2.1   National Electrical Code (NEC)

The primary section of the NEC that covers grounding is Article 250, with additional requirements outlined in other articles that cover specific areas, such as Article 800 for communications circuits. Within this section, the general requirements for grounding and bonding are discussed along with: specific requirements on the location of grounding connections; types and sizes of grounding conductors; methods of grounding and bonding; and conditions in which isolation or insulation may substitute for grounding. Section III in NEC Article 800 covers protection for communications cabling. It defines primary protectors and secondary protectors, and their places within the system. Additionally, Section IV covers grounding methods for cable and primary protectors, including outlining different situations such as one- or two-family dwellings, mobile homes, and buildings or structures with or without grounding means.

10.2.2   American National Standards Institute/National Fire Protection Association (ANSI/NFPA) 780

NFPA 780, titled "Standard for the Installation of Lightning Protection Systems," is another code used for the installation of electrical protection systems for communications cabling. Although the NEC and NFPA 780 seem to have definitions that conflict with each other, in reality, each one covers different systems and both are useful when designing and installing systems. We cover these differences where applicable throughout this chapter.

10.2.3   J-STD-607-A

J-STD-607-A, titled, "Commercial Building Grounding (Earthing) and Bonding Requirements for Telecommunications," supersedes ANSI/TIA/EIA-607-A. Jointly prepared and copyrighted by the Telecommunications

Industry Association/Electronic Industries Alliance (TIA/EIA) and the Alliance for Telecommunications Industry Solutions (ATIS), its purpose is to provide a uniform grounding and bonding infrastructure to support telecommunications systems and equipment within commercial buildings. As stated earlier, conformance with this standard does not override compliance with the NEC or local codes; rather, compliance with those documents is required to conform to this standard.

In the next section, we explore the topic of electrical exposure, how the codes and standards deal with electrical exposure, and how to minimize the affects of exposure.

10.3   Electrical Exposure

Electrical exposure can come in many forms and, as we discuss in this section, has many definitions. The primary consideration that the designer or installer must contend with is that electrical exposure may not always be accurately determined. Knowledge of local codes and standards along with examination of any available site records can all assist in the assessment of a site for exposure. According to NEC Article 800.2, electrical exposure is "a circuit that is in such a position that, in case of failure of supports and insulation, contact with another circuit may result." Interpretation of this definition shows us that it only accounts for protection from exposure to other live circuits; however, further examination of the section reveals coverage for other types of exposure as well. Perhaps the definition we should use is that **exposure** refers to a telecommunications cable's susceptibility to introduced electrical currents not used to transmit the telecommunications signal itself.

10.3.1   Lightning Exposure

Let us examine where we find exposure risks by starting outside of the building, where telecommunications cable is at the most risk. It is here that NFPA 780 is relevant. NFPA 780 defines exposure as anything above the ground and outside the **zone of protection**, which is an area that is under or nearly under a lightning protection system. A representation of a zone of protection is shown in Figure 10.1.

As you can see in the diagram, anything outside the dashed lines is unprotected. This is due to the potential for a lightning strike to use the telecom-

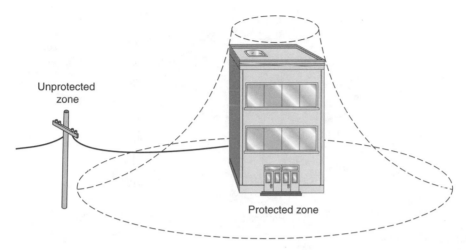

Unprotected
zone

Protected zone

Figure 10.1
Zone of protection

munications cable as its path to ground, which significantly decreases within the zone because there is a better path for it to use within the building and its grounding system. Does this mean that we can forego protection within the building? Absolutely not, as indicated by the NEC. Within Article 800.90 (A), installation of a listed primary protector is required on each circuit run aerially, in whole or in part. This prevents any exposure of the telecommunications circuit to light or power conductors operating at over 300 volts from harming anyone. One additional requirement for areas with potential exposure to lightning is that primary protectors must be present on each end of the cable. According to the NEC, lightning exposure always exists unless one or more of the exceptions listed in Fine Print Note (FPN) Number 2 within this section exist. The pertinent section of FPN Number 2 is as follows:

Interbuilding circuits are considered to have a lightning exposure unless one or more of the following conditions exist:

(1)   Circuits in large metropolitan areas where buildings are close together and sufficiently high to intercept lightning. [These circuits are within zones of protection.]

(2)   Interbuilding cable runs of 42 m (140 ft) or less, directly buried or in underground conduit where a continuous metallic cable shield or a continuous metallic conduit containing the cable is bonded to each building grounding electrode system.

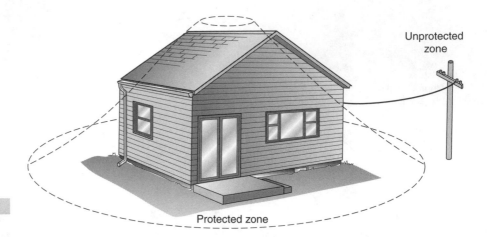

Figure 10.2

Cone of protection

(3)   Areas having an average of five or fewer thunderstorm days per year and earth resistivity of less than 100 ohm-meters.

The zone of protection, as defined, covers larger buildings such as city high rises. However, smaller structures are represented using the same type of philosophy, called the *cone of protection*, which is illustrated in Figure 10.2.

In both the zone of protection and cone of protection, the use of aerial cable is evident. However, do not think that if you use a buried cable that you are immune to lightning exposure. Aerial cable typically has high-voltage cable strung above it that can intercept and divert direct lightning strikes. This can help, but does not counteract the need for protectors. Additionally, a buried cable can attract ground strikes within certain distances depending upon soil resistance. This distance is typically 7 to 20 feet. For situations using underground cable, we can extend the zone of protection concept below the zone although it is a smaller area. Figure 10.3 illustrates this concept.

As you can see, depending on the situation, the underground cable may fall within the zone of protection of the building. More often, this is not the case and other factors, such as metal fences and underground shield conductors, can affect the zone. Other conductive structures can help or hurt the zone boundary as the case may be.

NFPA 780 includes a risk assessment guide that you can use to help determine the loss estimate due to lightning. This assessment includes factors such as type of structure and construction, location of the building, the fre-

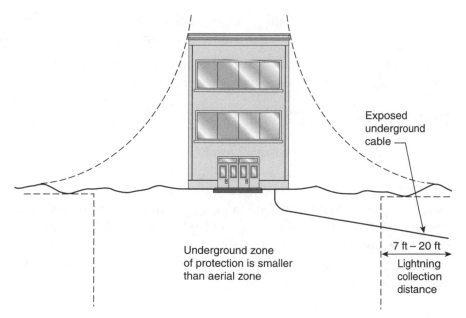

Exposed
underground
cable

Underground zone
of protection is smaller
than aerial zone

7 ft – 20 ft

Lightning
collection
distance

Figure 10.3
**Underground zone of
protection**

quency and consequences of lightning strikes, and the building's occupancy and contents. The assessment that is directly related to the site, its surroundings, and environment accounts for items such as thunderstorm activity, structural materials like steel, and any taller adjacent buildings. You should also research the area for any historical information and records of damage by lightning. You can consult with experts in the area and, above all else, refer to the local jurisdictional authority. Whenever in doubt about a site's potential exposure, err on the side of caution and consider it exposed.

10.3.2 In-Building Exposure

The danger of exposure does not end with protection from exterior elements. There are dangers inside the building that can cause just as much damage if left unprotected. The largest potential issue here is static buildup. Static electricity has the potential to release a large charge, which can damage circuits. There are preventative measures you can take to help mitigate this danger, such as a good grounding system and enforcing the use of electrostatic discharge (ESD) wrist straps when working on equipment. Another exposure risk is **induction**, which is stray voltage introduced onto telecommunications cables from higher voltage lines that run in parallel.

The mitigation strategies for this is to keep cables separate from each other as far as possible, allowing only perpendicular crossing.

In the next section, we discuss electrical protectors—primary, secondary, and enhanced—and how to install primary protectors.

10.4   Protectors and Terminology

In the best of conditions, electricity travels at the speed of light, which is roughly one foot per nanosecond. In ordinary conditions, the speed of an electrical surge is closer to 60% to 80% the speed of light. Protectors are used to "protect" equipment, buildings, etc. from electrical surges while allowing the flow of electricity. The NEC defines several layers of protection and the protectors involved. In the following sections, we cover these items along with other terminology, as you will need to know these in order to understand proper electrical protection.

10.4.1   Primary Protectors

Primary protectors come in different forms, such as carbon blocks, gas tubes, and solid state. Each are briefly discussed, as follows:

- **Carbon blocks:** One of the original protection arrestors. A carbon block has two electrodes. When a certain voltage is reached (in the 300 to 1,000 V range), an arc occurs in the air gap between the electrodes. The overvoltage is brought to a ground conductor. Once this condition is no longer present, carbon blocks return to their original protective state; however, they will cease to function with certain conditions. If a surge occurs for an extended period of time or if a fault overheats them, they will permanently short requiring replacement. Being the least expensive option, carbon blocks wear out quickly and can allow voltage leakage inducing noise especially on voice circuits. Carbon blocks allow a surge to travel nearly two miles on a cable, past a protector.

- **Gas tubes:** Similar to the way carbon blocks work, gas tubes are filled with an inert gas, which replaces the air gap and actually allows for a wider gap. This gas allows them to be more sensitive and can be set to arc at lower voltages giving them better protection capabilities than carbon blocks. Gas tubes allow a surge to travel up to one mile on a cable, past a protector.

- **Solid-state:** The fastest, highest-power technology available. Solid-state protectors use semiconductors, limiting the surge distance to only 1 to 5 feet past the protector due to the speed at which they react. Solid-state technology is the most expensive of the three primary protection options, but is the longest lasting.

NEC Article 800.90 recognizes two categories of primary protectors: fused and fuseless. It also specifies the conditions under which each is used. When installed, both types essentially operate in the same way when encountering excessive current. The exposed side causes the fuse or fine gauge fuse wire, in the case of the fuseless type, to break the connection open between the exposed circuit and the indoor wiring or the grounding conductor, thereby stopping the over voltage from causing damage.

10.4.2   Secondary Protectors

The primary protector's function is to prevent an excessive amount of current or overvoltage from passing along the telecommunications cable and causing damage; however, there are other dangers that primary protectors are not designed to catch. This is where secondary protectors come into play. Secondary protectors are not required by the NEC, but are commonly used for additional protection beyond the primary protector. These devices are usually designed to not only protect from an overvoltage but also to **sneak current**. Sneak current typically comes in two forms: (1) voltages that are too low for the primary protectors to catch and (2) a current drawn by faulty equipment that overheats the premises cabling. The following are examples of secondary protectors:

- **Heat coil:** A metallic device that resides inside the primary protector (overvoltage) device that reacts to the heat caused by an extended low current situation. When high enough, the heat melts the coil. At that point, a spring-loaded shorting bar shorts to ground, protecting the circuit. However, excessive overcurrent can result in fire. Because this is a one-time use design, inspect the protector regularly and replace once blown.

- **Sneak current fuse:** Similar to fuses found in automobiles and older construction homes, a fuse opens a circuit when a constant overcurrent situation occurs. Once the circuit is open, the load is removed and the danger of fire is avoided. Once a fuse blows, however, it must be replaced. Do not confuse this fuse with a primary protector fuse since they are designed for different functions and one cannot replace the other.

- **Positive temperature coefficient (PTC) resistor:** A resettable resistor that works similar to a sneak current fuse. However, PTC resistors do not have to be replaced after an overcurrent situation is removed.

Some manufacturers of telecommunications equipment recommend the use of these protection devices and may specify ones for their equipment. Some of these devices are designed to be installed anywhere within the system, while others reside behind the primary protector. Additionally, some primary protectors have options for adding secondary protection for reasonable cost.

10.4.3   Enhanced Protection

Other types of protection available include specifically designed devices for data circuits. The following are examples of these components and their uses:

- **Clamping diode:** Fast-acting voltage limiters that operate in low-voltage environments (for example, less than 50 V).

- **Isolation transformer:** Devices that provide line isolation, noise filtering, and enhanced common-mode surge suppression. These transformers, often found in data equipment interfaces, help to reduce interference and permit high-speed data transmissions.

- **Filter:** Can limit a circuit's bandwidth and improve impedance with regard to surges.

Some manufacturers of these devices include primary and secondary protection as an integral part of their product. If this is the case for the product you want to use, be sure that it is listed per the Underwriters Laboratories (UL) 497 requirements for primary and secondary protection.

10.4.4   Primary Protector Installation

NEC Article 800.90 and 800.100 cover the installation requirements of primary protectors. Primary protectors need to be located within or adjacent to the building they are serving and as close to the point of entrance as possible. A conductor listed for this purpose must ground the device, be made of copper or some other corrosive-resistant material of a size no smaller than 14 AWG, and be either stranded or solid in construction. This conductor, connected to the grounding electrode by appropriate means, complies with Article 250.70. The NEC requires that the grounding conductor be not more than 20 feet in length and run in as straight a line as possible. These requirements are for a typical installation and there are additional requirements that you may need to follow in special cases. Always be sure to check with your local jurisdictional authority to ensure compliance.

In the final section of this chapter, we cover the various components of an electrical grounding system and tie them together into a representation of a complete telecommunications grounding system.

10.5   Grounding Telecommunications Equipment

Buildings are required to have an electrical grounding system as specified by the NEC and a lightning control system as per the NFPA 780. Although these systems are required for safety considerations, a separate system designed for telecommunications equipment is good practice in order to ensure the equipment's reliability and performance. Such systems are additional protection for telecommunication systems and can provide for any special needs required by current cabling practices and high-speed sensitive equipment in use today that standard safety does not cover. Keep in mind that the intention of a telecommunications grounding system is not to replace an electrical system, but rather to supplement it. The best reference to use for this is the J-STD-607-A standard.

10.5.1   Ground Source

The preferred source for the telecommunications ground is the building's electrical ground. This is because the equipment gets its power from the electrical system. In cases where no system exists, install an acceptable ground source in accordance with NEC Article 800.100. This is in the form of a rod that is a minimum of 1/2 inch in diameter and at least 5 feet in length, driven into the ground. The connecting conductor to this source is a copper conductor, size 6 AWG or larger, and may be insulated. If insulated, the insulation must have labeling specifying this purpose. If placed within a metallic conduit, bond the conductor to the conduit at each end using either a 6 AWG bonding conductor or a grounding bushing. Label all grounding points and structures in their characteristic green color with clearly readable non-metallic labels as close as possible to the termination point. Figure 10.4 shows the required information on the label, as required by the J-STD-607-A standard. There may be additional requirements outlined in ANSI/TIA/EIA-606-A.

IF THIS CONNECTOR OR CABLE IS LOOSE OR MUST BE REMOVED, PLEASE CALL THE BUILDING TELECOMMUNICATIONS MANAGER

Figure 10.4

Example of a grounding label

10.5.2   Telecommunications Main Grounding Busbar (TMGB)

The telecommunications grounding system begins with the **telecommunications main grounding busbar (TMGB)**. This is where the separation between the electrical grounding system and the telecommunications grounding system begins. The J-STD-607-A specifies the TMGB to be a pre-drilled, copper busbar capable of using standard sized ground lugs. It should have space for present requirements as well as future expansion. The TMGB is a minimum of ¼ inch thick, 4 inches wide, and may be variable in length. The bar should be electro-tin-plated to reduce resistance, but if it is not, you should clean it and apply an anti-oxidant to the contact area to reduce resistance and corrosion. The TMGB is ideally located in the building's telecommunications entrance facility or placed to minimize the length of the grounding conductor as a rule. It serves as the grounding point for any telecommunications equipment within the room and as the central attachment for the telecommunications bonding backbone, which is discussed in the next section. Ordinarily there is only one TMGB, but there can be extensions of it called telecommunications grounding busbars (TGBs). Figure 10.5 shows a TMGB.

When the TMGB is in the same room as an electrical panel, the standard requires that the panel be bonded to the TMGB, and that the TMGB be placed as close to the panel as possible while maintaining NEC-required clearances. The standard requires that bonding to the TMGB consists of exothermic welding, listed compression two-hole lugs, and suitable one- or two-hole non-twisting lugs or other untwistable compression lug. All metallic raceways (pathways in which cabling is run) located in the same area not already bonded to the grounding conductor must bond to the TMGB. Insulated from its support, the standard recommends a minimum of 2 inches of clearance from the wall to allow access. Its location must be

Figure 10.5
Typical TMGB. Courtesy of
Chatsworth Products Inc.
2005.

accessible to all telecommunications personnel, and its height should accommodate both underfloor and overhead cabling.

10.5.3   Telecommunications Bonding Backbone (TBB) and Grounding Equalizer (GE)

The **telecommunications bonding backbone (TBB)** and the **grounding equalizer (GE)**, formerly known as the telecommunications bonding backbone interconnecting bonding conductor, are the next piece in the telecommunications grounding system. The TBB connects the TMGB to all the TGBs and, if there are multiple TBBs in the design, they are connected via a GE. The standard requires the TBB in a multi-story building be connected with a GE at the top floor and at every third floor inbetween at a minimum. Both the TBB and GE are sized depending on their length, with the minimum size being a 6 AWG. Table 10.1 shows the recommended size of the conductor based on length.

The standard recommends not having splices within any run of the TBB; however, if one is necessary, the splice must be accessible and located within the telecommunications spaces or rooms. Use the same methods of bonding as when bonding to the TMGB.

10.5.4   Telecommunications Grounding Busbar (TGB)

The **telecommunications grounding busbar (TGB)** is simply a miniature version of the TMGB. Its dimensions are the same, with the exception being that the width is 2 inches instead of 4 inches. Figure 10.6 shows a typical TGB.

Everything else about the TMGB applies to installation and bonding requirements. All of the conductors must be of the same gauge wire when

TABLE 10.1   TBB Size Chart

TBB Length linear meters (ft)	TBB Size (AWG)
Less than 4 (13)	6
4 to 6 (14 to 20)	4
6 to 8 (21 to 26)	3
8 to 10 (27 to 33)	2
10 to 13 (34 to 41)	1
13 to 16 (42 to 52)	1/0
16 to 20 (53 to 66)	2/0
Greater than 20 (66+)	3/0

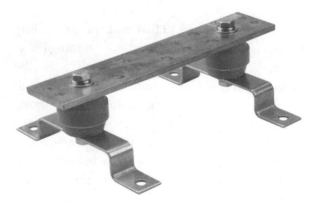

Figure 10.6

Typical TGB. Courtesy of Chatsworth Products Inc. 2005.

bonding conductors such as the TBB and GE to the TGB. Additionally, if there are multiple TGBs within a room, bond them together with the same sized conductor as the TBB or with a splice bar.

10.5.5   Other Grounding Connections

The TGB and TMGB can be the connecting point for other telecommunications-related items as well. As mentioned earlier, any raceway not already bonded to the grounding conductor bonds to either the TGB or TMGB. The primary protector ground bonds to the TMGB, with this bonding conductor maintaining a minimum 1-foot separation between it and any DC power, switchboard, or high frequency cables even if encased in electric metallic tubing (EMT). If the building's steel frame is accessible, either directly in the room or adjacent to where the TMGB or TGB is located, it is also bonded to it. Any incoming exterior cable feeds with metallic sheaths must bond to the TMGB including antenna cables, fiber-optic cables and housings, and inter-building backbone cables such as riser cables. Figure 10.7 shows an illustration of a complete telecommunications grounding system.

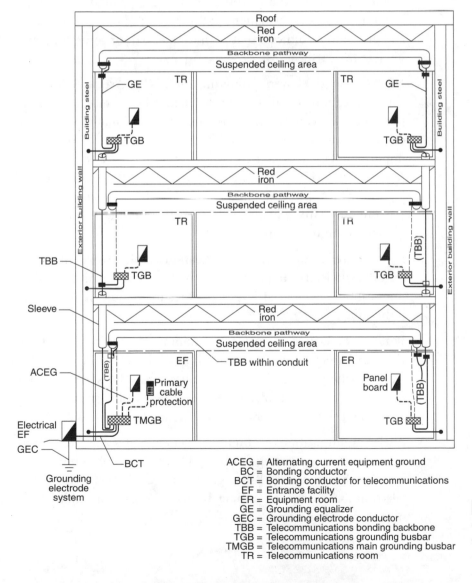

ACEG = Alternating current equipment ground
BC = Bonding conductor
BCT = Bonding conductor for telecommunications
EF = Entrance facility
ER = Equipment room
GE = Grounding equalizer
GEC = Grounding electrode conductor
TBB = Telecommunications bonding backbone
TGB = Telecommunications grounding busbar
TMGB = Telecommunications main grounding busbar
TR = Telecommunications room

Figure 10.7

Complete telecommunications grounding system.
Courtesy BICSI 2005.

10.6 Chapter Summary

- A ground is a conductive path or connection, typically a grounding rod, which carries an electrical charge to the earth or some other body that absorbs the charge. Earthing is the creation of a conducting connection to the earth or a body that serves as the earth. Bonding is the creation of a relationship between two devices that need to seek ground, and the method used to produce a good electrical contact between multiple metallic parts for the purpose of sharing ground. Therefore, bonding creates a permanent pathway for the current to travel, and grounding connects that path to the earth or something acting like the earth.

- NEC Article 250 covers grounding, and NEC Article 800 covers communications circuits. Section III in NEC Article 800 covers protection for communications cabling. Additionally, Section IV covers grounding methods for cable and primary protectors. Other important standards and codes include National Fire Protection Association (NFPA) 780, titled "Standard for the Installation of Lightning Protection Systems," and J-STD-607-A, titled "Commercial Building Grounding (Earthing) and Bonding Requirements for Telecommunications."

- Exposure refers to a telecommunications cable's susceptibility to introduced electrical currents not used to transmit the telecommunications signal itself. Types of exposure include lightning and in-building.

- Protectors are used to "protect" equipment, buildings, etc. from electrical surges while allowing the flow of electricity. Primary protectors include carbon blocks, gas tubes, and solid-state semiconductors. Secondary protectors include heat coils, sneak current fuses, and positive temperature coefficient (PTC) resistors.

- A telecommunications grounding system includes a telecommunications main grounding busbar (TMGB), telecommunications bonding backbones (TBBs), grounding equalizers (GEs), and telecommunications grounding busbars (TGBs), in addition to other grounding connections.

10.7 Key Terms

bonding: The creation of a relationship between two devices that need to seek ground, and the method used to produce a good electrical contact between multiple metallic parts for the purpose of sharing ground.

earthing: Another term for grounding, the creation of a conducting connection to the earth or a body that serves as the earth. Earthing is a commonly used term outside of the United States.

exposure: A telecommunications cable's susceptibility to introduced electrical currents not used to transmit the telecommunications signal itself.

ground: A conductive path or connection, which is typically a grounding rod that carries an electrical charge to the earth or some other body that absorbs the charge.

grounding equalizer (GE): Formerly known as the telecommunications bonding backbone interconnecting bonding conductor, connects multiple TBBs together in a multi-story building.

induction: Stray voltage introduced onto telecommunications cables from higher voltage lines that run in parallel.

sneak current: Voltages that are too low for the primary protectors to respond to, or a current drawn by faulty equipment that overheats premises cabling.

telecommunications bonding backbone (TBB): Connects the TMGB to all the TGBs. If multiple TBBs are included by design, they are connected via GEs. The TBBs in a multi-story building connect with a GE at the top floor and at every third floor inbetween, at a minimum.

telecommunications grounding busbar (TGB): A miniature version of the TMGB.

telecommunications main grounding busbar (TMGB): Represents the separation between the electrical grounding system and the telecommunications grounding system. The TMGB is a pre-drilled copper busbar, ideally located in a building's telecommunications entrance facility or placed to minimize the length of the grounding conductor. The TMGB serves as the grounding point for any telecommunications equipment within the room and as central attachment for the TBB.

zone of protection: An area that is under or nearly under a lightning protection system.

10.8   Challenge Questions

10.1 What is another term for grounding?

a. Bonding

b. Earthing

c. Insulating

d. Inducting

10.2 What article of the NEC primarily covers grounding?

a. 250

b. 569

c. 700

d. 800

10.3 Which NEC article covers protection for communications cabling, such as primary and secondary protectors?

a. 250

b. 569

c. 700

d. 800

10.4 What is bonding?

10.5 Standards are _____ oriented whereas codes are _____ based.

10.6 Which standard, jointly prepared and copyrighted by the TIA/EIA and the ATIS, provides a uniform grounding and bonding infrastructure to support telecommunications systems and equipment within commercial buildings?

10.7 What is the name of an area that is under or nearly under a lightning protection system?

a. Demilitarized zone

b. Zone of induction

c. Zone of protection

d. Ordinance zone

10.8 True or False: Buried cables are immune to lightning exposure.

10.9 What are the three main types of primary protectors?

a. Solid-state

b. Gas tubes

c. Heat coils

d. Carbon blocks

10.10 How does a zone of protection differ from a cone of protection?

10.11 What are the three main types of secondary protectors?

a. TMGBs

b. Heat coils

c. Sneak current fuses

d. PTC resistors

10.12 What is a TMGB?

10.13 How does a TMGB differ from a TGB?

10.14 What is the purpose of a TBB?

10.15 According to the NEC, a primary protector that is located within or adjacent to the building it is serving, and as close to the point of entrance as possible must adhere to which of the following requirements? (Choose all that apply.)

a. Capable of grounding the device

b. Be made of copper or some other corrosive-resistant material

c. Stranded or solid construction

d. Cannot exceed 30 feet in length

10.9   Challenge Exercises

Challenge Exercise 10.1

In this exercise, you familiarize yourself with a grounding system that incorporates J-STD-607-A, titled "Commercial Building Grounding

(Earthing) and Bonding Requirements for Telecommunications." To complete this exercise, fill in the labels represented by blank lines in Figure 10.8, and then answer the questions that follow:

Label 1: _____

Label 2: _____

Label 3: _____

Label 4: _____

10.1.1 Why is the GE located where it is?

10.1.2 If the length of the TBB is 35 feet, what is the AWG size?

10.1.3 What is the width of the TMGB?

10.1.4 What is the width of the TGBs?

Figure 10.8

Facility with a telecommunications grounding system

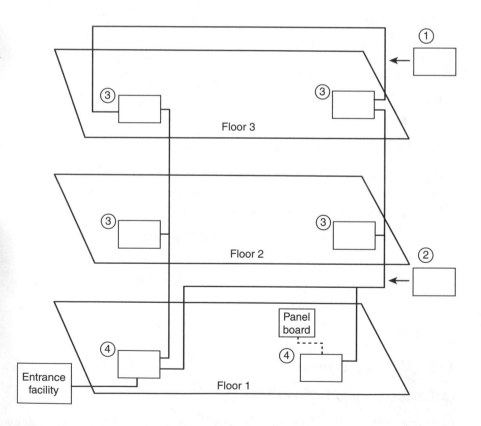

Challenge Exercise 10.2

In this exercise, you learn how to find and review local electrical codes, and compare them to NEC 250, NEC 800 Sections III and IV, J-STD-607-A, and NFPA 780. Conformance with the national standards and codes does not override compliance with local codes. To complete this exercise, research and summarize your local codes, and then compare your findings to NEC 250, NEC 800 Sections III and IV, J-STD-607-A, and NFPA 780.

Challenge Exercise 10.3

In this exercise, you learn the importance of grounding and how to assess the need for grounding with high-performance equipment. As a cable installer, you will need to understand (and be able to talk about) the importance of grounding for many reasons, safety being the most important. It's also helpful to be able to determine if a system is grounded properly for your own safety and the safety of others.

The requirements for ground resistance levels are becoming more stringent because of the lower operating voltages and higher operating speeds in today's electronics. In addition, electronic devices come out each year that raise the demand for the electricity that feeds them. High-performance grounding is more than achieving a low earth resistance. When using a high-performance ground, you should expect to get long-lasting performance that is stable.

Just like a well-performing cabled network, the only way to get high-performance grounding is to design it. To complete this exercise, use the information from the chapter and other sources that you find on the internet and in print at your local library to design a high-performance grounding system.

Challenge Exercise 10.4 (Optional)

In this exercise, you use an ammeter (a device that measures the flow of electricity in amperes) to try to detect wiring errors. Your instructor will have an incorrectly wired structure set up for you to practice on. With the instructor's guidance and supervision, attach the ammeter clamp to a cable (or conduit if the cable is hidden) and watch the results on the ammeter display. What might a reading of 1 to 10 amps indicate?

10.10   Challenge Scenarios

Challenge Scenario 10.1

You are one of three associate project managers on a large-scale, multi-story building project. You are responsible for the structures' electrical, grounding, and bonding specifications/requirements as they relate to the physical communications cabling plant. What steps should you take to properly plan and design the cable plant?

Challenge Scenario 10.2

You are a technician responding to a call to investigate why a recently installed Category 6 cable plant is not providing the expected bandwidth and performance levels that the final documentation indicated. You need to determine if the cable is damaged and needs to be replaced, if faulty connecting hardware is the problem, or if it's a poor installation.

Your investigation found that the reduction of system bandwidth and performance is due to electrical and communications cable being run in the same common 4-inch conduit. You know that this is not standards compliant and that it violates codes specified within the NEC, NFPA, and established local codes. What do you do to correct the problem?

CHAPTER 11

SOHO and Residential Infrastructure Technology

Learning Objectives

After reading this chapter you will be able to:

- Discuss HomePNA networking systems
- Discuss Powerline Carrier (PLC) networking systems
- Understand wireless systems, including wireless networking standards, wireless network equipment and layouts, wireless technologies, and security
- Understand SOHO options with broadband coaxial cabling, IEEE 1394 (FireWire), plastic optical fiber (POF), and direct broadcast satellite (DBS)
- Understand the basics of home automation systems

What is SOHO? **Small Office Home Office** (SOHO) is a term that is not easily pinned down by a simple definition. SOHOs are generally home offices that are used for work, either for entrepreneurial reasons or for work performed for a company. The people involved in a SOHO range from self-employed individuals to employees of a major corporation. Many terms are used to describe who utilizes a SOHO. These terms range from home-based business people, free agents, independent contractors, telecommuters, freelancers, and consultants, to any other independent professionals. This doesn't exclude those who work from home for their companies via a virtual private network (VPN). The definition is broad, so to sum it up, a SOHO is just that—an office in your home that is used for doing work.

Because a SOHO is located in the home, it is essential that you as the cable technician have a firm understanding of what this means in terms of wiring in the home. In this chapter, we cover SOHO wiring in detail. Home computer networking design and installation involves connecting computers and peripheral equipment with some type of media, such as copper cable or wireless network interface cards (NICs). In most business environments, the office buildings are already supporting cabling. Conduits, ducts, and in most cases, a wiring system exist. Most homes were not built with the need for a SOHO in mind. Because many homes are 25 years or older, you will not find much in the way of residential wiring that will facilitate data transfer within the home.

Beyond the SOHO is the recent trend towards home automation. *Home automation* is the process of connecting most of the electronic devices in your home with a central control system. This allows you to more efficiently use your electricity; it can actually lower your heating, cooling, and lighting costs. It also makes managing these devices easier. As more and more items in the home become networked, controlling them may be very cumbersome. Home automation solves this and many other issues. This chapter covers this in detail along with what you, as the cable installation technician, need to know about residential wiring.

11.1   HomePNA Systems

When discussing SOHO technology, the first stop is on some of the systems currently used in the home to achieve networking. **Home Phoneline Networking Alliance** (HomePNA) is one such technology. HomePNA is the standard that was adopted for using copper phone lines within the home as a way to connect network devices. The HomePNA standard is based on a set of standards that enables voice and data transmissions to utilize bandwidth that already exists in the home without a need for rewiring. The bandwidth utilized is on the home's current telephone cabling. Because most homes today have telephone cabling, HomePNA is a perfect fit when new cable installations are not possible or unwanted. The way it works is by using **frequency division multiplexing (FDM)**. FDM divides the current bandwidth on the cable so multiple solutions can use it. This means that you can use HomePNA while still using your phones, faxes, and anything else connected to the phone system cabling. A very good example is that someone could talk on the phone while another person is using the same line to access the internet or share files on networked computer systems.

Figure 11.1 simplifies the HomePNA network. For example, you have a SOHO environment in which two upstairs PCs need to connect to the

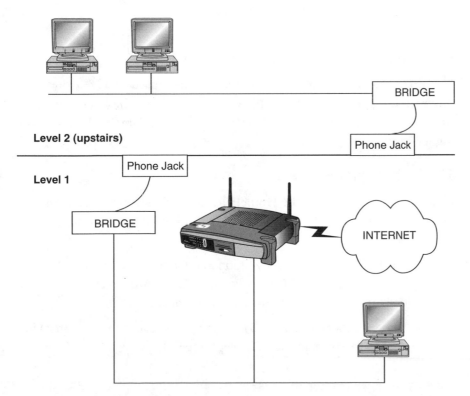

Figure 11.1

A HomePNA system

internet. If you don't use wireless and need this connectivity, you need to run a new cable from the downstairs (Level 1) portion of the home to the upstairs (Level 2) portion of the home.

When connecting devices to a HomePNA network, it is imperative to remember the following rules:

- The phone line (cable) you use does not have to be an active phone line. The phone and the phone number associated with the line have nothing to do with the cabling used. HomePNA only uses a portion of this cable for other purposes.

- Ensure that all HomePNA devices are connected to the same telephone wire pair.

- As seen in Figure 11.1, you need bridging devices so that your HomePNA networks can bridge your current Ethernet network to your telephone cabling, which normally uses RJ-11 termination.

HomePNA is a standards-based technology. Currently, it has evolved through three different versions.

- **HPNA 1.0:** Operates at 1 Mbps

- **HPNA 2.0:** Operates at 10 Mbps

- **HomePNA 3.0:** Operates at 128 Mbps

In 2002, the HomePNA organization (*http://www.homepna.org*) announced the approval of HomePNA 3.0, which is the next generation of this specification. The main advantage of this update is obviously the higher bandwidth.

11.1.1   HomePNA Hardware

Now that you know what the HomePNA network is, how do you connect devices to it? HomePNA networks use two different types of adapters to interface the network:

- **NIC:** This uses RJ-11 termination to interface with the current phone system in the home. Generally, the NIC has two RJ-11 connections: one for the wall phone outlet jack and one for a telephone.

- **Universal Serial Bus (USB) adapter:** This uses the computer USB port to interface to the wall telephone jack.

As a cable technician, you want to consider HomePNA technology as an alternative to those customers who may not want new cable installed in their homes. They can use the pre-existing cabling from their telephone cabling. This solution does not require powering of the cable, it is low voltage, and it can be set up quickly and easily.

11.2   Powerline Carrier (PLC) Systems

Another form of connecting devices in a residential home is to use existing power wiring (60 Hz/120 volt AC) in the home. When you need to get a SOHO up and running, there may be instances where it's either too costly to run new cable or it's not worth the effort of tearing up a home to install it. You may be able to use pre-existing wiring instead. As with HomePNA, the **Powerline Carrier (PLC)** system also uses the pre-existing telephone wiring in the home. You can use the power wiring in the home if you use special interface adapters and software to connect the devices. One thing to consider is that power lines in the house can be very noisy media. That means this noise will cause disruption with the systems in the form of interference. The HomePlug Powerline Alliance is a group of companies that developed the standard for power line networking. The standard you hear about most often is called HomePlug. PLC systems use pre-existing wiring in the SOHO to connect network devices, sometimes referred to as PLC devices. PLC devices connect to pre-existing home wiring via Ethernet or USB Powerline adapters.

HomePlug power-line networks are very easy to set up, because you only need some spare power outlets and whatever devices you want to connect using the special Ethernet or USB adapters with the HomePlug certification logo on them.

You cannot use surge protectors or power strips in this system because they interrupt the signal.   **NOTE**

11.2.1   PLC Protocols

When HomePNA and HPLA standards are implemented, no new cabling can be installed in the SOHO. Even without the new cabling, there is a lot going on under the hood here. Network devices use protocols to communicate and they communicate over transmission media. That media happens to be the pre-existing home wiring, but there is a new form of technology flowing over it: the **X-10** protocol. This control protocol is designed to use existing home wiring to send relatively low-speed control signals to home network devices as well as other home automation equipment, such as home lighting and security systems.

Pico Electronics, Ltd. developed X-10 in the late 1970s; now it is an industry standard, and X-10 has developed into a brand name. X-10 is the leading home automation protocol used today. With X-10 controllers, which send digital signals over home wiring systems to receivers that are plugged into the SOHO wall outlets, X-10 is transmitted over the home wiring at 60 bps (bits per second).

Another standard of note is Consumer Electronic Bus (CEBus). Developed by the Electronic Industries Alliance (EIA) in the 1980s and released in 1992, the CEBus standard, like X-10, defines how to make products communicate over power lines, coaxial and fiber-optic cabling, etc. You can learn more about CEBus by visiting, *http://www.smarthomeforum.com/ start/cebus.asp?ID=5.*

11.3   Wireless Systems

As a cable installer, it is important to understand a future world without wires. Wireless is a popular technology, and although the question of whether wireless will replace wired networks and systems remains, the answer is simple: Wired networks are cheaper and faster. That doesn't mean that wireless technology is not catching up. Every day, wireless systems are getting better, faster, and more efficient. In this section of the chapter, you discover how wireless systems fit into a SOHO and learn about the different types of wireless systems you are likely to come across in the SOHO environment. In addition, the wireless fundamentals and how wireless systems interact with wired networks are examined.

There are three standards of wireless technologies that you will most likely see: Institute of Electrical and Electronics Engineers (IEEE) 802.11; Home Radio Frequency (HomeRF); and Bluetooth. The following sections examine the capabilities, terminology, design considerations, and important performance issues of these standards.

NOTE ▪ IEEE 802.11, HomeRF, and Bluetooth are incompatible with each other.

11.3.1   IEEE 802.11 Wireless Networking

In June 1997, the IEEE completed its initial work on 802.11, which became the first standard for wireless local area networks (WLANs). This standard opened the door for wireless operating at the 2.4 GHz frequency, operating at speeds of 1 and 2 Mbps. Once this standard was ratified, the IEEE 802.11 working group began on newer standards in three different areas: 802.11a, 802.11b, and 802.11g.

802.11a Standard

In 2001, wireless equipment using the 802.11a standard started hitting the shelves for home owners wanting to run speeds up to 54 Mbps in their homes. The 802.11a standard operates in the 5 GHz radio band using orthogonal frequency division multiplexing (OFDM). Because 802.11a operates in the 5 GHz band, it is not compatible with other 802.11 standards that operate in the 2.4 GHz band such as 802.11 b and g.

802.11b Standard

In 2000, wireless equipment using the 802.11b standard became available for users wanting to run speeds of up to 11 Mbps in their homes. The 802.11b standard operates in the 2.4 GHz radio band using complementary code keying (CCK), a modulation technique that allows for the efficient use of the radio spectrum. This is how it achieves higher data rates over the 802.11 standard operating at 1 and 2 Mbps. 802.11b is compatible with 802.11 and 802.11g standards, but not with 802.11a, which operates in the 5 GHz band.

802.11g Standard

The newest standard to be released is the 802.11g standard. 802.11g is unique in that it brings the speed level up to 54 Mbps (matching that of 802.11a) and uses the same frequency band as the 802.11b standard, which is 2.4 GHz. 802.11g is not compatible with 802.11a; however, it is compatible with 802.11 and 802.11b.

11.3.2   How Does Wireless Work?

In this section, we look at the most common standard, 802.11b, and explain how it works. Instead of having traditional wired networks as seen in

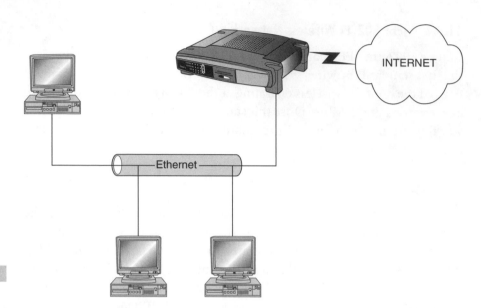

Figure 11.2

Viewing a wired network

Figure 11.2, we have a network that consists of far fewer cabling requirements because the data are sent over radio waves much like the music you hear in your car while driving is sent over radio waves. Data are sent the same way.

Figure 11.3 shows you the layout for devices in a wireless network. These devices must access a central wireless access point in order to have connectivity to resources on the network, such as files or internet access.

As you can see from looking at both setups (Figure 11.2 and Figure 11.3), far less cabling is needed to quickly establish a wireless SOHO network.

When working with wireless technology (such as 802.11b), you need two pieces of equipment. In Figure 11.3, notice that some computers have associated radio waves. This is because each of those computers is equipped with a wireless NIC, which is the first piece of equipment needed. The other piece of hardware needed is the wireless access point (WAP), which is also shown in Figure 11.3. The access point is combined with the router gateway to the internet. You can acquire access points that are separate from the router, but for this example (and what you will most likely be deploying in most SOHOs), the combination of the two (access point and router) is completely acceptable. The access point is fitted with antennas, which is what allows the wireless NICs to communicate with it.

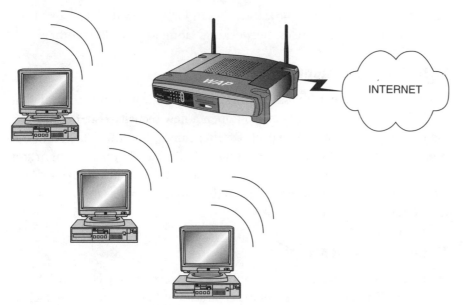

Figure 11.3
Viewing a wireless network

There are two operational modes in wireless networking with 802.11b: infrastructure mode and ad hoc mode. The infrastructure mode consists of at least one access point connected to a wired portion of a network, also known as a distribution system (DS), and can be configured two ways: **basic service set (BSS)** and **extended service set (ESS)**. They communicate via a Service Set Identifier (SSID), which is nothing more than a common name used between the devices so they all know they are part of the same service set.

- **BSS:** Uses one SSID. An access point acts as the central point for all devices wanting to participate on the network.

- **ESS:** The BSS access points communicate among themselves to forward traffic from one BSS to another. This allows devices to communicate with other devices via the ESS.

Ad hoc mode also uses a service set; it is called an independent basic service set (IBSS). When using ad hoc mode, the wireless stations communicate directly with each other; therefore, no access point is used in an ad hoc mode wireless network. One of the biggest downsides of using this technology is that all devices need to be in range of each other. If they are not, they may not have enough power to communicate. Now that you have a clear understanding of wireless, how it works, and its standards, let's look at the other two options for wireless networking: HomeRF and Bluetooth.

11.3.3   Wireless Security

As you learn about wireless networking, it's critical to understand that while forgoing the use of wires, you open a new security risk. Because the data do not traverse a cable, a packet sniffer can capture the data in mid-air. A packet sniffer is a capture and analysis tool. Because of this, it's important to consider security in the SOHO when using wireless technologies. Make sure you use the encryption options in the wireless systems to ensure a better level of security for the WLAN.

11.3.4   HomeRF Networking

HomeRF is also a WLAN standard and it is supported by the HomeRF Working Group. This group consists of many large, well-known companies such as Microsoft, Intel, HP/Compaq, and Motorola. HomeRF is not interoperable with the 802.11 family of wireless LANs or Bluetooth. HomeRF is in its current specification (Version 2.0) and supports up to 10 Mbps in the 2.4 GHz band. HomeRF uses Frequency Hopping Spread Spectrum (FHSS) modulation. HomeRF competes directly with 802.11 standards.

11.3.5   Bluetooth

Bluetooth is yet another wireless technology, but this one is different from 802.11 and HomeRF standards. Bluetooth is designed to connect one device to another device in close proximity with a short range radio link. This is perfect for use within an office, but probably not a great technology to use in a three-story home. Bluetooth was designed to be streamlined, using very small chipsets in the devices that needed to be networked, such as a wireless keyboard and mouse for the home. The range of each radio is approximately 10 meters or so.

Bluetooth also operates in the 2.4 GHz band and uses FHSS modulation. There are two ways you can set up a Bluetooth network: through a point-to-point connection and through a multipoint connection. Point-to-point means that one device is the master and the other is the slave. The multipoint connection refers to a *piconet*. A piconet is like the ad hoc network mentioned earlier in the chapter. In a piconet, up to seven slave devices can be set to communicate with a single master radio. Piconets can be connected to create scatternets. You create a personal area network (PAN) when you network using Bluetooth.

In summary, as a cable technician working within the SOHO environment, it's very likely you will come across at least one if not all of these wireless network technologies. It's important to understand their existence and how they operate.

11.4 Broadband Coaxial Cabling

Coaxial cabling will also show up in SOHO environments. Many internet service providers (ISPs) use cable TV to provide homes with internet access. As a result, many SOHOs use it for internet connectivity. It is important for you to realize coaxial cable's place in the home as well as what types of coaxial cable you may see there. Because coaxial cable was primarily used to carry television broadcasting over long distances, it only makes sense that if the power wiring inside your home and the telephone wires could be used for home networking, so could the cable TV feed.

Coaxial cable is manufactured in a very wide range of types, sizes, qualities, and cost. The SOHO environment typically uses the low-cost coaxial cable most commonly found in cable TV systems worldwide. Coaxial cable used for home cabling applications is most commonly type RG-6. There are a few major differences between the residential-based coaxial wiring and some of the more commonly seen legacy Ethernet (RG-58/Thinnet) networks used in many computer networks prior to the widespread use of twisted pair cabling. RG-6 has a 75-ohm impendence value, whereas RG-58 type coaxial cable has a 50-ohm impendence value. You must use hardware that matches the impendence or you will have installation problems.

Coaxial cable is used for video signal distribution in most modern residential structured wiring designs, so make sure you are familiar with its placement and use within that SOHO environment. **NOTE**

11.4.1 Coaxial Cable Connectors

The most common type of connector used with most coaxial cable deployments is the Bayonet Neill-Concelman (BNC) connector. Other types of adapters are also available for BNC connectors, including a T-connector, barrel connector, and terminator. Coaxial connections are covered in detail in Chapter 8.

11.5   IEEE 1394 (FireWire)

The IEEE 1394 standard is also known as FireWire. It is an extremely fast external serial bus standard used to support transfer rates up to 400 Mbps. Many modern computer devices (PCs, routers, cable modems, and so on) come with this technology installed. Apple originally developed and trade-marked the technology, and uses the name FireWire.

Other names exist, such as i.Link and Lynx, to describe their IEEE-1394 standard used with non-Apple products. You can connect 63 external devices to a single 1394 port.

You may find this technology in SOHO environments.

11.6   Plastic Optical Fiber (POF)

Plastic optical fiber (POF) is a type of optical fiber. POF typically uses poly-methylmethacrylate (PMMA, which is acrylic), a general-purpose resin, as the core material and fluorinated polymers for the cladding material. Most other fibers use glass or quartz; POF uses acrylic to keep costs down. POF is primarily used in SOHOs, digital home automation appliances, and auto-mobile networks. Although it may be used in place of any other optical fiber, POF is used primarily in these situations, which is why you should be famil-iar with it. POF does have some unique advantages besides its lower cost over other fiber cabling. POF uses a large diameter in its core to transmit data. This is good because even if the ends of the fiber become slightly damaged, the transmission most likely will be successful, provided that the damage is contained to the exterior cladding region and not the core region. However, the core of a POF is more sensitive than that of a glass optical fiber because of reflectance issues and refractive properties.

A major drawback with POF is that its larger core size prevents it from transmitting over large distances like traditional glass optical fiber. In the networking world, fiber is often used to connect multiple buildings in a campus environment. In the home, large distances normally are not cov-ered, so POF is a good solution.

11.7   Direct Broadcast Satellite (DBS)

Direct broadcast satellite (DBS) is another typical SOHO solution you may come in contact with as you work in the SOHO environment. DBS is an alternative to most current cable-based television systems. If you have ever seen a dish on top of someone's home for satellite TV, you know exactly what DBS is. DBS provides television programming directly from satellites using small, home-mounted satellite dishes.

11.8   Home Automation Systems

Home automation is becoming very popular; however, it has not been widely adopted due to its cost and complexity. Home automation can actually save money if implemented correctly.

Before we get deep into home automation systems, it's important for you to know why you need to understand these systems. As a cable installer working in SOHOs, you essentially are laying the foundation for future home automation devices by simply networking the home with cable, or setting the home up with PLC systems or X-10. As you learn how to use pre-existing cabling in the SOHO environment (or run new cable), you will realize this cabling is all you need to begin building home automation systems. You may also come across homes that are already set up with home automation systems, and you should understand them so you do not disturb them when you are running new cable.

So what exactly is home automation? It is the process of putting most of your home-based systems (environmental, lighting, computing, etc.) on a central control system. What does this do for you? Say you have your lighting systems set up with home automation. You can program the central control system to turn the lights on only when someone enters the room. Another example is to have the lights turn on each night at a certain time when you leave the house to give the appearance that someone is home when no one may be there.

The beauty of home automation is that your options are virtually unlimited. The examples just provided are two of hundreds of different scenarios you can program into home automation systems. You can control environmental systems as well; your heating and cooling can be controlled so that when a door opens, the air conditioner turns off and energy is not wasted.

Normally, home automation systems pay for themselves over time because the home uses energy more efficiently.

Another nice thing about home automation systems is that they normally do not require any new wiring or cable installation; you can use pre-existing cable. Sometimes, however, you may still have to install new cabling. The home automation field runs outside the scope of this book and this chapter, so the next few sections cover only some of the most common areas of home automation that you will come in contact with (as you will most likely be running cable parallel to these systems). We already covered home networking with wired and wireless systems as well as PLC and X-10. In the following sections, we cover lighting and security in detail.

11.8.1   Security and Monitoring Systems

You will most likely encounter security systems or security system cabling in the SOHO. A home security system is one of the largest portions of a home automation system, because it is generally the largest system in the house in terms of **closed-circuit TVs (CCTVs)**, sensors, microphones, and so on.

In addition to perimeter and interior protection offered by a security system, surveillance monitoring includes features that enable the home owners to observe environmental conditions inside and outside the home when at home or away.

Security, monitoring, and surveillance systems are comprised of many components, but the most common of these are:

- **A camera monitor:** Accepts the raw video signal from the camera and displays the image in a location inside the home.

- **Connecting cables:** Used to connect the monitor to the VCR or to connect sensors or the control pad.

- **A VCR:** Provides a way to view recorded activity for a current or later review if needed.

Home surveillance systems are used with CCTV to provide viewing of locations such as a day care center or a child's nursery inside a SOHO, or the outside perimeter such as the SOHO's external yard or grounds.

Surveillance control systems can use wired or wireless components, and therefore, you have many options for connecting them. X-10 and PLC are the most commonly used in the SOHO environment. Refer to PLC and the X-10 protocol earlier in this chapter.

11.8.2 Lighting Control and Energy Management Systems

Home lighting and energy management systems are also widely seen in SOHO environments. You need to understand how they connect to a central control system for home automation. Home lighting systems and energy management systems can be a little trickier than the other systems discussed in this chapter. In home automation, you need to be concerned with designing the overall system so that the total load on the current power system in the home is appropriate. Most homes have a maximum amount of power to work with, so you can't just keep adding new power-drawing devices without a penalty. You may have to upgrade the power in the home. This is why you need to consider your design first.

Now that you know design is critical to setting up lighting control and automation, let's look at setting the plan in motion so that you can evaluate and properly install them.

Estimation of Power Load

As a cable technician working in the SOHO environment, you may not have to worry about power load estimation, because you are simply running the physical wiring in a house. The power load estimation is the responsibility of the person installing the systems. Don't connect these systems until you understand the ramifications. Understanding power load is critical when working with a PLC system because if you plug too much in, you will trip breakers or damage a home's power system. Calculating total load is not easy and takes some time and detailed math. You need to analyze what is currently running in the SOHO. This is generally done by an electrical contractor, so consider finding one before you install new components in the residential home. When considering total power load requirements, you should include the following areas:

- **Computing systems:** Home computer equipment such as PCs, laptops, home broadband routers, modems, printers, Uninterruptible Power Supply (UPS) systems, and any other office equipment, such as shredders, fax machines, and copiers

- **Home entertainment and hobbies:** Radios, TVs, VCRs, DVD players, and musical equipment

- **Kitchen:** Microwave, refrigerator, coffee pot, can opener, etc.

- **Environmental systems:** Heating and cooling systems (such as HVAC), ventilation equipment, humidifiers, etc.

- **Home security systems:** The security system itself, control pads, cameras, VCRs, sensors, etc.

- **Lighting systems:** Lights, lamps, and other fixtures

- **Other systems:** Home power tools, battery chargers, garage door openers, washing machine, dryer, dishwasher, ceiling fans, and anything else that either has a plug or is hard wired into the high voltage system running through the house

If it draws power, consider it. After you know the total power load, you know what you will be dealing with as the cable installer.

Wired Runs

When you install a new wiring system for home lighting, you need to use wiring that passes local and national codes. The codes require that you use insulated wire that is the appropriate size for the application. The most commonly used cabling (which you should become very familiar with if you are doing residential wiring) is called Romex. Romex is a brand name for a type of plastic, insulated wire with a non-metallic (NM) sheath. The cable gauge is 12- or 14-gauge NM sheathed cable.

NOTE There are wireless systems available that allow you to easily connect new lighting systems to the central controller for home automation.

The Home Run

The **home run** is a term that describes the one central point (usually in a centralized closet) where all of the wiring in the SOHO terminates. This is also called home wiring and external telecommunications wiring.

Power Line Controls

Home lighting can be in your control with PLC as well. As mentioned previously, you can use PLC and X-10 controllers to control home automation using the existing high voltage wiring in the SOHO. As long as the power plugs are used, the lighting units can get and receive signals from the control unit. Once they are connected, they will utilize either X-10 or CEBus, two common power line protocols. Once installed and deployed, you can use this system to control the lighting fixtures in the SOHO.

Conduits

You can use conduit to enclose wire installed in locations that may be dangerous if exposed. What this means is that you need to protect the cable that you are running in hazardous areas, such as places that are damp (moisture), outdoors (which has a host of dangers), or places that have poor air quality. Whatever the hazard, run the wiring in conduits for protection. Conduits come in many shapes, sizes, and materials. Some specific types of conduit you should be familiar with are:

- Metal (rigid)
- Intermediate (flexible)
- Thin wall (more flexible)
- Nonmetallic PVC (polyvinyl chloride)

Thin wall conduits—what you are likely to see or frequently use in SOHO and residential wiring—is also called electrical metallic tubing (EMT). This conduit is made up of metallic tubing and can be used in either open or concealed electrical installations. It is important to consider it because it is the best choice for most conduit needs in the SOHO environment, is less expensive than rigid tubing, and is easy to install.

11.8.3 Other Home Automation Areas

To spark your interest in home automation, take a look at all the areas home automation has to offer. This field is emerging as more and more systems become available and the price drops. Lighting, phone systems, networking, security, and X-10 have been covered. In addition, you can control your home theater via home automation. You can replace a pile of remote controls with one controller that operates via a computer terminal. You can record all of your television shows based on programming, limit which channels can be seen at specific times, and so on. The limitations are endless if you spend the money to get the top-of-the-line equipment.

Other solutions are in the environmental arena. As mentioned before, you can reduce your home heating and cooling costs using home automation systems. You can also set up your irrigation systems via home automation. For example, you can have a sprinkler system run in the morning before you leave for work based on a computer program that you configure with the days of the week and the times you want this system to run.

11.9   Chapter Summary

- We discussed the fundamentals of working as a cable technician in the SOHO environment. A SOHO is a term that describes a Small Office Home Office that is used for doing work, and it's up to you to help achieve connectivity through wired and wireless systems. To do this, you must be concerned with the residential infrastructure technology.

- We covered HomePNA systems, PLC systems, wireless systems, broadband coaxial cabling, IEEE 1394 (FireWire), POF, DBS, and the recent trends in home automation systems. Home automation is the process of connecting most of the electronic devices in your home with a central control system that allows you to have centralized control.

- Because a SOHO is located in the home, you must have a solid understanding of what this means in terms of wiring in the home, all the other residential technology available, and how to work with or around it.

11.10   Key Terms

basic service set (BSS): Uses one SSID. An access point acts as the central point for all devices that participate on the network.

Bluetooth: A wireless protocol used to communicate from one device to another in a small location where the devices being connected are in the same general proximity from each other. Bluetooth uses the 2.4 GHz band to communicate.

closed-circuit TVs (CCTV): Doesn't broadcast TV signals, but transmits them over a closed circuit through a cable or wireless transmitter and receiver. Used often in security solutions and home security.

direct broadcast satellite (DBS): An alternative to most current cable-based television systems. DBS provides television programming directly from satellites using small, home-mounted satellite dishes.

extended service set (ESS): In an ESS, the BSS access points communicate with each other to forward traffic from one BSS to another. This allows devices to communicate with other devices via the ESS.

frequency division multiplexing (FDM): A technique used to allow for the transmission of data across the available bandwidth of a circuit to be divided by frequency into narrower bands, each used for a separate voice or data transmission channel. FDM allows for more than one conversation to be carried on a single circuit.

Home Phoneline Networking Alliance (HomePNA): The standard that was adopted for using copper phone lines within the home as a way to connect network devices. The HomePNA standard is based on a set of standards that enables voice and data transmissions over a home's existing telephone cabling.

Home Radio Frequency (HomeRF): A WLAN standard that supports up to 10 mbps in the 2.4 GHz band. HomeRF is not interoperable with 802.11 or Bluetooth.

home run: The one central point (usually in a centralized closet) where all of the wiring in the SOHO terminates. This is also called home wiring and external telecommunications wiring.

plastic optical fiber (POF): A type of optical fiber that uses polymethyl-methacrylate (PMMA)—a general-purpose resin—as the core material and fluorinated polymers for the clad material.

Powerline Carrier (PLC) systems: Another way of connecting devices in a residential home using the existing high-voltage power wiring (60 Hz/120 Volt AC). PLC systems normally use the X-10 protocol to communicate.

service set identifier (SSID): Specifies which 802.11b network you are attempting to join. This is how you participate in a wireless network.

Small Office Home Office (SOHO): A home office used for doing work. It is up to you to help achieve connectivity through wired and wireless systems.

X-10: The leading home automation protocol used today. With X-10 controllers, which send digital signals over home wiring systems to receivers that are plugged into the SOHO wall outlets, X-10 is transmitted over the home wiring at 60 bps (bits per second).

11.11 Challenge Questions

11.1 You are a cable technician assigned to help deploy a wired network in a client's home. If this area in the home is used as an office for work, what would the correct terminology be to describe this office?

a. ROBO

B. SOHO

c. LOGO

d. Client office

11.2 Which of the following is the standard adopted for using pre-existing copper telephone cabling in the house to connect and network devices within the home?

a. SSID

b. DBS

c. POF

d. HomePNA

11.3 Which technique allows the transmission of data across the available bandwidth of a circuit to be divided by frequency into narrower bands, where each is used for a separate voice or data transmission channel?

a. FDM

b. TDM

c. FQDN

d. HPNA

11.4 Which of the following terms describes the technique of using pre-existing high-voltage wiring in the SOHO to connect and network devices?

a. PLC

b. FDM

c. POF

d. HomePNA

11.5 You are deploying a solution in the home with X-10. Your client asks how fast X-10 operates over pre-existing high voltage wire. What is the correct speed?

a. 90 Kbps

b. 120 bps

c. 60 bps

d. 20 Mbps

11.6 You need to set up a security system with transmitters and receivers. Which protocol would you use that is the leading home automation protocol in use today?

a. X-12

b. X-120

c. X-100

d. X-10

11.7 You are deploying a wireless solution in the home. You need to connect two workstations to one central access point. What is the name of this type of wireless network?

a. SSID

b. IBSS

c. BSS

d. ESS

11.8 You are deploying a wireless solution in the home. You need to connect two BSSs so that workstations from the other BSSs can communicate with each other and within both BSSs. What is this type of wireless network called?

a. SSID

b. IBSS

c. BSS

d. ESS

11.9 You are deploying a wireless solution in the home. You need to set up a unique identifier so all of your wireless devices can communicate among themselves through a single access point. What is this unique identifier called?

a. SSID

b. IBSS

c. BSS

d. ESS

11.10 You are deploying a wireless solution in the home. You are setting up a small peer-to-peer Bluetooth-based network. What is the name of this type of wireless network?

a. SSID

b. IBSS

c. BSS

d. ESS

11.11 Bluetooth technology operates within what radio band?

a. 5.0 GHz

b. 2.4 GHz

c. 4.4 GHz

d. 4.2 GHz

11.12 You are deploying new cable in a SOHO. You need to keep a specific run dry (it may get wet from leaks from the roof). What can you install to make sure that the new cable is protected?

a. A conduit

b. A home run

c. A POF

d. A DBS system

11.13 You are installing new fiber into a client's home. The client would like to keep the cost down. There is no need to run this fiber over a long distance. What will meet the customer's requirements?

a. POF

b. BSS

c. ESS

d. PVC

11.14 What is the only issue with using POF in the home?

a. POF is not standardized.

b. POF can't be used over very long distances.

c. POF is very expensive in comparison to other types of fiber (glass).

d. POF is very weak and breaks easily.

11.15 When running cable in the home, which term is used to describe the location where the wiring in the SOHO is laid to one central point and terminates?

a. A conduit

b. A home run

c. A POF

d. A DBS system

11.16 You received a request from a client who operates a SOHO to include the TV programs coming from a satellite. What is the proper name for such a solution?

a. A BSS

b. A home run

c. An IBSS

d. A DBS system

11.17 You are deploying a new security system in a client's home. What are the three basic components of a security and surveillance system?

a. Camera monitor

b. Connecting cables and hardware

c. Sensors

d. VCR

11.18 You are recommending a wireless solution in a SOHO. The client requests a 54-Mbps transmission rate in the 5 GHz band. Which solution should you install for this client?

a. 802.11

b. 802.11a

c. 802.11b

d. 802.11g

11.19 You are recommending a wireless solution in a SOHO. The client requests an 11-Mbps transmission rate in the 2.4 GHz band. Which solution should you install for this client?

a. 802.11

b. 802.11a

c. 802.11b

d. 802.11g

11.20 You are recommending a wireless solution in a SOHO. The client requests a 54-Mbps transmission rate in the 2.4 GHz band. Which solution should you install for this client?

a. 802.11

b. 802.11a

c. 802.11b

d. 802.11g

11.12   Challenge Exercises

Challenge Exercise 11.1

In this exercise, describe how you will plan and deploy a wireless network in a SOHO. To complete this exercise, you need a computer with a word processing application in which to type your planning document, and a Web browser and internet access with which to perform research. Your document should include details for the following phases:

11.1.1 Plan what you want to do. Before deploying a wireless solution, plan and design what you hope to accomplish, and know what you want the wireless systems to be doing for the client. For example, you may want five computers in the SOHO to access files on a server as well as print and access the internet (all in the same SOHO). You want to do this wirelessly, and you already have everything but the wireless equipment. You have a partially wired segment of this network; it has a router and a switch that provides you with internet access as well as connectivity to the file server and a printer. What else do you need? You need an access point and NICs for the five computers.

11.1.2   Select a standard. What bandwidth are you looking for? Because wireless systems are selected by standard, you have to select a standard based on what you need. 802.11b runs at 11 Mbps, 802.11a and 802.11g run at 54 Mbps.

11.1.3   Acquire the appropriate hardware. If you only need 11 Mbps, then you can select 802.11b wireless equipment. If you have an 802.11a access point and five PCs with 802.11b NICs, you will not be able to use the wireless system. All radio frequencies have to match, so select the proper equipment for the job. A good rule is that 802.11a, which operates in the 5 GHz range, only operates with 802.11a hardware. 802.11, 802.11b, and 802.11g all operate in the 2.4 GHz range so they are all interoperable.

11.1.4   Configure the equipment properly. You have to install the access point correctly. Make sure you connect it via Ethernet to the wired network switch that connects to the router. That's how you have to do it in this scenario, but you can buy devices today that contain the router, switch, and access point. Configure the five PCs with wireless NICs.

11.1.5   Test the solution and use it. You have to configure all of the devices via software applications to get the logical configurations set up so the computers can access network resources, and so on.

11.13   Challenge Scenarios

Challenge Scenario 11.1

You are a cable installer/technician servicing a client in a SOHO environment. You need to assess whether you need to run new cabling or use pre-existing cable. You are working for a client who would rather not run new cabling, but needs to network four computers in the SOHO. Because the client does not own the building, he can't have new cabling installed in the walls. Provide a solution for your client.

CHAPTER | 12

Structured Cabling

Learning Objectives

After reading this chapter you will be able to:

- Understand the purpose of structured cabling and the modular concept behind structured cabling

- Adhere to structured cabling standards

- Understand terminology related to structured cabling

- Understand backbone cabling

- Recognize the various types of raceway systems

- Understand horizontal cable, consolidation points, and work areas

- Differentiate between various structured cable types and mediums

- Discuss lighting and its effects on cabling

The term **structured cabling** has evolved from what was once known as local area network (LAN) and/or premise cabling. These terms refer to the inside plant world, or the cabling that connects our voice-, data-, and video-driven communication systems of today. The term structured cabling will likely morph again in the future. In fact, you will likely come across the term "enterprise cabling." At the same time, the term **structured cabling** has been steadily migrating from the commercial communications side of the industry to the residential side.

In this chapter, we cover the basics of structured cabling as it pertains to the commercial cabling industry. What it is, the terminology involved, standard types of raceways (pathways for cabling), horizontal cabling (which ties all cabling to the backbone), and the harmful affects of lighting fixtures on cabling in general are examined.

12.1   The Purpose of Structured Cabling

The theoretical purpose of structured cabling is to unify and provide vendor nonspecific architecture to an industry that has been restricted by the limitations of proprietary, or **vendor-specific**, cabling and connecting components for years. In the early years of vendor-specific cabling, the end user was limited to the options provided by the vendors as well as to the specific network protocol of those products. These limitations included, but were not limited to:

- A single network protocol concept

- Non-interoperability with more than one vendor's product throughout the entire system

- Upgrades or network protocol changes requiring that the entire Physical layer/cabling infrastructure be replaced due to multi-vendor incompatible components and/or protocol inflexibility

- Multiple cable infrastructures within a company, based on specific department or divisional needs

Significant reconfiguration of the Physical layer infrastructure occurs for even the simplest changes or upgrades (also known as **moves, adds, and changes [MACs]**), and proprietary troubleshooting techniques are required to locate Physical layer problems.

Structured cabling has provided the communications industry with the ability to use multi-vendor and **vendor-neutral** solutions; the ability to select and then change network protocols without having to replace cable infrastructures; and the opportunity for joint cooperation between passive telecommunication manufacturers, active telecommunication manufacturers, and the applicable industry-standard organizations to develop a common solution.

12.2   Structured Cabling Standards Organizations

Prior to 1985, the concepts and practices for the design or installation of any communications system were based on the chosen vendor and, therefore, that vendor's proprietary requirements and practices. In 1985, due to the growth and complexities that networks were beginning to take, several industry organizations found it necessary to coordinate a standardization of the fundamental infrastructure design and installation methodologies of the cabling world.

In 1991, the initial document that spelled out the standards for generic customer cabling, called TIA/EIA-568 (short for Telecommunications Industry Association/Electronic Industries Alliance-568, which you should be quite familiar with at this point in the book!), was published. In 1995, the document was revised to reflect the rapidly changing environment of the industry that had taken place over the previous four years. This document was released as the TIA/EIA-568-A industry standards.

In the fall of 2000, the document was again revised to reflect and keep pace with an industry that had grown more in the previous five years than it had in the past decades combined. This document is currently known as TIA/EIA-568-B. It is scheduled to be revised again in 2006 and will be released as TIA/EIA-568-C.

The following groups are the primary organizations that generate the generic guidelines and recommendations for the structured cabling industry.

- **American National Standards Institute (ANSI):** Founded in 1918 to serve as the impartial national coordinator in standards development for the United States.

- **International Organization of Standardization (aka ISO):** Formed in 1947 and based in Geneva, Switzerland, and formerly known as the International Telephone and Telegraph Consultative Committee. ISO is not actually an acronym for the organization, but it's how the organization is commonly known.

- **International Electrotechnical Commission (IEC):** Founded in 1906, publishes international standards associated with electrical, electronic, and other related topologies.

- **Telecommunications Industry Association (TIA):** Formed in 1988 after a merger of USTSA and the information and telecommunications technologies group EIA.

12.2.1   Structured Cabling Standards Documents

Many documents detail the various standards to which the cabling industry must adhere. We've already mentioned TIA/EIA-568-A and TIA/EIA-568-B, and they'll come up throughout this chapter. The following are other pertinent national and international documents that you should be familiar with as a cable technician:

- **AS/NZS-3080:** The Australian/New Zealand document developed for the performance and specification criteria regarding the Australian/New Zealand commercial cabling industry.

- **AS/NZS-3084:** The document that covers the specifications for the pathways and spaces within the telecommunication industry for Australia and New Zealand.

- **AS/NZS-3085:** The standards document that covers the applicable criteria for labeling practices of the Australian and New Zealand telecommunication commercial cabling markets.

- **CSA-527:** The standards document that covers the criteria for grounding and bonding within the Canadian telecommunications industry.

- **CSA-529:** The Canadian document that covers the performance and specification criteria regarding the Canadian commercial cabling industry.

- **CSA-530:** The document that covers the specifications for the pathways and spaces within the telecommunication industry for Canada.

- **EN50173:** Produced by the European Committee, CENELEC, the document that covers specifications for generic cabling for a customer's premises for many parts of Europe.

- **ISO/IEC-11801:** The document that describes international generic cabling for a customer's premises, including performance and specification information regarding the commercial cabling industry. TIA/EIA-568-B is the U.S. document that covers the performance specifications for the commercial cabling industry. The basis of this document is derived from the ISO/IEC-11801 global document.

- **ISO/IEC-14763:** The global document that covers the specifications for the pathways and spaces within the telecommunication industry.

- **TIA/EIA-569-A:** The document that covers the specifications for the pathways and spaces within the telecommunication industry for the United States.

- **TIA/EIA-606-A:** The standards document that covers the applicable criteria for labeling practices within the U.S. telecommunication commercial cabling industry.

- **J-STD-607:** The standards document that covers the criteria for grounding and bonding within the U.S. telecommunications industry.

 NOTE

J-STD-607 is not a life safety document; rather, it's a document the covers the grounding and bonding practices for telecommunications equipment.

Other documents of note include:

- National Electrical Code (NEC) article 770

- NEC article 800.52

- Institute of Electrical and Electronics Engineers (IEEE) 802.11a is the standard that covers the wireless technology that operates at the 5.8 Ghz/54 Mbs frequency.

- IEEE 802.11b is the standard that covers the wireless technology that operates at the 2.4 Ghz/11 Mbs frequency.

- IEEE 802.11g is the standard that covers the wireless technology that operates at the 2.4 Ghz/54 Mb frequency.

- 802.11n is the standard that covers the wireless technology that operates at the 100 Mb frequency.

- 802.3ae is the standard that covers the fiber-optic media used in the Ethernet protocol.

- 802.11af is the standard that covers the power over Ethernet technology.

Now that the standards have been covered, we move on to commonly used terminology within the cabling industry, especially with regard to horizontal structured cabling, which we cover at length in this chapter.

12.3   Structured Cabling Terminology

Before we discuss the nuts and bolts of Physical layer cabling infrastructure, we need to define some common industry terminology and acronyms that you will often hear as a cabling technician. The acronyms IDF, HC, HCC, TR, and FD are references to the horizontal cabling section of the structured cabling hierarchy. These are the generational terms used as each revision of the U.S. standards evolved and were used by each group of industry technicians as they entered the cabling business.

Intermediate data frame/facility (IDF) was commonly used prior to the 1984 divestiture of AT&T, when the phone companies were overseeing all of the installations within the telecommunications industry. IDF is still used by many today because cable installers passed it down from one generation to another. IDF became horizontal closet or cabling (HC), which then became horizontal cross-connect (HCC). This term was followed by the acronym **TR**, which is an abbreviation for telecommunications room. This brings us to the current and most accurate term **FD (floor distributor)**. As you will come to realize, it properly depicts and defines the horizontal area and its purpose. Ironically, the ISO/IEC-11801 has always used the term FD when discussing the facility/location that houses the horizontal cable, its applicable connecting hardware, and other specific equipment. An additional term that's often thrown into the discussion about horizontal section is work area (WA). We discuss the term in more detail later in this chapter.

The next set of acronyms reference the backbone section of the Physical layer/cabling infrastructure. The term main data frame/facility (MDF) is a carryover from the pre-1984 divestiture of AT&T. As industry standards evolved, MDF changed to main closet (MC), which eventually became known as main cross-connect (MCC). This change was made in the interest of better describing the specific area. The current term, **campus distributor (CD)**, reflects the identical terminology that's used in the ISO/IEC-11801 global "Generic Customer Premises Cabling" document. If a structure is either large enough and/or unique in design, there may also be a secondary level to the backbone cabling. The applicable terms are intermediate cross-connect (ICC) and, as of the most recent revision to the industry standards, **building distributor (BD)**.

The purpose of all of the revisions or changes to the acronyms and terms within the telecommunications industry over the years has been to better describe the specific locations within a structured cabling infrastructure. The original terms referred to the various locations as data frames/facilities. The industry then defined the prospective locations as closets. The terminology next described the various locations as cross-connects, thereby keeping up with the state of the industry and reflecting its internal structure. The current and most likely permanent acronyms reflect the term *distributor* rather than closet or cross-connect within each acronym to reflect unification with the ISO/IEC-11801 global document and its terminology. Now that we have discussed the terminology, let's look at backbone cabling in more detail.

12.4   Backbone Cabling

Backbone cabling is the main cable and connecting hardware, which includes the main and intermediate cross-connects, in a proposed telecommunications environment. Backbone cabling includes the cabling that runs between telecommunications closets and equipment rooms. Backbone cabling centers around two cabling infrastructure design types: **distributed network architecture (DNA)**, the traditional concept that deploys both voice and data cable from the campus distributor to the floor distributor over various cable types and **central network architecture (CNA)**. CNA is the newer of the two design types. It deploys fiber optics for data transmission, and then either Category 3 copper cable, if it's an actual physical cable for voice transmission, or Voice-over IP (VoIP) technology for voice transmission. These architectures provide connectivity between intrabuilding (within a building) and interbuilding (from one building to another) designs as well as vertical connections between floors and connections between closets/distributors.

To break this down further, intrabuilding cabling connects the primary or first level of backbone cabling from the MDF, MC, MCC, CD, ICC, or BD locations to the IDF, HC, HCC, or FD facilities. The primary purpose of the campus distributor is to bring the outside services together at the demarcation point. This is located on the inside of the building's outer wall, traditionally at ground level in a DNA and on a mid-level floor in a CNA, thereby integrating them within the rest of the physical structure.

The interbuilding cable provides connections between separate structures that are within the premise and/or campus environment. Backbone cabling also has been referred to as riser cabling, intercloset/interconnect cabling, and vertical cabling.

Backbone cabling is inclusive of, but is not limited to, the cable medium types that connect one building to another or the cables that connect a campus distributor to a floor distributor within a single structure. They can be a copper-based medium such as Category 3, 5e, 6, Augmented 6, or Category 7, or they can be an optical medium such as multimode or single-mode fiber-optic cables.

Backbone cabling is also inclusive of cross-connect cables that tie one rack together with another in the floor distributor location. In most applications, there is a floor-mounted seven-inch rack or, on occasion, a wall unit that houses the backbone cabling from the campus distributor location. This unit is normally positioned next to an identical unit housing the horizontal cabling within the floor distributor location. These two racks are linked using cross-connect cabling in defined lengths set by the industry standards documents.

A **cross-connect** can be defined as a location within an infrastructure that allows the connection of new or existing cables. This is inclusive of, but not limited to, optical jumpers, copper patch cords, and termination blocks such as 110-style devices. Additionally, TIA/EIA-568-B specifies other applicable design requirements that pertain to backbone cabling, as follows:

- Applicable grounding and bonding requirements that can be found in the J-STD-607 document as they relate to both system electronics and the physical cable plant.

- Avoidance of electromagnetic interference (EMI)/radio frequency interference (RFI) sources when installing the actual cable plant.

- Backbone cabling distances for both DNA and CNA collapsed backbone network architecture as they relate to both fiber-optics and copper mediums.

- The use of multi-pair cable types, providing they meet or exceed the minimum performance criteria of the established standards document.

- The maximum allowable amount of conductor untwist as it relates to installation practices and performance criteria for the various cable categories. The specifications include: (1) no more than three inches of conductor untwist for the termination of Category 3, (2) no more than 1/2 inches of conductor untwist when terminating Category 5 or 5e, and (3) no more than 1/4 inches of conductor untwist when terminating Category 6.

- The use of proper connecting hardware, that it must be made of at least the same grade as the installed cable plant.

- The maximum allowable bend radius (20×) when cable is being pulled, the outside diameter of the cable that's being pulled, and 10× the diameter of the cable once it's been placed or is static/dormant.

- The maximum fill rate capacity for the systems raceway. This includes specific fill ratios for conduit when used. Those fill rates are not to exceed 40% when running multiple cables through conduit or 53% if one larger pair cable is being run through the conduit. These percentages are fixed for all sizes of available conduit.

- The maximum allowable scalability and/or hierarchical design of backbone cabling levels. This is defined as two options: (1) the backbone cable can be installed and directly run from the campus distributor to the floor distributor without exceeding any of the allowable distances specified in the TIA/EIA-568-B document; and (2) if the maximum distances were going to be exceeded due to the design of the structure, a second level of backbone cabling with a building distributor would be required. This is required if any copper media is going to exceed the 90-meter maximum allowable distance between the campus and floor distributors. Optical cable distances are far greater at 2,000 meters for multimode and 3,000 meters for singlemode, and are not the primary causes of this option.

- The maximum number of cross connects that can exist within the cable plant between the campus distributor and floor distributor is one.

- The maximum allowable lengths of all applicable patch cords and/or jumpers. This varies per location. Refer to the TIA/EIA-568-B document for the specific requirements.

- The specific temperature of 57°F to 64°F and percentage of humidity for controlled environments such as campus and floor distributors. Refer to the TIA/EIA for other specified criteria.

- The specifications for maximum allowable door width and height.
- The recognized cable types for backbone cabling applications are Category 3, Category 5e, Category 6, Augmented Category 6, and Category 7 in either shielded twisted-pair (STP) or unshielded twisted-pair (UTP) and multimode or single-mode optical fibers.

In the next section, we discuss the various types of structured cabling raceways, through which backbone and other cabling are run.

12.5   Structured Cabling Raceway Types

A raceway is a pathway in which cabling is run. Structured cabling raceway systems are critical to the successful deployment of all cabling infrastructures. There are many different types and the considerations for which type or types are appropriate are project-specific. The options are:

- Underfloor duct
- Cellular floor
- Ladder rack style
- Basket/cable tray styles
- Raised/access floor
- Ceiling pathways
- Perimeter raceway systems
- Modular furniture raceways

Primary considerations start at the point of the building's structural integrity to determine if the appropriate raceway system needs to be a part of that integrity. This is not only because of the facility's integrity, but also a consideration based on what types, size, and cumulative total of installed cables there will be within the scope of the project.

If it is determined that the raceway system is critical to the structural integrity of the building due to issues such as structural bracing, seismic activity, or other imposed/overriding codes/laws, then the appropriate choice might be the underfloor duct option.

12.5.1   Underfloor Duct Systems

An underfloor duct, or cell-based system, is embedded into the structure's concrete in the initial construction phase of the building. This system has a network of distribution and feeder ducts, or cells. The distribution ducts are used to distribute the assigned cables from the floor distributor to the work areas.

The feeder ducts distribute the cables from the floor distributor to the distribution ducts. Most of the ducting has factory-preset access points that allow technicians to gain access to the cabling being installed. This in turn allows them to connect to work area outlets.

Prior to the actual installation of the cables, the technician needs to verify and locate the factory preset inserts. On occasion, these products do not include the inserts, requiring the installer to acquire an after-market product that serves the same function. You can also obtain access or handhole devices that allow accessibility to the intersections of the feeder and distribution ducts.

12.5.2   Cellular Floor Systems

A cellular floor system, in concept, closely resembles the underfloor duct system; however, there are differences. For example, if you need to run multiple types of cabling such as copper and fiber as well as power cable through a common duct, you might need to consider physical separation issues, especially when considering EMI, RFI, or redundancy. There may be multi-level pathway issues as well as grounding and bonding issues related to metallic devices.

If it's deemed unnecessary to embed the raceway of the cable plant into the structure's concrete, a very common choice is the deployment of a ladder rack-based system. However, that is a costly option.

12.5.3   Ladder-Rack Raceway Systems

A significant advantage of a ladder-rack raceway system is that you do not have to embed the entire system into the concrete of the building, as you do with underfloor duct systems and cellular floor systems. However, you do have to sink anchors into previously existing concrete, after the route of the raceway system has been determined. You can sink the anchors using a tool known as a Hilti gun, which sinks an anchor into concrete with the force of a 22-caliber handgun. You must be certified by the manufacturer to use this device legally on job sites.

The next step is to attach all-thread, which is nothing more than a threaded tubular meter that threads into the installed anchors. At this point, you can install the associated hardware that comes with the system, which is designed in a clamp form to secure the actual sections of the ladder rack.

This raceway system is primarily designed for cable infrastructures with a significant amount of cabling, thereby requiring adequate load bearing and weight displacement ratios.

12.5.4   Basket/Cable Tray Option

If you determine that the installation of a ladder rack-based system is overkill, the ideal choice may be to install a basket/cable tray-based system. These systems are modular in design and are often much less expensive to purchase, thereby providing increased flexibility to the customer.

A large variety of these types of raceway systems are available on the market. The installation process for these systems is similar to that of the ladder rack-based system, with noted differences. Basket/cable tray systems use a variety of hardware. They can be mounted in the ceilings, against walls, or under modular floor systems. The physical tray is flexible, scalable, and moldable in many cases. This feature provides both the installer and customer with tremendous flexibility and options.

12.5.5   Raised/Access Floor Systems

These types of raceway systems are normally found and/or used in large data facilities, computer rooms, or in advanced office environments. These systems are designed with modular floor tiles that are connected to pedestals supported on the subfloor by steel stations/footings.

Typically, these systems are designed with a minimum of 6 inches of space underneath in an office environment, or 12 inches of space if they are in an environment intended for plenum usage. A plenum space in this application may be intended for the distribution of conditioned air to maintain the recommended temperature of 57°F to 64°F specified in the TIA/EIA-568-B standard for campus and floor distributors. Containment should be considered for the cables that will be running under the modular floor tiles. An ideal choice would be the deployment of a basket-type cable tray system. Another option would be some type of duct-based system such as interduct, which is nothing more than rated hollow plastic tubing. Whichever option is chosen, you should make sure to position it in such a way that it does not create length violations that exceed the 90-meter maximum, and that it adheres to all related grounding and bonding requirements and practices.

12.5.6   Ceiling Distribution Systems

This raceway system has taken many different forms over the years and thus has been used to describe many different scenarios. Therefore, for the purpose of this book we limit its definition to the following: the usage of this concept is most often in open warehouse architectures with material known as *unistrut* as the central component.

This type of system typically has been one of the least expensive to install. However, its application is specific. Its components include hardware to mount it to the concrete ceiling, pre-cut sections of unistrut, and connecting hardware to support the installed bundles of cable. Refer to TIA/EIA-569-A for placement and spacing specifics.

Conduit

Conduit-based raceway systems are an option; however, they should only be considered in a permanent environment. The following types of conduit are considered acceptable for communication raceways:

- Electrical ridged-metal tubing, EMI rated
- Ridged non-flammable polyvinyl chloride (PVC)
- Fiber glass-based material

However, the use of metal flex tubing is not allowed due to abrasion characteristics. If you are deploying conduit as the raceway of choice, it must be grounded on at least one end, and no section of conduit should be longer than 100 feet or 30 meters in length without the use of pull points and pull boxes.

All conduits should be reamed and fitted with insulated bushings to guard against abrasions. Every section of conduit should be installed with jet line/pull line inside of it for the future installation of cable. Care should be taken not to install or run conduit in high-risk areas such as crosswise to cellular floor raceway systems, hot water lines, steam limes, boiler areas, or through areas of stored flammable materials.

When installing any type of conduit as a raceway system for communication cables, refer to TIA/EIA-568-B and 569-A for appropriate fill rate capacities and pulling lubricants.

12.5.7 Perimeter Raceways

This raceway is ideal for educational facilities, libraries, or other work area environments. These systems are designed to be mounted on the exterior of walls, at an average waist-level height for easy access to allow for MACs. Traditionally, they are made with a plastic design, although metal is used occasionally. It's important to make sure that the installation is level around the entire room and that the cable is pulled in tightly, so that the panels can close upon completion of your installation.

12.5.8 Furniture Raceways/Pathways

Furniture raceways and pathways are the most challenging of the various raceways. There are no options when it comes to furniture raceway systems. They are part of an existing prefabricated office structure that has been purchased as one complete entity.

Therefore, this is an "as is" environment. You will be working in very limited space and around many sharp sections of metal raceway. It is imperative that you exercise extreme caution when installing the cable within these raceways.

Keep in mind that, although you are already working in a limited space, TIA/EIA-569-A specifies that you are to utilize only 33% of the available space when installing communications cable and to reduce that percentage at corners. Refer to the proper standards document for applicable grounding and bonding practices.

12.5.9 Vertical Raceway/Pathway Options

For this chapter, this option is limited to the typical methodology. That application is for the running of backbone cabling from various distributor locations to each other, with the understanding that the great majority of distributors sit above each other from floor to floor and there is an absolute necessity to secure that cable.

This is achieved by either securing sections of ladder-rack style of raceway to the wall or with the use of D-rings. At this point, the individual bundles of cables that have traveled from location to location through the 4-inch penetration points in the floors are gathered and secured with Velcro to either the installed ladder-rack tray or with the use of D-rings bolted to the wall.

This also controls the weight of the cables over time, which is commonly referred to as the vertical rise or the long-term tensile strength that a cable can withstand over time. This is critical to the long-term performance of communication cable.

Additional requirements also could come from the NEC or the National Fire Protection Association (NFPA) documentation as well as any local/regional requirements that may exist or be applicable.

12.6 Horizontal Cabling

Horizontal cabling ties to the backbone cable that has joined all of the closet/distributors together. This is most often achieved through the concept of a horizontal cross-connect methodology.

The traditional scenario is a 7-foot floor-mounted rack that houses the backbone cabling, which has transitioned from the campus distributor into the specific floor distributor. At this point, there is an identical rack located adjacently in the specific floor distributor that houses all of the horizontal cabling that the floor distributor is intended to service in the overall system design.

The method used to link the backbone cable to the horizontal cable is with either copper patch cords or optical jumpers, depending on the installed cable plant. It's this practice that allows individuals to have access to today's shared networks. This concept allows individual users the access needed to share, transmit, or save information to common drives located on network servers or to other work stations on a network.

The primary purpose of each and every floor distributor is to house the horizontal cable. However, it's not the only purpose. Although nearly everything that you read within this industry indicates that the telecommunications room/floor distributor and equipment rooms are different, in fact they are one and the same facility in the overwhelming majority of applications.

Additionally, the floor distributor houses all applicable electronics and other connective devices, such as punch-down blocks for traditional voice applications.

The concept behind horizontal cabling is to route cables from the floor distributor to the individual work areas that will be served by that specific floor distributor. This is to be achieved without the introduction of any splices or taps at any point in any cable run.

Additionally, several industry organizations, including the ISO, IEC, ANSI, TIA, EIA, and the International Telecommunication Union (ITU), have placed stringent requirements for the hierarchy of horizontal cabling. These include using the same grade of cable throughout the process, and the use of a matching grade of connecting hardware. For example, the same grade or better of an eight-position **insulation displacement contact (IDC)** contact copper jack must terminate an end of the installed cable in a work area application.

One of the most important requirements set forth has been the requirement to use the same grade or better of factory constructed/terminated patch cords. There are a number of important issues about this topic, because for many years it was common for installers to build their own patch cords.

However, as with the development of Category 5e cables, the practice of building your own cables has been highly discouraged due to the higher transmission speeds, bandwidths, and intended applications of the new generation of networks. This issue has been taken a step further with the introduction of Category 6 cable.

WARNING ⚠ Although it is recommended that you do not use handmade patch cords when installing Category 5e cable, manufacturers and various standards organizations require that only factory-constructed patch cords be used when installing a Category 6 cable infrastructure.

The requirement has been imposed to the point that if factory-constructed patch cords are not used, the system will not be certified or carry a warranty by the manufacturer. The reason for this may not seem that obvious; however, it has everything to do with system performance, system certification, and vendor warranties.

Handmade cables have been the root cause of a substantial amount of network-related problems and/or failures over time, and it is the intention of the industry to eliminate any avoidable scenarios such as connectivity issues to prevent costly network downtime and/or failures. These practices, if followed, have significantly reduced both the time and cost incurred when it becomes necessary to troubleshoot network problems or failures.

Horizontal cabling is the connecting station cable that links the floor distributor to your individual work station. Another term that is often used within the industry is *home run cable*. This term has enormous importance. It indicates a specific application for the cable. This also means that there are specific types and grades of cable that can be used for this application.

Horizontal cabling specifications are also inclusive of a consolidation point, if implemented, as well as the actual physical work area location where people will interact with variety of devices, such as computers, phones, and fax machines. We look at the consolidation point first.

12.6.1   Consolidation Points

A **consolidation point** (**CP**), which is the location of the interconnection between horizontal cables that extend from building pathways and horizontal cables that extend into work area pathways, traditionally has been used where modular furniture is serving as work areas. This is also known as zone cabling. In this scenario, one of the primary considerations made during the design phase was to limit the amount of traffic and access allowed to enter the floor distributor by nonessential personnel.

This is achieved by deploying a consolidation point at a minimum of 15 meters or 45 feet away from the floor distributor for the purpose of being able to run a long jumper that doesn't exceed 22 meters (or a maximum of 72 feet) in length from the consolidation point to the physical work areas. This enables nonessential personnel to perform MACs (moves, adds, and changes) for individual work areas while minimizing the need for nonessential people to have access to the actual floor distributor. This minimizes potential risk to unrelated active components and/or other cabling located in that facility.

12.6.2   Work Areas

We now look at work areas. As defined earlier, the horizontal cable is the cable that connects to the floor distributor designated to service your specific work station. Beyond the number of drops that each work area is to have, which is defined by the TIA/EIA-568-B standard, the primary issue here is the design.

As discussed in the previous section about consolidation points, if the design is to include the use of modular furniture, you need to consider using a **multi-user telecommunications outlet assembly (MUTOA)**. Each MUTOA can serve up to 12 work area locations. MUTOAs are normally installed in a central location within each modular furniture cluster to equalize the cabling distances to each work area, and to uniformly locate each unit for troubleshooting and overall system documentation purposes.

If the work area design consists of actual hard offices and modular furniture located next to a physical wall, the hardware will likely consist of individual mud rings to secure the wall plates at each location, loaded with the recommended amount of IDC-type jacks, and installed 18 inches from the floor and within 1 meter (or 3 feet) of any electrical outlets to be compliant with the Americans with Disabilities Act (ADA).

Additionally, TIA/EIA-568-B specifies other applicable design requirements that pertain to horizontal cabling, as follows:

- Applicable grounding and bonding requirements that can be found in the J-STD-607 document as they relate to the requirements of the grounding bus bar within each facility, the racks, the connecting hardware if applicable, the system electronics, and the physical cable plant.

- Avoidance of EMI/RFI sources such as fluorescent lights and any other motorized and/or electrically charged device when installing the actual cable plant.

- Maximum horizontal cabling distances for both DNA-distributed network architectures and CNA-collapsed backbone network architecture as they relate to both fiber-optic and copper mediums.

- The use of multipair cable types providing they meet and/or exceed the minimum performance criteria of the established standards document. Multipair cable such as 25-pair cable is not considered a standards compliant cable at this time.

- The maximum allowable amount of conductor untwist as it relates to installation practices and performance criteria for the various cable categories. These include: no more than 3 inches of conductor untwist for the termination of Category 3; no more than 1/2 inches of conductor untwist when terminating Category 5 or 5e; and no more than 1/4 inches of conductor untwist when terminating Category 6.

- The use of proper connecting hardware. It needs to be at least the same grade as the installed cable plant.

- The maximum allowable bend radius when cable is being pulled: ($20\times$) the outside diameter of the cable that's being pulled, and ($10\times$) the diameter of the cable once it's been placed and/or is static/dormant.

- The maximum fill rate capacity for the system's raceway. This includes specific fill ratios for the conduit when it is used. Those fill rates are not to exceed 40% when running multiple cables through conduit or 53% if one larger pair cable is run through the conduit; these percentages are fixed for all sizes of available conduit.

- The maximum number of cross-connects that can exist within cable; between the campus distributor and floor distributor the number is one cross-connect.

- The maximum allowable lengths of all applicable patch cords and/or jumpers. This varies per location, so refer to TIA/EIA-568-B for the specific requirements.

- The specific temperature of 57°F to 64°F and percentage of humidity for controlled environments such as campus and floor distributor. Refer to TIA/EIA-568-B for other specified criteria.

- The specifications for maximum allowable door width and height.

12.7 Structured Cable Types

The minimum rated cable and drops to each work station from their assigned floor distributor are: (1) Category 3 UTP for voice applications; (2) Category 5e or better drops for data applications; and (3) per the TIA/EIA-568-B standards revision, there is also a recommendation to have multimode optical cable drop to each work area location for future, yet to be determined applications.

In addition, both TIA/EIA-568-B and TIA/EIA-569-A specify parameters for the actual square footage and location of floor distributors, building distributors, and campus distributors. The grade of applicable hardware, recommended raceways, and their appropriate applications as well as the requirements for both plenum and non-plenum environments are also specified. A plenum environment is defined as an air duct or return air duct system that utilizes a forced air circulation, such as from heating or air conditioning (HVAC) system. Any air space used for forced air circulation, other than an actual common area scheduled to be occupied by people in a work area, is a plenum. (This means that the area above the drop in ceiling tiles is designated as plenum space if separate ducted air returns are not used.) An example of a non-plenum space would be where backbone cable runs vertically from floor to floor, connecting the campus distributor to each floor distributor. Although this is not a plenum area by definition, it must meet or exceed all established safety codes, including using proper fire-stopping putties or other protective materials in these areas to prevent the spread of fire.

These types of cables must be developed with certain materials for the insulation that covers each individual wire pair as well as the overall cable jacket. These selected materials must also meet or exceed all of the established safety requirements set forth by UL, NFPA, NEC, the ISO, the CSA, and the European Consortium (EC).

In closing out the discussion on horizontal cabling, there is one last significant topic to discuss. The recognized cable types for horizontal cabling applications are: Category 3, Category 5e, Category 6, Augmented Category 6, and Category 7 in either STP or UTP as well as multimode optical fibers.

12.7.1 Copper Patch Cord Types

The copper patch cord types include crossover cables, rollover cables, and straight-through cables. A *crossover cable* is a cable that is wired with the T568A wiring pattern at one end and the T568B wiring configuration at the opposite end. This means that the orange and green pairs are switched. An application for this type of cord is for connecting two computers to one another, such as in a point-to-point network scenario.

A *rollover cable* is one in which the wires are literally "rolled over," that is, pin 1 on one end becomes pin 8 on the other end, pin 2 on one end becomes pin 7 on the other end, and so on. You use a rollover cable to connect hubs and as a console cable on a router so that the router may be configured.

Straight-through cable is identical on both ends of the cable and is used in everyday common applications.

12.8   Lighting

Many of the most common failures within cabling plants can be traced back to lighting products. One of the most common sources of EMI is the lighting fixtures within a communications environment. The following are a few examples of failures generated by cabling installed too close to a lighting fixture:

- Using long rows of installed fluorescent light fixtures as a cable raceway system, thereby inducing EMI.

- Failing to allow for at least a 5-inch clearance between a communication cable and the ballast of a fluorescent light fixture to avoid EMI.

- Multiple test failures relating to various types of cross-talks.

- A negative impact to a network in many different ways, once it's gone active, that is, operational. The key here is that it has not always been an immediate problem. It may not show up until the bandwidth limits of the network are being approached or upgraded with newer electronics, such as Fast Ethernet devices upgraded to Gigabit Ethernet devices over Category 5e cable infrastructures.

12.9   Chapter Summary

The fundamental areas of structured cabling include: the purpose of structured cabling; industry standards and organizations; industry terminology; backbone cabling; structured cabling raceway systems; horizontal cabling; consolidation points; work areas; facility locations; cabling types; patch cord types; and lighting and its effects on the Physical layer of communication cabling infrastructures.

12.10   Key Terms

backbone cabling: The cable and connecting hardware that include the main and intermediate cross-connects in addition to cabling that runs between telecommunications closets and equipment rooms.

building distributor (BD): A secondary level to backbone cabling. Also called intermediate cross-connect (ICC).

campus distributor (CD): The main communications center in a large cabling infrastructure. Also called a main data frame/facility, main closet, or main cross-connect.

central network architecture (CNA): A cabling concept that deploys fiber-optics for data transmission, and then either Category 3 copper cable, if it's an actual physical cable for voice transmission, or Voice over IP (VoIP) technology for voice transmission. Like DNA, this architecture provides connectivity between intrabuilding and interbuilding designs as well as vertical connections between floors and connections between closets/distributors.

consolidation point (CP): Used in zone cabling, the location of the interconnection between horizontal cables that extend from building pathways and horizontal cables that extend into work area pathways.

cross-connect: A location within an infrastructure that allows the connection of new or existing cables. This is inclusive of, but not limited to, optical jumpers, copper patch cords, and termination blocks such as 110-style devices.

distributed network architecture (DNA): The traditional cabling concept that connects both voice and data cable from the campus distributor to the floor distributor over various cable types. Like CNA, this architecture provides connectivity between intrabuilding and interbuilding designs as well as vertical connections between floors and connections between closets/distributors.

floor distributor (FD): The horizontal area that houses the horizontal cable as well as its applicable connecting hardware and other specific equipment.

horizontal cabling: The connecting station cable that links the floor distributor to an individual work station.

insulation displacement contact (IDC): The design characteristics on an RJ-45 data jack.

intermediate data frame/facility (IDF): Another name for horizontal closet or cabling, horizontal cross-connect, telecommunications room, and floor distributor. It refers to the horizontal area that houses the horizontal cable and its applicable connecting hardware and other specific equipment.

International Electrotechnical Committee (IEC): Founded in 1906, an organization that publishes international standards associated with electrical, electronic, and other related topologies.

moves, adds, and changes (MACs): Upgrades or modifications made to an existing cabling infrastructure.

MM fiber: Abbreviation for multimode optical fiber.

multipair cable: Twenty-five pair cable.

multi-user telecommunication outlet assembly (MUTOA): A device that can service 12 or more work area locations, equalizing the cabling distances to each work area and uniformly locating each unit for troubleshooting and overall system documentation purposes.

SM fiber: Abbreviation for single-mode optical fiber.

structured cabling: The methodology used for cabling in the commercial and residential cabling industries.

telecommunications room (TR): A room housing telecommunications wiring and wiring equipment usually containing one or more cross-connects.

vendor neutral: A design, unrelated to a specific vendor, that enables unilateral interoperability between an industry's vendors.

vendor proprietary: A design related to a specific vendor; usually incompatible with other vendor's designs.

12.11 Challenge Questions

12.1 _____ is the methodology used in today's commercial industry.

12.2 The _____ document specifies the generic cabling requirements for customer premises cabling, specifically written for the United States.

12.3 What does EIA stand for?

12.4 Exactly how many pairs of cables does multi-pair cable have?

a. 4

b. 8

c. 12

d. 25

12.5 What does the term IDC stand for?

12.6 What does the term MCC stand for, and which industry-related terms has it replaced?

12.7 Which type of cabling connects the floor distributor to the work area?

   a. Vertical

   b. Horizontal

   c. Inter-raceway

   d. Residential vertical

12.8 The _____ is the connection point where backbone and horizontal cabling come together.

12.9 What is the name of the location that houses the horizontal cable?

   a. Equipment room

   b. Underfloor cell

   c. Floor distributor

   d. Floor point

12.10 What document describes international generic cabling for customer premises, including performance and specification information regarding the commercial cabling industry?

   a. EN50173

   b. ISO/IEC-11801

   c. TIA/EIA-568-B

   d. CSA-530

12.11 What document describes the U.S. performance specifications for the commercial cabling industry, and is derived from the ISO/IEC-11801 global document?

   a. EN50173

   b. ISO/IEC-11801

   c. TIA/EIA-568-B

   d. CSA-530

12.12 Per TIA/EIA-568-B, what is the maximum allowable amount of conductor untwist when terminating Category 6 cabling?

a. 1/4 inch

b. 1/2 inch

c. 3/8 inch

d. 1 inch

12.13 True or False: Using matched grades of cable and connecting hardware are not recommended.

12.14 TIA/EIA-569-A specifies that you are to utilize only what percentage of available space when installing communications cable and to reduce that percentage at corners?

a. 23%

b. 30%

c. 33%

d. 36%

12.15 What does the term MAC refer to with regard to cabling?

a. Moves, adds, and changes

b. Moves and changes

c. Move access control

d. Medium access cable

12.16 Describe the configuration of a crossover cable.

12.17 Category 3 cable is a _____ grade cable.

12.12   Challenge Exercises

Challenge Exercise 12.1

In this exercise, you provide details for the steps that are required to install horizontal cabling on an upper floor of a multi-story building. The cabling must meet today's standards and be as "future-proof" as possible. To complete the exercise, you need a computer with a word processing application. You also should have a Web browser and internet access to research background information for the steps to provide as much information as possible.

12.1.1 *Plan all details of the horizontal structured cabling before beginning deployment.* Discuss various aspects of the planning process, including information you need to get from the customer.

12.1.2 *Review all applicable standards and related documents.* Which documents should you review and why?

12.1.3 *Order the correct grade of hardware.* Discuss matching grades of hardware.

12.1.4 *Configure the components properly.* Discuss the importance of vendor training to ensure the highest level of installation quality.

12.1.5 *Test the solution.* At a minimum, which documents should you review for information on appropriate testing measures?

12.13   Challenge Scenarios

Challenge Scenario 12.1

You are a cable installer/technician who has been sent out on a call to investigate why a bundle of recently installed horizontal cables have suddenly started to experience failures within the customer's newly installed network. You must determine if new cables need to be installed or if the existing cables can be used.

Through investigation you have found that the failures are due to the cables being installed on top of eight fluorescent light fixtures. You know that this is not compliant with the standards and that it also violates a locally enforced code. What should you do?

CHAPTER 13

Building Your Cabling Toolkit

Learning Objectives

After reading this chapter you will be able to:

- Understand which tools make the job of pushing and pulling cable in an existing structure easier

- Choose from a variety of cable labeling systems

- Understand the purpose of stripping and cutting tools

- Use termination tools to install connectors on the ends of cable

- Understand splicing tools and methodology

- Identify crimping tools

- Choose which diagnostic tools to use for testing cables and reporting compliance

- Identify other miscellaneous tools to add to your toolkit

- Understand material safety data sheets

Just as a carpenter cannot build a house without the proper tools, no cable installer or troubleshooter can effectively perform their duties without the proper tools. In addition to installation tools, a variety of tools are available to assist an installer with troubleshooting new or existing cable plant infrastructure. This chapter outlines many of these tools as well as any safety concerns associated with their use.

13.1   Cable Pushers and Pullers

A **cable caster** is similar in operation to a fishing rod and reel. A weight is attached to the end of a light, flexible line similar to fishing line. The weight is then cast across an area, such as a drop ceiling, taking the flexible line with it. This process is repeated until the desired length is reached. The lightweight line is then used as a pilot to pull a bundle of cable across the drop ceiling. The flexible line is easier to throw than the individual cable that is to be run, and can be cast much further (Figure 13.1).

A **fish tape** is similar in function to a cable caster, but the way it operates is significantly different. A fish tape is either a stiff nylon tape or a flat metal tape approximately 1/4- to 1/2-inch wide and 1/8-inch high. The length of the fish tape can vary depending on the need. The fish tape is kept on a spool to make it more manageable. A fish tape is usually used to make the job of *fishing* a cable or bundle of cables through a hollow wall an easier task (Figure 13.2).

A **push/pull rod** is a series of narrow lightweight rods in which you thread your cable and then push/pull it through hard-to-access areas such as wall and ceiling drops. The pull rod offers greater stiffness than a fish tape, and most pull rods have a wisp head that is helpful in moving the rods past obstructions and preventing snags (Figure 13.3).

Figure 13.1

Cable caster. Courtesy of Greenlee Textron Inc. 2005

Figure 13.2
Fish tape. Courtesy of Klein Tools 2005

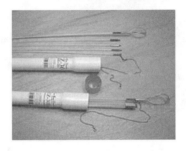

Figure 13.3
Pull rod. Courtesy of Labor Saving Devices Inc. 2005

A **gopher pole** is a lightweight telescoping pole with a hook on one end, used for pulling or moving cabling in hard-to-reach places; it is especially useful if you don't have a ladder handy (Figure 13.4).

13.2 Labeling Systems

Cables are usually run in bundles. It becomes difficult to identify which ends of a cable belong to each other. The easiest way to avoid this problem is by labeling both ends of the cable with the same identifying number or label. Several devices are available to label the cables, from a method as simple as using a permanent marker on each end of a cable, to write-on mylar labels, to elaborate printers that print tubular labels to affix over the both ends of a cable.

Wire and cable labelers are commercially available from companies such as Brady, Brother, and Panduit. These labelers are hand-held, electronic, thermal transfer printer units that you use to print cable descriptions onto tape. The tape is usually of high quality and sticks to a variety of materials, including wire and textured plastic.

Figure 13.4

Gopher pole

You can also purchase labeling software that can be installed and run in most operating systems, with the Windows operating system being the most common. These applications generally include a vast database of standard cable types and descriptions, symbols, bar codes, and label sizes (templates), which you select and/or customize and then print onto labels.

13.3   Stripping and Cutting Tools

A variety of cutting tools and tools to strip away protective sheathing are available to assist cable installers in performing their duties. Stripping tools function to remove the protective sheathing on a cable without damaging the inside of the cable. This is especially true when handling fiber-optic cables, although damage can also be caused to a copper cable because copper is a fairly soft metal. Damaging the cladding on a fiber-optic cable will prevent it from reflecting the maximum amount of light back into the core, thus disabling it. Nicking the core of a copper cable will make it more susceptible to breakage,

Figure 13.5
Typical stripping tool. Courtesy of IDEAL Industries Inc. 2005

crosstalk, possible shorting, and increased attenuation. Figure 13.5 shows an example of a typical stripping tool.

You also can acquire round cable strippers that allow you strip STP, UTP, or fiber-optic cabling. The round cable strippers are fully adjustable, allowing you to set the blade depth to strip insulation without nicking the core.

Cable knives, which often feature forged steel blades and dielectric handles, can be used to strip cables as well. Some cable knives include a hooked blade for greater flexibility and control while stripping, and adjustable handles with locking mechanisms to hold the handle and blade in place.

13.4 Crimping Tools

You use crimping tools to attach connectors to the ends of cables. As with most other tools covered in this chapter, you have a wide variety to choose from.

13.4.1 Coaxial

To apply a terminator or to splice a coaxial cable, in many cases it is very efficient to simply crimp the new end onto the cable. A coax crimper is similar in function to a pair of pliers. Because coaxial cable has two main parts—the core and the shielding—two crimps are required. First you must crimp a bullet-shaped end connector onto the core by placing the connector only over the tip of the coaxial core, and second, insert it into the coax crimper and squeeze the handles of the crimper. (Although we cover the standard two-piece style connectors in this section, be aware that one-piece style connectors are also available in addition to compression-fitted connectors.) Most crimpers have a ratchet-style action that requires a certain amount of force to ensure the end has been crimped adequately. After the

Figure 13.6

A crimper has differently sized openings in the die to accomplish different tasks. Courtesy of Paladin Tools 2005

proper amount of pressure has been applied, the crimpers will release, allowing you to remove the crimper from the end of the cable.

The jaw of the crimper usually has two sizes of openings in the die: one for the core and the other to crimp the *ferrule* to the end of the cable (Figure 13.6). Crimping the ferrule uses the same procedure as the core; you insert the coaxial terminator over the grounding braid, slide the ferrule over the braid and connector combination, and insert into the coax crimper. Apply pressure to the handles of the crimper until the crimp is completed. After completing a crimp, gently pull on the terminator to ensure a solid fit.

TIP
The most common problem when crimping coaxial cable is having the sheathing touch the core of the cable, resulting in a short circuit. A simple way to test for this problem is to place one probe of a multimeter on the core and one on the ferrule to perform a continuity test. The test should indicate an open, or infinity impedance. If there is any resistance, it is possible that you grounded some of the braid to the core. Multimeters are discussed in the "Multimeters" section (13.7.5).

13.4.2 Twisted Pair

Putting an end on a new twisted-pair cable involves placing all pairs of individual wires into a connector and using a twisted-pair crimping tool to press the connector onto the end of the cable. The die on the crimping tool will be oriented so the tab on the RJ-45 or RJ-11 connector lines up with the tab on the die. You apply pressure to the crimper to push the pins from the end onto the cable. A typical RJ-11/RJ-45 crimper is shown in Figure 13.7.

TIP
An average person should be able to pull on the connector with a moderate level of force and the connector should not come off.

Figure 13.7

RJ-11/RJ-45 crimper.
Courtesy of IDEAL Insustries
Inc. 2005

13.4.3 Fiber SC

Crimping fiber-optic cable is not that different from crimping other styles of cable but has one notable exception: the fiber-optic cable must be properly cured in a specialized heated oven after the end has been crimped onto it. All of the same preparation and precautions are necessary whether you will be utilizing a traditional epoxy connector or a crimp-style connector.

13.4.4 Fiber ST

Fiber-optic ST style crimp-on tools are quite similar in use to an SC style connector crimping tool. The main difference is the physical characteristic of the cable itself; the SC style is a square connector and the ST is a BNC style.

13.5 Termination Tools

Cable termination is covered in Chapter 5. In this section, we discuss the tools used to terminate various cables.

13.5.1 Coaxial

To terminate coaxial cabling, you use a coax stripper to strip back the ends of a coaxial cable, leaving the desired amount of core, grounding, and shielding exposed. Crimping tools, new style compression tools, and other special tools specifically designed to assist in pushing the standard F-type connector onto the cable can be used.

13.5.2   Twisted Pair

Tools used to terminate twisted-pair cables include: cable cutters (such as a round cutter, ratcheting cutter, or diagonal flush cutter); a sheath remover; a punchdown tool; and a crimping tool. For punchdown tools, you can use a standard, inexpensive punchdown tool that requires you to apply adequate pressure to make a good connection between the cable and punchdown block. Another punchdown tool is the impact tool, which is spring-loaded. You compress the head of the tool, and upon release, the tool exerts force on the connector, firmly seating it in the punchdown block. Impact tools generally have many blades to accommodate 66, 110, Krone, and Bix termination blocks. Multi-pair termination impact tools that include proprietary heads for Category 6 products are available from companies such as Siemon, Panduit, and Leviton.

13.5.3   Balun

A cable with a balun style connector on it usually has a 75-ohm BNC connector on one side and a twisted-pair connector on the other. Although you can terminate the ends yourself, baluns are usually purchased and not made in the field because the success rate is low. Additionally, there are other resistors in the internals that are not easily field terminable. If you must terminate a cable with a balun style connector, terminate the coaxial (BNC) end of the cable using the same tools that we discussed in the "Coaxial" section (13.5.1). Terminate the twisted-pair side using the tools discussed in the "Twisted Pair" section (13.5.2).

13.5.4   Fiber

Fiber-optic cable is more difficult than the other types to terminate. Special termination equipment has to be used, and the materials must be perfectly clean during the process. In addition, many problems can arise while terminating and splicing fiber-optic cable, such as end gaps, concentricity, uneven ends, and air gaps, as discussed in Chapter 8, "Cabling System Connections and Termination." For those reasons, we discourage new cable technicians from terminating their own fiber cabling, and instead recommend that you consult an expert for those services. If you plan to work with fiber cabling exclusively or as a major part of your career, consider taking a course in fiber-optic cabling and termination or learn from an expert. In this section, we provide a brief overview of tools commonly used to terminate fiber cabling.

Popular cable strippers for fiber include No-Niks, Millers, and Clauss, among others. You will also use Kevlar cutters and scissors, and different styles of cleavers, such as hand scribes, staple cleavers, precision cleavers, and proprietary tools such as the Leviton Threadlock. Additionally, you'll need to be familiar with ovens, ultraviolet lamps, hypodermic needles, and syringes. Hand-polishing equipment and materials include polishing pucks (or polishing disk), polishing pads, lapping film (polishing cloths), dry wipes, and alcohol.

13.6 Splicing Tools

Splicing coaxial cable involves using a coax stripping tool to strip away the protective sheathing and prepare the end similar to the method used for terminating coaxial. In many cases, two coaxial cables are connected using a barrel connector. All you need to do is connect the standard female ends of the BNC coax connector to the barrel connector and twist to lock in place.

Splicing category-rated cable (for example, Category 5, 5e, and 6) is forbidden by the standards and as a rule should not be done.

Fiber-optic cable requires the highest degree of precision to splice due to its demanding requirements for tight tolerances. Remember from the fiber-optic discussions in Chapters 7 and 8 that the glass core of the fiber-optic cable is very small, and the splicing must be exact. In the case of single-mode cable, there can be as little as 8 microns to work with, so the individual performing the splicing also must be well trained in the use of fiber-optic splicing tools. The two main methods of splicing fiber-optic cables are mechanical and fusion. A mechanical splicer is simply a device that allows you to perfectly align two fiber cables and maintain that alignment. A fusion splicer perfectly aligns two fiber ends and then fuses the glass ends together using heat or electricity. Mechanical splicers are lower in cost ($1,000 to $2,000) than fusion splicers ($15,000 and up), but the per-splice cost is lower with fusion splicers. In addition, fusion splicers offer a much reduced loss of light.

13.7 Diagnostic Tools

As a cable technician, you will inevitably have to address cabling problems. This section covers many diagnostic tools that can help you resolve cabling communications problems. Many of these tools are also routinely used to test newly created cabling to ensure quality and compliance.

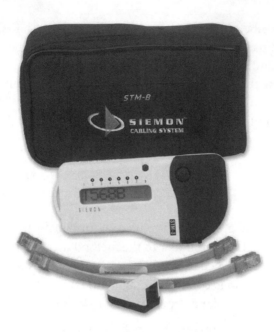

Figure 13.8

Cable tester. Courtesy of
Siemon 2005

13.7.1   Cable Testers

A **cable tester** has the ability to test several characteristics of a cable depending on the cable type being tested. Most standard cable testers can identify opens, shorts, reversals, split pairs, and miswires, among other problems. A typical cable tester is shown in Figure 13.8.

Twisted-pair cabling, for example, can be tested for pair mapping, in other words, making sure pin 1 on side A is the same as pin 1 on side B.

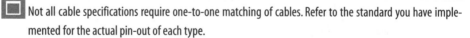

NOTE   Not all cable specifications require one-to-one matching of cables. Refer to the standard you have implemented for the actual pin-out of each type.

Another common test for twisted-pair cabling is **near-end crosstalk (NEXT)**. NEXT occurs when the pairs on one end of a cable are bleeding over to another pair. This is caused by either insufficient twisting of the pairs or twisting the wrong pairs together.

A cable tester is also designed to help identify the length of a cable. A cable tester is placed on one end of a twisted-pair cable, and a signal is generated from the cable tester to the other side where another device is connected that will reflect back the signal. By performing a calculation based on the amount of time it takes to receive the returned signal, the distance of the cable can be determined.

A popular cable tester, the STM-8, is manufactured by Siemon. This tester is actually two hand-held units: a master unit and a remote. The system features an LCD display and can test cable runs from 1 to 900 meters; it also can detect and identify several wiring configurations, such as T568A, T568B, USOC, 10BaseT, and Token Ring.

13.7.2   Certification Meters

Most cable specifications have certain requirements they must meet to be considered in compliance with the specification standard. A useful tool that can be used to prove compliance is called a **certification meter**. This device is like a cable tester, with an added feature, it has a database of most cabling standards. It can generate a report of all cable runs that have been tested to validate their compliance with the specifications. Many cable vendors offer warranties on their products to provide a comfort level to the end user that the products will support a particular speed with minimal data loss. In many cases, it is a requirement to provide a hard copy of the published results from the certification meter to activate the warranty.

Different cable specifications require different parameters as to what are acceptable values for signal loss, NEXT, and length. Be sure your cable certification meter is capable of supporting the cable standard you are working with. For example, be certain your cable certification meter is capable of certifying Category 6 cable if that is the media you are working with.

13.7.3   Time Domain Reflectometer (TDR)

A TDR is a device that is used to measure length of a cable or the distance from one end of a cable to a fault, using a principle that is similar to sonar. A signal is transmitted down the length of a cable. Once the signal reaches either the end or a fault in the cable, part of the signal returns to the transmitter. The time it takes to receive the signal back as well as the strength of the returned signal identifies to the TDR several useful pieces of information. The length of the cable can be determined, or in some cases the length to a fault in the cable, either an open circuit or a short circuit. Whenever two metal cables are in proximity to each other they will form some measurement of impedance. This impedance can be measured for its absolute value, in other words what the actual impedance value is numerically, or the TDR can measure a variation in impedance.

Many TDRs have variable pulse or signal strength settings, so it is possible to use a TDR to artificially increase a load on a cable to determine its capacity to handle bandwidth.

NOTE 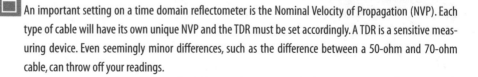 An important setting on a time domain reflectometer is the Nominal Velocity of Propagation (NVP). Each type of cable will have its own unique NVP and the TDR must be set accordingly. A TDR is a sensitive measuring device. Even seemingly minor differences, such as the difference between a 50-ohm and 70-ohm cable, can throw off your readings.

When TDRs first appeared on the market, they were very expensive and only a few companies were able to purchase them. Recent advancements in technology have allowed the cost of the manufacturing process to be reduced significantly, making a TDR a necessity in most cable installer's and/or troubleshooter's tool kits.

13.7.4   Optical Time Domain Reflectometer (OTDR)

An OTDR operates in a similar manner as a TDR, only over fiber-optic cable instead of copper-based cable. Instead of looking for shorts and opens or differences in impedance, the OTDR looks for variations in attenuation. It is just as important for an OTDR to be calibrated properly, based on the characteristics and type of fiber-optic cable being tested, as it is for a TDR to be set correctly. An OTDR can measure and report light loss at a connector, a splice, or along the length of a cable.

13.7.5   Multimeters

A **multimeter** can perform several tests. It can test for the presence or the absence of a signal and the voltage of that signal. It can also reveal shorts or opens between the core of a coaxial cable and the grounding sheath as well as for twisted-pair cabling. The main advantage a multimeter has over other styles of test meters is its relative low cost. Multimeters can be purchased for as little as a few dollars and can provide a lot of data for their investment. A multimeter has two leads connecting to the meter: one red and one black. The red lead is usually connected to the positive terminal or post being tested, and the black is usually connected to the negative terminal or ground. When testing continuity, it does not matter which connector is used. It is common to use different connectors on the leads of a multimeter that have, say, a coaxial type connector on them to make connecting them to a cable end easier.

13.7.6   Tone and Probe Sets

A **tone and probe set** is a tool with two separate pieces: the tone generator and the probe that helps you listen for the tone. The tone generator is connected to one end of a cable and injects an audible tone down the length of the cable. The probe is held against many cables until the one that has the tone generator connected to it has been identified. A tone and probe set is most commonly used by telephone cable engineers who need to identify a particular pair of telephone (copper) wires. The engineer places the tone generator on one end of the pair of wires, and by placing the probe on various cables in a bundle can locate the other end of the cable he is trying to find. The tone will be transmitted down the entire length of the cable pairs, and can be detected anywhere down the length of the cable right through the plastic sheathing surrounding the cable.

13.7.7   Telephone Test Sets

A **telephone test set**, sometimes referred to as a butt set, looks similar to a telephone hand set. It features a numeric dialing keypad that is capable of generating dual-tone multi-frequency (DTMF) signals like a standard telephone, with some added functions. Four keys, labeled A, B, C, and D, are used for signal and control tones in a PBX. In addition to having the ability to test a telephone circuit by placing and receiving telephone calls, many telephone test sets have digit grabber capability. A *digit grabber* is a device that connects in-line and can identify the exact characters that are being transmitted. The DTMF tones that are generated are displayed on a text screen on the test set. This can be useful when trying to isolate problems between a PBX and another communication device.

13.7.8   Stud Finders

A **stud finder** is a device that is used to find wood frame uprights (studs) located in walls. Most stud finders are electronic devices that use a procedure similar to sonar to find densities under the surface of a floor, wall, or ceiling. Some stud finders are simply magnets attached to a lever that swings out when the magnet passes over the metal nails or screws used to attach the drywall material to the wood studs that make up the framing of a wall. The more elaborate stud sensors are electronic devices that can sense the density of the material they are being passed over. To operate a stud sensor, simply lay it flat against the wall, ceiling, or floor where you want to find a stud, press the calibration button, and move it slowly back and forth. The stud sensor works by responding to the difference in densities between the hollow spaces between the wall studs and the studs themselves. As you pass over a stud, the sensor will notify you by either turning on a light (or a series of lights to indicate proximity of the stud) or emitting an audible tone to indicate the presence of a stud.

TIP Most buildings are constructed on either 16-inch or 24-inch centers. That means the center of one stud to the center of another is usually separated by 16 inches or 24 inches, depending on the length of the wall and local building ordinances. After you identify the spacing between studs, it is likely that the remaining wall studs are spaced apart at the same measurement.

13.7.9   Wire Finders

A **wire finder** is a device that is used to trace energized circuits. It can be used to trace the path a cable takes underground, even up to depths of several feet. The wire finder operates by having a transmitter send a signal down the length of a cable; a receiving unit is then passed along the length, or where you think the cable runs may be. The receiver will pick up the signal as it passes over the cable, making it easy to identify in a drop ceiling, down a hollow wall, or in the ground.

13.8   Other Tools

There are a variety of other tools that can be used by a cable installer, some of which are specialized tools to be used in special situations. However, common tools and equipment used in other industries that also can be used by cable installers and troubleshooters include:

- Ladders

- Electric and hand-operated drills and drill bits

- Hammers

- Wrenches
- Screwdrivers
- Socket sets
- Drywall saws
- Flashlights

13.9   Tool Usage and Material Handling Data Sheets

Each tool and piece of test equipment should include a safety data sheet that should be referred to for safe operation. Proper precautions should be taken anytime you are working around electrical circuits.

! WARNING

Any electrical circuit should be assumed to be a live circuit with voltage unless you can verify the absence of voltage with a meter.

13.9.1   Hazardous Materials and Material Safety Data Sheet (MSDS)

MSDS is a generic term that is widely used to identify material safety data sheets. These sheets identify the hazards associated with the use of chemicals, and how to safely handle them. In 1986, the Occupational Health and Safety Administration (OSHA) began requiring MSDSs for all hazardous materials. OSHA has a suggested format for MSDSs, Form 174 (OMB #1218-0072), which is available for download from the U.S. Department of Labor's Occupational Safety and Health Administration Web site (*http://www.osha.gov*). Although the use of the form itself is not mandatory, all MSDSs must contain the following information:

- **Chemical identity:** The identity of the substance as it appears on the label.

- **Manufacturer's name and telephone number:** The name, address, telephone number, and emergency telephone number of the manufacturer. It can also include the date the MSDS was prepared and an optional signature of the preparer.

- **Hazardous ingredients:** A list of the hazardous components by chemical identity and other common names.

- **Physical/chemical characteristics:** A description of the boiling point, evaporation rate, vapor pressure, physical appearance, and odor.

- **Fire and explosion hazard data:** The flash point as well as the method used to determine the flash point. Flammability limits and the method to be used to extinguish the chemical are stated.

- **Reactivity data:** Includes stability, conditions to avoid, incompatible materials to avoid, and hazardous decomposition, or other by-products.

- **Health hazard data:** Includes routes of entry (inhalation, skin, ingestion), whether the hazard is acute (immediate) or chronic (built up over time), symptoms of exposure, medical conditions that can aggravate reaction, and emergency first aid procedures.

- **Precautions for safe handling and use:** Includes measures to be taken in case material is spilled, or other precautions to be taken for safe handling.

- **Control measures:** Includes respiratory protection required, hand and eye protection required, any other protective clothing requirements, or other hygienic safety considerations.

13.9.2 Repetitive Strain Injury (RSI)

Repetitive strain injury (RSI), is a term used to define a variety of soft tissue disorders caused by repetitive motion, bad posture, or stress. Some of the more commonly known RSIs are carpel tunnel syndrome and tendonitis. Although it is not generally life-threatening, an RSI can decrease your overall quality of life by introducing chronic pain and suffering. The key to dealing with an RSI is not to just handle the symptoms, but to reduce the actions that lead to the injury. Some of the symptoms of an RSI are: pain, stiffness; swelling and/or numbness in the joints; discomfort caused by performing repetitive tasks that reduce or disappear when no longer performing the task; and discomfort that migrates from one part of the body to others. If you suspect that you might have an RSI, stop performing the task(s) you believe may be causing the problem and seek medical advice as soon as possible.

WARNING ❗ If you believe that you have a problem with repetitive stress injuries, it is imperative that you seek competent medical advice as soon as possible, since early diagnosis of the problem will result in a more rapid recovery.

Treatment of RSIs can be difficult because the very thing that caused the injury in the first place is most likely work-related, and in most cases seeking alternate employment is impractical. There are treatment methods that can be utilized to minimize the impact of RSI, such as wearing protective equipment (braces, knee wraps, etc.) modifying posture, and even reducing stress wherever possible, since stress leads to tension that can exacerbate the injury. In some cases, something as simple as adding an exercise routine to your lifestyle can dramatically reduce the effects of repetitive stress injuries. Also, if your job requires physical labor, stretching your muscles before beginning physical tasks can help to reduce the effects of RSI; remember to take regular breaks.

13.10   Chapter Summary

- Cable pushers and pullers help the cable technician run cabling efficiently. These tools include cable casters, fish tapes, push/pull rods, and gopher poles.

- Labeling systems are used to label cables and include simple methods such as marking each end of a cable with a permanent-ink marker, to using write-on mylar labels, to investing in electronic cable labelers that print tubular labels that you affix over both ends of a cable.

- Stripping and cutting tools help remove the sheathing from cables to prepare it for crimping or termination. These tools include ordinary strippers, round cable strippers, and cable knives, among many others. Crimping tools are used to connect connectors to the ends of cables.

- Termination tools include cable cutters (such as a round cutter, ratcheting cutter, or diagonal flush cutter), a sheath remover, a punchdown tool, and a crimping tool. Impact punchdown tools greatly reduce the amount of pressure required by the technician to firmly seat cables in punchdown blocks.

- Diagnostic tools help test cables for quality and compliance with codes and standards, and resolve communications problems. These tools include cable testers, certification meters, time domain reflectometers (TDRs), optical TDRs, multimeters, tone and probe sets, telephone test sets, stud finders, and wire finders.

- OSHA requires a materials and material safety data sheet (MSDS) to be filed when handling hazardous materials.

- To diminish injury, know the signs and symptoms of repetitive stress injuries (RSIs) and how to seek assistance in the event you suspect you have been afflicted with RSI. You can prevent RSIs by knowing what causes RSIs and avoiding those habits.

13.11   Key Terms

cable caster: A device that is similar in operation to a fishing rod and reel that is used to pull cable over a long open span like a drop ceiling.

cable tester: A multi-function cable testing device that can provide information regarding the length of a cable and faults in a cable (short circuits or open circuits) plus the distance to a fault.

certification meter: A device that is used to measure compliance with published cabling standards.

coax stripper: A tool that is used to expose the core, grounding, and protective sheathing of a coaxial cable.

ferrule: A metal sleeve with a barrel-like appearance that is used when terminating coaxial and some fiber-optic cables.

fish tape: A tool that is used to pull cable through a hollow wall, floor, or ceiling.

gopher pole: A lightweight telescoping pole with a hook on the end; used for pulling or moving cabling in hard-to-reach places.

multimeter: An inexpensive device that measures resistance and voltages on a cable.

near-end crosstalk (NEXT): The leaking of an electrical signal between two pairs of copper cabling.

optical time domain reflectometer (OTDR): A device that measures the length of a fiber-optic cable as well as any faults in the cable and the distance to the fault.

push/pull rod: A series of narrow lightweight rods in which cable is threaded and then pushed/pulled through hard-to-access areas such as wall and ceiling drops. The rods are attached to a base pole, which is held by the cable technician.

stud finder: A tool used to identify where wood studs are located in a wall, floor, or ceiling.

telephone test set: A tool used by telecom engineers to test phone circuits.

time domain reflectometer (TDR): A device that measures the length of a copper cable as well as any faults in the cable and the distance to the fault.

tone and probe set: A test tool that is used to identify a pair of copper wires; usually used in telecom circuits.

wire finder: A tool used to locate the path a cable follows through walls, floors, ceilings and underground.

13.12   Challenge Questions

13.1 Which cable types require the most amount of precision when splicing?

    a. Coax

    b. Balun

    c. Fiber-optic

    d. Twisted pair

13.2 What is the tool that is used by a telephone engineer to place calls, test phone circuits, and send DTMF signals including extended DTMF?

    a. Tone and probe set

    b. Multimeter

    c. OTDR

    d. Telephone test set

13.3 What tool can be useful for a cable installer who is required to pull cable through a hollow wall?

    a. Stud finder

    b. Multimeter

    c. MSDS

    d. Cable caster

13.4 Which of the following are on an MSDS? (Choose all that apply.)

    a. Chemical identity

    b. Manufacturer's telephone number

c. Hazardous ingredients

d. Reactivity data

13.5 What is the first step you should take in the event you suspect that you are afflicted with a repetitive strain injury?

a. Put ice on the affected part.

b. Consult your doctor.

c. Stop performing the act you believe led to the injury.

d. Apply heat to the affected area.

13.6 Which of the following cable types can require the use of a puck to terminate the cable?

a. Balun

b. Twisted pair

c. Coaxial

d. Fiber-optic

13.7 A cable stripper is used for which of the following? (Choose all that apply.)

a. Removing the protective outer sheathing on a coaxial cable

b. Removing the cladding from the core of a fiber-optic cable

c. Removing the protective plastic coating on a twisted-pair cable

d. Removing labels from any type of cable

13.8 What is the result of tying the core of a coaxial cable to the grounding braid?

a. Better data throughput

b. An electronic open circuit

c. An electronic short circuit

d. Less data throughput

13.9 What cable type is concerned with the issue of NEXT?

a. Fiber-optic

b. Balun

    c. Twisted pair

    d. Coaxial

13.10 Which of the following are issues to be concerned about when terminating or splicing fiber-optic cable? (Choose all that apply.)

    a. Air gap

    b. Cleanliness

    c. Grounding

    d. Dispersion

13.11 What is the tool that is used to help a cable installer pull cable through a drop ceiling?

    a. Stud finder

    b. Fish tape

    c. Cable caster

    d. Probe

13.12 On an MSDS, the *reactivity data* section refers to which of the following?

    a. Whether the hazard is acute or chronic

    b. Measures to be taken in case the material is spilled

    c. Stability, conditions to avoid, and incompatible materials to avoid

    d. Respiratory protection requirements

13.13 What is the tool that can be used to help a telephone cable installer identify a particular pair of copper wires?

    a. Tone and probe set

    b. Cable caster

    c. OTDR

    d. MSDS

13.14 Which of the following are considered RSIs? (Choose all that apply.)

    a. Carpel tunnel syndrome

b. Arthritis

c. Tendonitis

d. Fatigue

13.15 What is the most common problem encountered when crimping coaxial cable?

a. Pinching the cable, causing signal loss

b. Allowing strands of the sheathing to touch the core, causing a short circuit

c. Applying inadequate pressure to the crimpers, resulting in an intermittent connection

d. Applying too much pressure to the crimpers, causing a severed cable

13.16 Which of the following can be used to determine the length of a cable? (Choose all that apply.)

a. Cable tester

b. Multimeter

c. Wire finder

d. TDR

13.17 Where can you go to obtain additional information regarding MSDS?

a. *http://www.osha.gov*

b. *http://www.cisco.com*

c. *http://www.us.gov/msds*

d. *http://www.msds.org*

13.18 Which tools would you use to help identify a short or open circuit on a copper cable? (Choose all that apply.)

a. Multimeter

b. Cable tester

c. TDR

d. OTDR

13.19 Which tool would you use to identify the underground path a copper cable takes between two buildings?

    a. TDR

    b. Cable tester

    c. Wire finder

    d. Multimeter

13.20 Which tools operate in a similar way as sonar? (Choose all that apply.)

    a. TDR

    b. Multimeter

    c. OTDR

    d. Line tester

13.13   Challenge Exercises

Challenge Exercise 13.1

In this exercise, you learn about the various properties of copper coaxial cabling by using a multimeter. To complete this exercise, you need a multimeter and a length of copper coaxial cable of varied resistance (i.e., 50 ohm and 75 ohm).

13.1.1 Place the probes into the multimeter if they are not already connected.

13.1.2 Set the multimeter to the **OHM** setting. If you have several ranges to select from, choose the range that corresponds with the type of cable you are testing. In this case, select the range that includes 50 and 75 ohm.

13.1.3 Purposely short the core copper of the 50-ohm cable to the grounded sheathing on one end of the cable.

13.1.4 Place the red lead on the end (not the end you shorted) of the cable and clip the black lead to the grounding sheath.

13.1.5 What do you observe on your multimeter?

13.1.6 Remove the short circuit from the far end of your cable and repeat the test.

13.1.7   What did you observe after the short was removed?

13.1.8   Perform the entire exercise a second time with the 75-ohm cable and observe the results.

Challenge Exercise 13.2

In this exercise, you research material safety data sheets. To complete this exercise, you need a computer with a Web browser, internet access, and a printer.

13.2.1   Log on to your computer and open your Web browser.

13.2.2   In your Web browser, type: *http://www.osha.gov*.

13.2.3   In the search field, type **MSDS** and press **Enter**.

13.2.4   Research the format of an MSDS. Find some sample forms, and download and print them.

13.2.5   Fill out a form as completely as possible assuming you are handling asbestos.

13.2.6   What is the main goal of an effective hazard communication program?

Challenge Exercise 13.3

In this exercise, you use a crimping tool to put an RJ-45 connector on a twisted-pair cable using the TIA/EIA-568 cable specification. To complete this exercise, you need a length of twisted-pair cable, two RJ-45 connectors, an RJ-45 crimping tool, a cable tester capable of providing wire mapping, and the specification for TIA/EIA-568 cable.

13.2.1   Obtain a length of Category 5e cable and an RJ-45 connector.

13.3.2   Following the TIA/EIA-568 specification, place the individual strands of copper cable into the RJ-45 connector. With the cable end facing away from you and the tab of the RJ-45 cable facing down, pin 1 will be to the right.

13.3.3   After you are satisfied that the individual strands are in their proper slots and fully seated, place the end and cable into the crimper and apply pressure until the crimper releases. This ensures a solid crimp.

13.3.4   Perform steps 2 and 3 on the other end of the cable.

13.3.5 Place the cable ends into the cable tester and test the cable by performing a wire map.

13.3.6 Ensure all individual strands are in compliance with TIA/EIA-568.

13.14 Challenge Scenarios

Challenge Scenario 13.1

Your task is to supply a parts list for a cable installer's tool kit. Budget is not as important a factor as having a complete test kit. Your kit should provide the necessary tools to test and terminate copper as well as fiber-optic cable.

13.1 What tools would you include in your kit?

13.2 What tools would you leave out?

13.3 What tools can be marked as redundant in the event something needs to be removed from the budget?

CHAPTER 14

Planning and Implementation of Premise Wiring Installations

Learning Objectives

After reading this chapter you will be able to:

- Recognize flexible design considerations
- Understand the bid process including Requests for Proposals (RFPs) and Requests for Quotes (RFQs)
- Be aware of the CSI MasterFormat
- Comprehend the principles of project management as it relates to cabling projects
- Identify the phases of the construction process
- Recognize the tools of project management and their uses

Until recently, construction projects did not include telecommunications systems as a major part of construction. As these systems became more important to business operations, they moved up the priority list in construction projects. Now architects, planners, and construction designers include these systems in the design phase in the building process.

The importance of a well thought out design that has the needs of the customer and their performance requirements in mind is vital to ensure the reliability and life of the installed system. To make certain that the system delivered will meet the design, particular attention given to the construction specifications, plans, and careful project management must take place. These are the elements of an effective pre-construction planning stage and go a long way to assuring the successful completion of the project.

14.1   Planning Premises Cabling Systems

As the saying goes, prior planning prevents poor performance. While planning will not eliminate jobsite issues, having a good plan and sticking to it can minimize the impact of any obstacles that arise during the completion of the project.

14.1.1   Consultation

The planning stage of a cabling installation begins with a meeting with the customer. This meeting typically will be attended by the building architect, the telecommunications system designer, and any of the customer's key representatives like the Chief Executive Officer (CEO), Chief Operations Officer (COO), or the Chief Information Officer (CIO) as well as any subordinate personnel that are necessary. The topic of discussion centers on the customer's operations, including current systems, plans and goals, and the structure of the new or existing building where the installation is going to take place. Questions about these factors help you, the telecommunications system designer, to identify the type of system, its performance requirements, and the initial size of the installation.

Once the first consultation is complete, you can create an initial design based on the information gathered from the customer. At that point, you review the initial design with the customer, most likely in another meeting similar to the first consultation and including the same attendees. Then you incorporate any changes identified into the design and perform another review. This process may have several revisions depending upon the size and complexity of the project, or it might go through the process only once if the project is small.

14.1.2   Flexible Design Approaches

With the convergence of voice, data, and video systems as well as the conversion of analog signals into digital transmissions enabling the emergence of new technologies such as Voice over IP (VoIP), the distribution system needs to be robust, flexible, and have an extended lifespan. Similar to a commercial office building where the building structure represents the cable plant and the internal offices represent the computers and other equipment, the premise wiring structure must be able to handle the inevitable changes and upgrades to the equipment connected to it. Additionally, this convergence and flexibility has been a driving force in the development of a standards-based uniform method of providing the path for these technologies to interact.

One of the first items you need to consider is the cable pathway method you are going to use. Many options are available and each has pros and cons. The method you choose should expect moves, additions, and changes (MACs) to accommodate the needs of the customer.

Underfloor Ducts

Underfloor ducts are metal raceways, usually rectangular, that form a grid within the floor allowing for the distribution of the horizontal cable. Installed within the concrete floor of the building, like a commercial highrise for example, these raceways are in combinations of large-sized, also known as feeder, and standard-sized, also known as distribution, runs depending upon the needs of the floor area served. These ducts also have access points installed to allow for access both during and after installation of the cable. Because this type of distribution system becomes part of the building structure and is not expandable, carefully calculate fill capacity and account for expected growth as well. Also, placement of the access points is crucial because these will be the only way to gain entrance to and use the duct system.

Access Floors

Access floors are essentially raised floors placed on support structures on top of the normal or subfloors, with the cabling lying underneath the floors in the space created by it. This design allows for excellent access to the cable and generally provides unencumbered paths, since no services other than a small amount of electrical is usually permitted to use that area. It is because of this that the use of access flooring in the **telecommunications room (TR)** and **equipment room (ER)** is quite common. Using such a system requires an analysis of the type and weight of the equipment going on top of it to ensure that overloading the system does not occur. In addition, most areas of the country require that these systems be seismically braced and inspected prior to use, which usually adds time and cost to the project.

Conduit

Conduit is a system of pipes or tubes of different sizes that are connected together to provide the pathway for cable distribution. Conduits can be installed in the concrete floor, down walls, and even overhead in the ceiling. Access to a conduit system is made using **junction boxes**. These boxes provide cable-pulling points, points to allow change of direction, and can be termination points as well. As with underfloor ducting, sizing of the conduit system to accommodate current use and future growth is essential because changes to the system are difficult, if not impossible, once installed. Conduits offer physical protection of the cable and can also help with electromagnetic interference (EMI) if they are comprised of a metallic material and are properly grounded.

Cable Tray

A **cable tray** consists of a rigid, prefabricated support system that provides excellent access and greater flexibility for cable distribution. Cable tray systems are one of the most common distribution types typically suspended overhead above a false acoustical ceiling, like in commercial office buildings, or installed within an access floor space. Usually, main cable pathways within a TR, ER, and the first section of distribution to service areas use a cable tray. Beyond that, other methods of cable distribution are typically used. Seismic bracing of the cable tray is required in most areas, and is installed in accordance with local building codes.

Ceiling Distribution Systems

Ceiling Distribution Systems are not individual support structures; rather, they are systems comprised of many different methods of distribution such as cable trays, conduits, and another type of support called J-hooks. J-hooks are, as their name implies, metal brackets shaped in a J where the cable rests. These hooks have minimum standard dimensions in order to ensure they provide proper support to the cable while not contributing to performance issues such as pinching or deforming the cables. J-hooks are usually used for small cable bundles that shoot off from the main pathway. The most important consideration when using these systems in a design is how much space is available within the ceiling. As a designer, you must be aware that the ceiling space is shared by other building systems such as fire sprinklers, plumbing, HVAC (heating, venting and air conditioning), and electrical, and you should plan accordingly to avoid conflicts with these systems.

Surface Mounted Raceway

Surface mounted raceway is a distribution system mounted on the surface of walls and ceiling structures. It is available in many different sizes and styles as well as different materials. You use surface mounted raceways in areas where the walls and ceilings are either solid in construction, such as brick or concrete, or sealed up with drywall. Size considerations of surface mounted raceway also need to be accounted for when designing a system. Additionally, many customers do not like the aesthetics associated with these distribution systems and will only consider them as a last resort.

Media Type

Although choosing the media type is the second topic in this discussion, the type of distribution pathway can affect the media type and vice versa. This is due to the physical size difference between fiber-optic and copper cables, and the consideration of bend radius and other issues that affect the performance of each. In a cabling system design, backbone and inter/intra-building applications typically use fiber-optics while copper is the media of choice for the distribution to the work areas. The performance requirements of the customer will greatly determine what media type and grade are used.

As discussed in Chapters 5 and 6, copper cable comes in many different category ratings that denote performance level. When you are designing a system, keep in mind that most often these systems have been in place for several years while the electronic equipment attached to them change with time and technological advances. Therefore, it would serve you and your customer to ensure that the system you design will support their bandwidth needs both now and in the future. To illustrate this, let us assume that your customer currently has a Token ring network with a maximum data transmission rate of 16 Mbps (megabits per second) and would like to upgrade to an Ethernet 100 Mbps system. Once this upgrade is complete, they do not expect to perform another upgrade for at least five years and, since they own their building, they do not wish to upgrade their cabling infrastructure at that point; rather, they want to only upgrade the electronics in five years. The best system you could design for a new installation of copper cable would be Category 6. This system would support a minimum of one Gigabit data transmission rate with the expectation of being capable of supporting multi-Gigabit applications. By designing the cable infrastructure with this high level of performance capability, you can assure your customer that they will have a distribution system that will support their current goals as well as future upgrades.

Fiber-optic cable also has different performance capabilities, as noted in Chapter 7. You need to know these capabilities in order to design the right solution for your customer. Since fiber-optic use is prevalent in backbone as well as intra-building applications, accounting for the higher bandwidth needed is a must. Keep in mind that the backbone is like a highway of a city where all the side streets flow to the main thoroughfare; if there is not enough room for all those cars to travel, traffic will back up until there is gridlock. Fiber-optic has the benefit of being relatively immune to outside interference such as EMI and radio frequency interference (RFI). This is particularly useful when the only pathway you have between floors in a building travels through the mechanical room that houses the motors for the elevator, which can be a large potential source of EMI.

When deciding on the media type you will use, there is another aspect for consideration: safety. The **National Electrical Code (NEC)**, produced by the National Fire Protection Association (NFPA), is a document specifically intended to decree the minimum requirements necessary in order to protect people and property from electrical hazards. The NEC, revised every three years, covers the rules regarding electrical and telecommunications cabling installations in the United States in relation to safety. Although not law by itself, the NEC is the basis of almost all legal electrical code. Canada has its own version of the NEC called the Canadian Electrical Code (CEC), Part 1. As a designer, it is important for you to know local legal codes and regulations as well as standards requirements as they affect your design.

Other Spaces

TRs, ERs, and **telecommunications entrance facilities** are other areas that you will assist with or completely design. More often than not, networking equipment resides in the same vicinity as the head end of the cabling infrastructure. It becomes your responsibility to ensure that the space provided in these areas accommodates your proposed design and allows for future expansion. Due to the interaction between the cabling vendor and the information technology (IT) department of the customer, you will typically be asked to include their space layout and equipment as part of your design, so that any additional equipment cabinets, racks, or enclosures are included within the specifications sent out for the bidding process.

Understanding how much space is required for your termination equipment as well as any additional equipment you include is essential in the planning stage. The architect uses this information to incorporate these spaces into the building drawings. The total amount of space needed will vary with the size of the infrastructure required and the style of the termination point, whether it is wall-mount or rack-mount configuration. Depending on the size of the installation, these configurations can require large amounts of floor space, including the clearance surrounding them for accessibility.

Unfortunately, having the ability to affect the size and shape of the room available for the TR, ER, or entrance facility does not happen often. Most of the time, you will be given a print designating the area that you have available and you must design around that space. This is where your creativity and ingenuity come into focus. Being flexible and having close working knowledge of the products available for your use will make designing these rooms a much easier process.

14.1.3 Plans and Specifications

Once the customer approves the design, the architect incorporates it into the building plans and specifications. This requires a round or more of meetings and reviews by the designer and the architect to ensure that everything the designer wanted to incorporate was included in the plans.

Most commercial building projects follow guidelines set by the Construction Specifications Institute (CSI) called the **MasterFormat**. The MasterFormat is essentially an organizational outline for a building and everything that resides within it. It provides a way for all of those involved in the construction project to quickly find the information they are looking for and to standardize this outline across the industry. The CSI is an association of many members of the construction community including specifiers, contractors, architects, engineers, suppliers, and building owners. Their combined contributions have made the MasterFormat a success in the commercial building market.

The MasterFormat consists of several sections in a building specifications document called divisions. Each one of these covers specific areas within the construction process such as general information, foundations, electrical, and plumbing. Within each of these divisions are subdivisions that break down each section even further. In 2001, the CSI began reviewing and updating the MasterFormat to reflect the changes within the construction industry.

One part of the changes was the separation of telecommunications, life safety, integrated automated services, and others that had been integrated into other sections under the 1995 MasterFormat. With these changes, components that are vital to business operations today but had been included only as an afterthought are now part of the preplanning stage of construction, giving them the attention they deserve. Also, with their inclusion into the MasterFormat, these services now have validity never experienced before when vying for space in an already crowded environment.

14.1.4 Construction Specifications

Now that you have a basic understanding of the CSI MasterFormat, let us look at an example of it. In Figure 14.1, you see a sample of Division 27—Communications from *MasterFormat 2004 Edition*. There are many levels of detail, each progressively deeper as you move into the outline. You also should note that there is a detailed list of all aspects of construction including

DIVISION 27-COMMUNICATIONS

27 00 00   COMMUNICATIONS
27 01 00   Operation and Maintenance of Communications Systems
27 01 10   Operation and Maintenance of Structured Cabling and
              Enclosures
27 01 20   Operation and Maintenance of Data Communications
27 01 30   Operation and Maintenance of Voice Communications
27 01 40   Operation and Maintenance of Audio-Video Communications
27 01 50   Operation and Maintenance of Distributed Communications
              and Monitoring
27 05 00   Common Work Results for Communications
27 05 13   Communications Services
27 05 13.13 Dialtone Services
27 05 13.23 T1 Services
27 05 13.33 DSL Services
27 05 13.43 Cable Services
27 05 13.53 Satellite Services
27 05 26   Grounding and Bonding for Communications Systems
27 05 28   Pathways for Communications Systems
27 05 28.29 Hangers and Supports for Communications Systems
27 05 28.33 Conduits and Backboxes for Communications Systems
27 05 28.36 Cable Trays for Communications Systems
27 05 28.39 Surface Raceways for Communications Systems
27 05 43   Underground Ducts and Raceways for Communications
              Systems
27 05 46   Utility Poles for Communications Systems
27 05 48   Vibration and Seismic Controls for Communications Systems
27 05 53   Identification for Communications Systems
27 06 00   Schedules for Communications
27 06 10   Schedules for Structured Cabling and Enclosures
27 06 20   Schedules for Data Communications
27 06 30   Schedules for Voice Communications
27 06 40   Schedules for Audio-Video Communications
27 06 50   Schedules for Distributed Communications and Monitoring
27 08 00   Commissioning of Communications

27 10 00   STRUCTURED CABLING
27 11 00   Communications Equipment Room Fittings
27 11 13   Communications Entrance Protection
27 11 16   Communications Cabinets, Racks, Frames and Enclosures

Figure 14.1

Sample *MasterFormat 2004 Edition*, Division 27—Communications.

The Numbers and Titles used in this product are from *MasterFormat*™ 2004 edition published by The Construction Specifications Institute (CSI) and Construction Specifications Canada (CSC) and is used with permission from CSI, 2005. The Construction Specifications Institute (CSI), 99 Canal Center Plaza, Suite 300, Alexandria, VA 22314, 800-689-2900; 703-684-0300, CSINet URL: http://www.csinet.org.

both products and activities. This list helps to facilitate communication between all personnel involved in a construction project.

A typical section within a division has the specific requirements for that item usually beginning with the "General Requirements" section. This continues with subsections such as "Products" and "Execution." Within each of these subsections are the details of the item or items listed, usually in a bulleted format. While regulating the overall layout, whoever is writing the specification decides what is included in the details. Generally speaking, the more detail included in the specifications, the more accurate the work will be.

For example, imagine that you are the specifier for a construction project called ABC Business Park. You job is to write the specification for the telephone and data structured wiring that will be contained within the buildings as well as the interconnection between them. As part of this specification, you will select the performance level required for the project as well as suggest manufacturers of products that meet your specifications. In our example, we focus on the horizontal cable infrastructure that is part of Building A within the business park. A sample specification is shown in Figure 14.2.

When looking at this example, one thing to notice right away is the detail in which you describe each product within Part 2—Products. With this amount of detail, there can be no question as to what you mean. Finding products that meet this specification will not be too difficult, especially because an example of an approved manufacturer is included with each product specification. By writing the specification in this way, the contractors who respond to the bid request and the ones who finally perform the installation will have a clearly defined standard with which to compare their chosen product. This also allows for a checks and balances system when the project is being verified prior to the hand off to the customer.

Other than specifying specific products that you require for a project, the construction specifications allow you to give specific instructions on processes, like how to respond to a bid, acceptance testing, how to handle substitutions for products, and what warranties you want the contractor to provide. The construction specifications may include the process undergone when finding erroneous items covered by the specification when discovering

Customer and Project Name
Project #
12/18/2001

SECTION 17160

HORIZONTAL CABLING

PART 1-GENERAL
1.1 WORK INCLUDED

  A. Provide all labor, materials, tools, and equipment required for the complete installation of work called for in the Contract Documents.

1.2 SCOPE OF WORK

  A. The horizontal portion of the telecommunications cabling system extends from the work area telecommunications outlet/connector to the horizontal cross-connect in the telecommunications closet/room. It consists of the telecommunications outlet/connector, the horizontal cables, and that portion of the cross-connect in the telecommunications closet/ room serving the horizontal cable. Each floor of a building should be served by its own horizontal cabling system.

  B. This section includes minimum requirements for the following:

    1. Horizontal cable from telecommunications closet/room to Workstation

    2. Category 5e Modular Jacks

    3. Category 3 Modular Jacks

    4. Work Area Equipment Cords

    5. Faceplates and Jacks

    6. Installation and Termination Methods

1.3 QUALITY ASSURANCE

  A. All cable shall be installed in a neat and workmanlike manner. All methods of construction that are not specifically described or indicated in the contract documents shall be subject to the control and approval of the Owner's Representative. Equipment and materials shall be of the quality and manufacture indicated. The equipment specified is based upon the acceptable manufacturers listed. Where "or equivalent" is stated. equipment shall be equivalent in every way to that of the equipment specified and subject to approval.

  B. Strictly adhere to all **Category 5e** installation practices when installing horizontal cabling.

  C. Materials and work specified herein shall comply with the applicable requirements of:

    1. ANSI/TIA/EIA-568-A Telecommunications Cabling Standard (including all the latest amendments and applicable addenda)

    2. ANSI/TIA/EIA-569-A Pathway and Spaces

    3. NFPA 70-1999

Figure 14.2

Horizontal cable specification. Courtesy of Richard Steiner 2005

Customer and Project Name
Project #
12/18/2001

      4.   BICSI Telecommunications Distribution Methods Manual

      5.   FCC 47 CFR 68

      6.   NEMA-250

      7.   NEC-Articles 770 and 800

      8.   ADA-Americans with Disabilities Act

      9.   ISO/IEC 11801 (International) Generic Cabling for Customer Premises Standard (including all the latest amendments and applicable addenda)

1.4   SUBMITTALS

    A.   Manufacturers catalog sheets, specifications and installation instructions for the following: cable. modular jacks media adapters, optical fiber connectors, faceplates and jacks if different from the Design Make (submit with bid).

    B.   If providing pre-standards manufacturer system solution, submit installer/contractor certification documentation and channel certification information and requirements from manufacturer.

PART 2-PRODUCTS

2.1   100 OHM UNSHIELDED TWISTED PAIR CABLE (UTP)

    A.   Physical Characteristics:

      1.   Be 100 Ω 4-pair, 22–26 gauge, Category 3 or 5e rated cable, as applicable

      2.   Be appropriate for the environment in which it is installed

      3.   Labeled and third party Verified Category 3 or 5e rated cable

      4.   White cable shall be provided for all cables terminated to 110 system blocks in the telecommunications closet/room that are designated for analog or digital telephone services

      5.   Yellow cable shall be provided for all cables terminated to patch panels in the telecommunications closet/room that are designated for data services

      6.   Be UL listed

    B.   Transmission Characteristics:

     The following cable specifications shall also be met by the cable manufacturer for 4-pair UTP Category 5e cables:

      1.   Attenuation

     Qualified Cables shall exhibit worst case attenuation less than the values derived using the equations shown in the chart below from 1 MHz to the highest referenced frequency value. Worst case qualified cable attenuation performance for selected frequency points of interest is also provided:

Figure 14.2 (continued)

Horizontal cable specification. Courtesy of Richard Steiner 2005

Customer and Project Name
Project #
12/18/2001

Cat-5e	
Frequency Range	**1-100 MHz**
Worst Case	$\leq 1.967\sqrt{f} + .023 \cdot f + \dfrac{0.05}{\sqrt{f}}$
Frequency Points of Interest	**MHz**	
	100	**22 dB**
	200	-
	250	-

2. Near End Crosstalk (NEXT) Loss

Qualified Cables shall exhibit worst case NEXT Loss greater than the values derived using the equations shown in the chart below from 1 MHz to the highest referenced frequency value. Worst case qualified cable NEXT Loss performance for selected frequency points of interest is also provided:

Cat- 5e	
Frequency Range	**1-100 MHz**
Worst Case Cable NEXT Loss	$\geq 67 - 15\log\left(\dfrac{f}{0.772}\right)$
Frequency Points of Interest	**MHz**	
	100	**35.3 dB**
	200	-
	250	-

Figure 14.2 (continued)

Horizontal cable specification. Courtesy of Richard Steiner 2005

Customer and Project Name

Project #

12/18/2001

3.   Power Sum Near-End Crosstalk (PSNEXT) Loss

     Qualified Cables shall exhibit worst case PSNEXT Loss greater than the values derived using the equations shown in the chart below from 1 MHz to the highest referenced frequency value. Worst case qualified cable PSNEXT Loss performance for selected frequency points of interest is also provided

Cat-5e		
Frequency Range	**1-100 MHz**	
Worst Case	$\leq 1.967\sqrt{f} + .023 \cdot f + \dfrac{0.05}{\sqrt{f}}$	
Frequency Points of Interest	**MHz**	
	100	**22 dB**
	200	-
	250	-

4.   Equal Level Far-End Crosstalk (ELFEXT)

     Qualified Cables shall exhibit worst case ELFEXT greater than the values derived using the equations shown in the chart below from 1 MHz to the highest referenced frequency value. Worst case qualified cable ELFEXT performance for selected frequency points of interest is also provided

Cat- 5e		
Frequency Range	**1-100 MHz**	
Worst Case ELFEXT	$\geq 66 - 20\log(\dfrac{f}{0.772})$	
Frequency Points of Interest	**MHz**	
	100	**23.8 dB**
	200	-
	250	-

Figure 14.2   (continued)

Horizontal cable specification. Courtesy of Richard Steiner 2005

Customer and Project Name
Project #
12/18/2001

5. Power Sum Equal Level Far-End Crosstalk (PSELFEXT)

Qualified Cables shall exhibit worst case PSELFEXT Loss greater than the values derived using the equations shown in the chart below from 1 MHz to the highest referenced frequency value. Worst case qualified cable PSELFEXT performance for selected frequency points of interest is also provided

	Cat- 5e
Frequency Range	**1-100 MHz**
Worst Case PSELFEXT	$\geq 63 - 20\log(\frac{f}{0.772})$
Frequency Points of Interest	**MHz**	
	100	20.8 dB
	200	-
	250	-

6. Return Loss

Qualified Cables shall exhibit worst case Return Loss greater than the values derived using the equations shown in the chart below from 1 MHz to the highest referenced frequency value. Worst case qualified cable Return Loss performance for selected frequency points of interest is also provided

	Cat- 5e	
Frequency Range	**1-100 MHz**	
Worst Case Return Loss	**Frequency (MHz)**	**Return Loss (dB)**
	$1 \leq f < 10$	$20 + 5 \cdot \log(f)$ dB
	$10 \leq f < 20$	22 dB
	$20 \leq f \leq 100$	$25 - 7 \cdot \log(f/20)$
Frequency Points of Interest	**MHz**	
	100	20.1 Db
	200	-
	250	-

Figure 14.2 (continued)

Horizontal cable specification. Courtesy of Richard Steiner 2005

Customer and Project Name
Project #
12/18/2001

7. Propagation Delay (ANSI/TIA/EIA-568-A-1)

Qualified Cables shall exhibit worst case Propagation Delay less than the values derived using the equations shown in the chart below from 1 MHz to the highest referenced frequency value. Worst case qualified cable Propagation Delay performance for selected frequency points of interest is also provided

		Cat-5e
Frequency Range		**1-100 MHz**
Worst Case Propagation Delay		$< 534 + \dfrac{36}{\sqrt{f_{MHz}}}$
Frequency Points of Interest	**MHz**	
	100	**538 ns**
	200	-
	250	-

8. Delay Skew (ANSI/TIA/EIA-568-A-1)

Qualified Cables shall exhibit worst case Delay Skew less than the values specified in the chart below per 100 m from 1 MHz to the highest referenced frequency value

		Cat- 5e
Frequency Range		**1-100 MHz**
Worst Case Delay Skew	**MHz**	
	100	**45 ns**
	200	-
	250	-

Figure 14.2   (continued)

Horizontal cable specification. Courtesy of Richard Steiner 2005

Customer and Project Name
Project #
12/18/2001

   C.   Design Make UTP: **Mohawk/CDT MEGALAN 400 or equivalent**

   1.   Yellow (for cables terminated to patch panels), 4 pair, Category 5e

   2.   White (for cables terminated to 110 system blocks), 4 pair, Category 3

2.2   CATEGORY 5e MODULAR JACKS

   A.   Physical Characteristics:

   1.   Be available in Orange

   2.   Have 50 micro-inches minimum of gold plating over nickel contacts

   3.   Be 8-position/8-conductor with 110 IDC termination

   4.   Provide universal application/multi-vendor support

   5.   Support T568A or T568B standards wiring options

   6.   Allow termination with standard termination tools

   7.   Be constructed of high impact, flame-retardant thermoplastic

   8.   Be made by an ISO 9001 Certified Manufacturer

   9.   Be able to except 100 Ω 22–24 AWG copper cable

   B.   Transmission Characteristics (as tested in accordance with ANSI/TIA/EIA-569A Annex and ISO/IEC 11801 Category 5e compliant)

   1.   The following requirements shall also be met (NEXT Loss and FEXT tested in both Differential and Common Mode:

Parameters	Performance @ 100 MHz
NEXT Loss	43.0 dB
FEXT	35.1dB
Insertion Loss (Attenuation)	. 4 dB
Return Loss	20 dB

   2.   Be UL VERIFIED (or equivalent) for TIA/EIA Category 5e electrical performance

   3.   Be UL LISTED 1863 approved or equivalent

   C.   Design Make: **Leviton GigaMax 5e or equivalent**

2.3   CATEGORY 3 MODULAR JACKS

   A.   Physical Characteristics:

   1.   Be available in Ivory

   2.   Have 50 micro-inches minimum of gold plating over nickel contacts

Figure 14.2   (continued)

Horizontal cable specification. Courtesy of Richard Steiner 2005

Customer and Project Name
Project #
12/18/2001

        3.   Be 8-position/8-conductor with 110 IDC termination

        4.   Provide universal application/multi-vendor support

        5.   Support T568A or T568B standards wiring options

        6.   Allow termination with standard termination tools

        7.   Be constructed of high impact, flame-retardant thermoplastic

        8.   Be made by an ISO 9001 Certified Manufacturer

        9.   Be able to except 100 Ω 22–24 AWG copper cable

    B.   Design Make: **Leviton GigaMax 5e or equivalent**

2.4   MODULAR EQUIPMENT CORDS: (CATEGORY 5e)

    A.   Physical Characteristics:

        1.   Be round. and consist of eight insulated 24 AWG, stranded copper conductors, arranged in four color-coded twisted-pairs within a flame-retardant jacket

        2.   Be equipped with modular 8-position plugs on both ends. wired straight through with standards compliant wiring

        3.   Use modular plugs, which exceed FCC CFR 47 part 68 sub-part F and IEC 60603-7 specifications, and have 50 micro-inches minimum of gold plating over nickel contacts

        4.   Be resistant to corrosion from humidity, extreme tempera-tures, and airborne contaminants

        5.   Utilize cable that exhibits Power Sum NEXT performance

        6.   Be made by an ISO 9001 Certified Manufacturer

        7.   Be available in several colors with or without color strain relief boots providing snagless design. Must meet the flex test requirements of 1000 cycles with boots and 100 cycles without boots

        8.   Be available in any custom length and standard lengths of 3, 5, 7, 10, 15, 20, and 25 feet

    B.   Electrical Specifications:

        1.   Input impedance without averaging 100 $\Omega \pm 15\%$ from 1 to 100 MHz

        2.   100% transmission tested for performance up to 100 MHz. Manufacturer shall guarantee cords are compatible with Cate-gory 5e links

        3.   Utilize cable that is UL VERIFIED (or equivalent) for TIA/EIA proposed Category 5e electrical performance

    C.   Design Make: **Leviton GigaMax 5e or equivalent**

2.5   FACEPLATES

    A.   Physical Characteristics:

Figure 14.2   (continued)

Horizontal cable specification. Courtesy of Richard Steiner 2005

Customer and Project Name
Project #
12/18/2001

  1. Be single gang faceplates for wall mounted locations
  2. Be 3.25"W 1.755"H .75"D for furniture faceplates
  3. Be applicable to both fiber and copper applications.
  4. Be available in Ivory (Wall mounted) and Black (Furniture mounted).
  5. Be made by an ISO 9001 Certified Manufacturer
  B. Design Make: **Leviton QuickPort series or equivalent**
  1. **Acceptable Styles:**
      1) **Wall Mounted-Leviton Part No. 41080-41P**
      2) **Furniture Mounted-Leviton Part No. 49900-EE4**
      3) **Equivalent Manufacturer Styles as necessary**

2.6   JACK ASSEMBLY (WALL PHONE)
  A. Physical Characteristics;
  1. Be constructed of stainless steel
  2. Be a 6 position, 4 conductor modular connector
  3. Have mounting lugs designed to mate with corresponding telephone base plate or adapter
  4. Mount to single gang outlet box or to wall directly
  5. Have 50 micro-inches minimum of gold plating over nickel contacts
  6. Be made by an ISO 9001 Certified Manufacturer
  7. Be able to except 100 Ω 22–24 AWG copper cable
  B. Design Make: **Leviton Type 630A Part No. 40223-S or equivalent**

PART 3-EXECUTION

3.1   INSTALLATION
  A. UTP Cable:
  1. All wiring shall be concealed in walls or above ceilings except where necessary to feed into furniture.
  2. All exposed furniture feed wiring shall be concealed with some type of spiral cable wrap or loom tubing so as to provide an aesthetically pleasing appearance.
  3. All wiring above ceilings shall be installed using J-hook style supports as specified in Section 17130 of this specification.
  4. Cable above accessible ceilings shall be supported a minimum of 4' on center from cable support attached to building structure.
  5. Do not untwist cable pairs more than 0.5 in. when terminating.

Figure 14.2 (continued)

Horizontal cable specification. Courtesy of Richard Steiner 2005

Customer and Project Name
Project #
12/18/2001

6. The Contractor shall be responsible for replacing all cables that do not pass applicable Category rating requirements.

7. Maximum length shall be 90 meters.

8. Cable shall have no physical defects such as cuts, tears or bulges in the outer jacket. Cables with defects shall be replaced.

9. Install cable in neat and workmanlike manner. Neatly bundle and tie all cable in closets. Leave sufficient cable for 90∞ sweeps at all vertical drops.

10. Maintain the following clearances from EMI sources.

   a) Power cable-6 in.

   b) Fluorescent Lights-12 in.

   c) Transformers-36 in.

11. Cables jackets that are chaffed or burned exposing internal conductor insulation or have any bare copper ("shiners") shall be replaced.

12. Firestop all openings where cable is installed through a fire barrier.

B. Inserts and Faceplates

1. All cables shall be terminated with modular jacks that snap into a faceplate mounted on a wall outlet box. surface race-ways or power pole.

2. Outlet boxes shall be secured to building with mechanical fasteners. Adhesive fasteners are not allowed.

3. All extra openings to be filled with blank inserts.

4. Terminate cable per EIA/TIA T568B standard pin assignments.

5. Locate so that combined length of cables and cords fro panel to phone or computer does not exceed 3m.

Figure 14.2 (continued)

Horizontal cable specification. Courtesy of Richard Steiner 2005

non-conforming work. You can see an example of this type of section in Figure 14.3.

With a well thought out and detailed specification, you can control many aspects of the installation. This is a benefit that you can take advantage of

---

Customer and Project Name
Project #
12/18/2001

SECTION 17190
SUPPORT AND WARRANTY

PART 1-GENERAL

1.1   WORK INCLUDED

    A.   Provide all labor, materials, tools, and equipment required for the complete installation of work called for in the Contract Documents

1.2   SCOPE

    A.   This section includes:

        1.   Cutover Support Requirements

        2.   Warranty Requirements

1.3   SUBMITTALS

    A.   Submit manufacturer warranty information with bid

    B.   Contractor certification to provide warranty with bid

PART 2 PRODUCTS

2.1   PRODUCT WARRANTY

    A.   The Contractor shall provide a manufacturer's warranty for the installed horizontal cabling solution that:

        1.   Will support and conform to TIA/EIA 568 standards

        2.   Will be free from defects in material or faulty workmanship

        3.   Will warrant the installation for a minimum of 25 years

PART 3-EXECUTION

3.1   CUTOVER SUPPORT REQUIREMENTS

    A.   Cutover support requirements will be provided by the Contractor on an as needed basis as determined by the Customer at the time of acceptance of the project and are not covered under this specification.

Figure 14.3
Warranty specification.
Courtesy of Richard Steiner
2005

in many ways. One way in particular is when you have a customer who favors a particular product. By writing the specification to fit a particular product, you can control the outcome of the product selection by the bid respondents. This works for just about anything else within the specification as well. How bidders respond to you, acceptable forms for the test results, and, as we have just seen, acceptable warranties are just a few of the ways you can ensure that the products and services delivered meet your expectations.

14.1.5   Drawings

While the construction specifications give the product, method of delivery, and other requirements, the drawings or plans enhance the overall picture of the construction project. By having the ability to read drawings, also known as prints, you gain further knowledge about the construction site and the environment where the contractor will be installing your system. The drawings convey in pictorial form what text cannot.

Drawings typically come in sets of several sheets. The set starts with a cover sheet, which will have the name of the project, the site address, and usually the architect's or general contractor's information. The second sheet normally contains abbreviations, definitions, and symbols along with general information that pertains to the whole set of drawings, such as a table of contents. The following sheets contain the information on specific sections of the project including: site plans, foundation, individual floor plans if the building has multiple floors, plumbing, electrical, HVAC systems, and telecommunications. Not all plans will have every sheet listed here, nor will they have the same information every time. Each set will have the information required for the project that it covers.

The telecommunications drawing of a project contains the information specific to that portion of the project (Figure 14.4). The most common

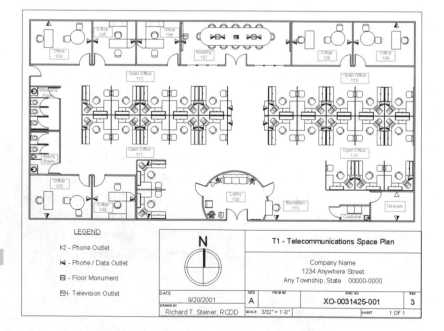

Figure 14.4

Telecom drawing.

Courtesy of Richard Steiner
2005

information that the drawing will show is the location of the telecommunication cable drops. Depending on the architect, the telecommunications drawing shows the number of cables at each location as well as the cable types, if there are multiple types at the given locations. Within the "Notes" section of the drawing, there is additional information regarding items such as the type of pathway, the location of the ends of conduits, and any other explanations that are necessary. A floor plan is the base of the drawing; it displays all of this and shows the walls, rooms, and even furniture, if necessary.

Another section of the set of drawings that contains useful and relevant information is the "Elevations" section. Here is where one or more graphics display the layouts of the telecommunications racks as seen from the front and side (Figure 14.5). This section may also contain a graphic showing a typical telecommunications outlet, including labeling (Figure 14.6). These graphics are especially helpful to the installer because they describe in a simple and effective manner the way the customer wants these items to appear. As a designer, you would provide these displays to the architect for inclusion into the drawings.

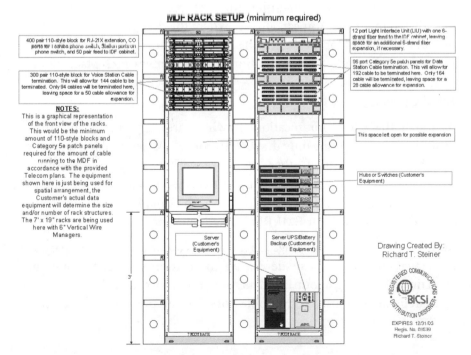

Figure 14.5

Telecommunications racks elevation. Courtesy of Richard Steiner 2005

FACEPLATE LAYOUTS

Colors represented here can be
determined at a later date and are just
being used for illustration purposes.

Drawings Created By:
Richard T. Steiner

6-Port Faceplate

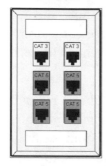

2-Port Faceplate

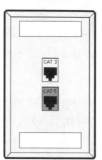

Figure 14.6

Faceplate elevation. Courtesy of Richard Steiner 2005

Sometimes the equipment rooms and telecommunication rooms house several types of equipment laid out in a specific design. This design shows up as a part of the drawings as well, and is usually placed in the Details section of the plan set. Figure 14.7 shows a sample equipment room layout.

The example shows an overhead view of a simple ER. You can see in the layout that there is an accounting for each piece of equipment that is included in the room. You can also see where the designer wanted to place the conduit sleeves for the cables, the ladder rack to carry them to the termination racks, and each termination and equipment rack placement. It is this level of detail that will ensure that the system delivered is the one you designed.

Although the telecommunication drawings are very important to the installer, they are not the only drawing to consult. When looking at a set of prints, keep in mind that other trades are working in close proximity to you. By looking at other sheets like the Reflected Ceiling Plan, Plumbing and Sprinkler System Plan, and HVAC Plan, you can plan your cable routes to avoid either physical or EMI/RFI-related interference.

Keep in mind that the drawings are not always accurate when compared to the actual site. Mistakes made during installation or the drawings not reflecting something about the structure you are working in (i.e., wrong measurements) are just a couple of factors that can present obstacles to the

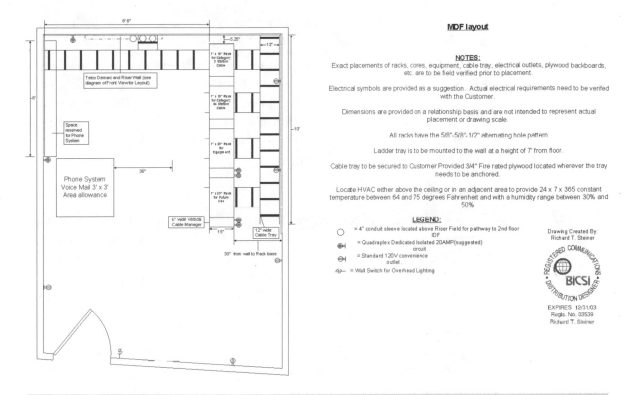

Figure 14.7

Equipment room layout. Courtesy of Richard Steiner 2005

installation. This is where having a good plan is extremely helpful. Having a good plan will prepare you for minor adjustments on site.

14.1.6   Working with Retrofit/Remodel Situations

Retrofit and remodel situations require an additional level of careful planning and preparation. These situations usually involve existing systems that need to stay functional during construction or something about the site that is unique, such as working with asbestos. For you to be successful in these projects, you need to have a high level of attention to detail.

The most common of these projects involves keeping one system functional while the new system installation is taking place. Most businesses are sensitive to having their telecommunications systems offline, because doing so means losing productivity and/or sales, which directly affects their profit. As a designer or project manager, you need to keep this in mind when planning one of these projects.

However important it is to keep the current system running, often there is not enough room either in the TR, ER, or in the cable distribution system to support adding the new system next to the old one. This is where creativity and careful preparation are great assets. Performing a **phased cutover** or "floating" the existing system are just two ways to facilitate the completion of a remodel project.

Phased cutover

A phased cutover is where completion of some sections of the project happens prior to moving on to other areas. This is the most common method of handling a remodel, and is effective with planning and careful execution. The number of phases depends on the size of the project or the total amount of work done, including other trades. By knowing the existing systems, what the removal order is, and when it should change, you can work out the details well in advance in an attempt to avoid mistakes made on the job site. Careful documentation and labeling these existing systems can go a long way toward successful completion of the project.

Floating

Floating the existing system while installing the new one carries its own set of considerations. When performing this type of remodel, decisions about how you will keep the existing system functional during installation are crucial and creative thinking is essential. The people performing this type of remodel project should be of the highest caliber, since one mistake can bring down the entire existing system. Because of the precariousness that occurs when doing this method, your level of planning and execution must be top-notch.

Other situations

Just keeping the current system alive is not the only situation you will encounter. Often the building itself is just as problematic. Special situations like dealing with asbestos or the retrofitting of a historical landmark carry an additional set of issues. Hazardous materials like asbestos require experienced people, specialized equipment, and site preparation. This type of issue will add to the time and the cost of the project.

Historical sites have extraordinary requirements as well. Typically, these buildings have additional rules that are followed when performing an installation, most of them dealing with how and when changes to the structure are made. This affects how you plan your pathways, what type of pathway you use, and how you will seismically brace the equipment among other things. Local codes and ordinances will govern how to handle these situations. By staying flexible when designing a system in this environment and knowing all your options regarding methods of distribution, you will be able to satisfy both the inspectors and your customer.

14.2 Installing Residential Cabling

Residential structured wiring has just begun to take a foothold in the residential market. While it has not entered into the mainstream of construction as yet, many of the more progressive building companies are embracing these installations. As demand for the services these systems support increases, so will the number of installations that take place. Digital subscriber line (DSL) and broadband cable service area expansion along with the higher bandwidth demands of streaming video, internet gaming, and other interactive products will help with this growth.

14.2.1 Residential Design Considerations

The biggest difference between residential and commercial cabling projects is the environment where they reside. Instead of removable tile ceilings, dedicated pathways, and individual rooms for equipment that are found in commercial buildings, residences consist of hard drywall ceilings, limited pathways, and shared spaces. With these factors in mind, you need to plan accordingly. The advantage that residential wiring has over commercial is that the project is usually on a small scale so these differences are not a big issue.

The most important consideration is to pre-wire, pre-wire, pre-wire. Because of the limited access and that to make an addition or change later requires a lot of work, the best mitigation strategy is to make sure you have enough cable in the walls to support any future requirements. If your customer tells you there is even a remote possibility of requiring a cable at a location, you should put it in the wall.

The TIA/EIA 570 standard recommends two levels of wiring designs. The first level recommends a minimum of one telecommunications outlet in

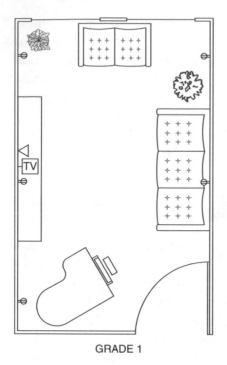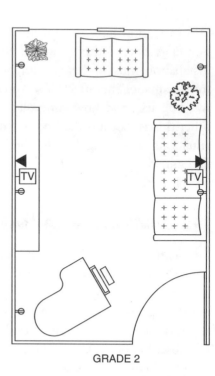

GRADE 1   GRADE 2

Figure 14.8

Grade comparison. Courtesy of Richard Steiner 2005

each room of the home, configured with one Category 5 and one RG-6 coaxial cable in the outlet. The second, and most recommended level, is to have two outlets configured with two Category 5e and two RG-6 coaxial cables in every room, preferably on opposite walls. You can see an example of this in Figure 14.8.

Beyond the voice, data, and video requirements, many customers in the residential market want whole-house audio as well. There are many different methods to accomplish this task; you need to be familiar with all of them so you can design the best and most cost-effective solution to accomplish your customer's needs. Many products are on the market that require different cable types to function. Planning for these differences and potential changes can make your design adaptable in future, thus avoiding locking the customer into one particular method or product.

As the designer of residential wiring, you will need a broad range of knowledge and understanding of the products and services available. Since you are the expert in the customer's eyes, they will ask you to recommend or specify the equipment that connects to the cabling infrastructure. High definition television (HDTV) is a perfect example of an emerging technology that is becoming widely available and sought after. If you did not know that some HDTV systems require up to six coaxial RG-6 cables to connect the rooftop equipment to the television set in the house, you would not plan for this, which would force your customer to add the required cables later, making them incur additional expenses and frustration.

Although pre-wiring is always a good strategy, many customers will balk at the cost for what they see as unnecessary at the time of installation. Even if you explain the cost advantage of putting the cable in while the walls and ceiling are open and accessible, some customers will still not see the advantage of planning for the future. In cases like this, the best thing you can do is to plan for as much additional pathway capacity as possible. Designing the cable routes to pass through attic or crawl spaces and providing pathways in the areas that are not accessible will assist your customer later on when they decide to make the additions or changes.

14.2.2   Residential Plans and Specifications

While commercial projects have standardized many items like drawings and specifications, residential projects do not have much of these standardized. The CSI MasterFormat that you saw earlier does not necessarily apply here. The codes covering electrical and telecommunications cabling are not as tight either. Regardless of these differences, when you look at a residential building plan, and a commercial building plan, you will see enough similarity that you should be able to interpret them.

Because the size of a residential project is quite small when compared to most commercial projects, the need to separate one or two trade's information onto multiple plan sheets would be overkill. Often you will see only one or two sheets in a residential plan set, so your ability to dig through all the commotion to get to what you need is useful. Residential specifications are usually the same, if they even exist at all. More often than not, a residential customer will have only a concept of what they want, leaving you to interpret the details and tell them what they need. As these systems become more complex and commonplace, this gap will close and become more like commercial projects with detailed specifications and plans.

14.3   Request for Quote/Proposal and Bidding

Regardless of what type of project, you will undergo a selection process to decide who is going to do the installation. This process, known as a **Request for Quote (RFQ)** or **Request for Proposal (RFP)** and bidding, is a structured and time-sensitive event. How potential installers respond to an RFP or RFQ depends on how these were written and what the customer would like to see. As a rule of thumb, you will find these requirements in the first division of the MasterFormat specification for the project.

14.3.1   Request for Quote versus Request for Proposal

Generally speaking, the difference between an RFP versus an RFQ is not that wide and most people use the terms interchangeably. Usually you use an RFQ when the total amount of the product or service requested is small and easily defined. You would use an RFP for larger projects that may not be so cut and dried. Both have clearly defined instructions for your respondents to follow and, in some cases, there are penalties assessed when these instructions are not followed. These penalties can range from simply asking for a rework of the quote or proposal to refusing to accept the paperwork.

When writing them, the more precise you are in your instructions the better the responses will be. The best RFPs and RFQs outline in specific form how the responder is to break out each item that you want to see written. Let us look at an example of this.

You are the consultant on a large cabling project for a business that is designing their headquarters office. As part of your services, you design the cabling infrastructure, write the specifications, and will assist the customer in evaluating the responses. The customer wants to see what it costs for the total project with the pricing structure as shown below.

- Total cost of labor with a breakout of the hourly cost per person based on skill level: installer, technician, crew leader, foreman, and project manager. They also want to specify the number of persons at each skill level who are expected to work on the project.

- Total cost of parts with a full bill of materials showing part number, part description, unit cost, and quantity of each.

- A per location cost for either additions or subtractions to the total location count of the project. (This is **per-drop pricing**. Some installation companies base all of their pricing on this format, which would include all materials and labor to install one location of a given configuration.)

- The hourly rate for onsite support on the day of the customer's occupation of the space.

As you can guess, if you did not clearly define what the customer was looking for in the response, you would see lump sum responses, separate labor and material responses, and anything other than what the customer wanted. By defining clearly what you expect of a respondent, you can decrease both the amount of time required for them to respond and the number of questions that they have for you. To facilitate communication of what you are looking for, it is a good idea to include an example of what you want or even create a blank form that they may use to respond to you.

14.3.2   Bidding

After the RFP/RFQ, specifications, and drawings are all complete, the project goes out to bid. This bid process can be open to the public, which is called a public bid, or it can be sent to preselected contractors, called a selected bid. When a call for a bid occurs, there is usually a time limit on how long the bidders have to respond. Additionally, as part of the bid package, there may be requirements such as submitting past project records, providing references from current or former customers, and posting a bond. All of these are methods used to narrow the field of bidders and find the most qualified contractor.

Public bids are the most cut-throat and often the less profitable of the two types of bids. Government projects, schools, and churches are the most common projects found in the public bid sector, along with general contractors looking for subcontractors for projects. You will find many companies vying to win these projects so your profit margins are slim if not nonexistent. Also, many of these contract awards go to the lowest bidder, so the winning contractor had better be accurate on the estimate or will lose money on the project. Change orders are common on these projects since they are so tight on profits.

Selective bids are more preferable to contractors. Here the customer or consultant will have a list of preselected contractors who can meet their criteria. These contractors receive the bid directly with, once again, a time frame for responding. Sometimes the customer or consultant holds a prebid meeting to distribute the bid packages all at once, thus avoiding the possibilities of any unfair advantage by favoring one company over another. Often the contractor awarded the bid is in the middle of the pack of bidders where the differentiating factors are not based on price alone. The profit margins in this type of bid are better, because the customer is probably concerned with the quality of the installation in addition to the price and they have already excluded those contractors they do not want prior to sending out the bid.

14.4   Construction Planning, Documentation, and Management

Once the bid award occurs, preplanning and design phase comes to a close and a new phase begins with the construction planning. Before any physical work begins, there is quite a bit of organization and coordination to be done. During this time, you sign the contract, create a project timeline, and meet with all the parties involved.

14.4.1   Contract

The contract is the most important document during a project. This document is the legal and binding description of the specific work performed, during what time frame, for what cost, and any of the penalties invoked should one or both parties break the agreement. This will be a very detailed document listing everything expected of both parties, and the terms and conditions under which they will be working for the duration of the project.

Most contracts will outline in the utmost detail exactly what the contractor does and does not provide during the project. This part of the contract, typically called a **Scope of Work** (**SOW**), will provide minute detail on every aspect of the work performed. It may also outline the hours of the jobsite and any other conditions required by both parties such as attendance at onsite weekly safety meetings, for example. Sometimes the SOW may even include specific exclusions to the contract, such as the contractor not being responsible for any issues with regard to computer equipment purchased from an outside source.

Another part of the contract is the "Terms and Conditions" section. It outlines the payment terms, such as a one-third payment to begin the project, one-third once fifty percent of the work is completed, and the final third upon satisfactory completion. Within this section are any clearly defined penalties, such as withholding $300 per day for every day that the project extends past the final due date from the last payment. This section may also cover any project deliverables required by the customer upon completion, such as test results placed in a binder.

Finally, a common section to be included in the contract is the Warranty and Service section. This section of the contract explains what warranty coverage the contractor provides. Typically, parts and equipment manufacturers will offer a standard warranty and may offer an extended one if the installer holds a certificate to install their products. Currently, manufacturers of telecommunications cable and connecting hardware offer these extended warranties under the condition that the installation contractor meets specific criteria. The length of these warranties varies; however, most are 10 to 15 years or more. In addition to a parts warranty, the contractor may offer coverage on labor, often providing repair service if the cause is faulty workmanship. The section will detail any added services such as support during a customer's occupation of the space after the handoff or cutover occurs.

14.4.2   Project Schedule

As the construction plans near finalization, the project manager (PM) or general contractor (GC) creates a preliminary project schedule. The preliminary schedule contains a high level listing of the multiple tasks that need accomplishing and the dates they expect to start, which will give you a rough idea of their order and their expected duration. Within this high level view will be individual trade tasks usually listed under sections that correspond to the phases of construction; generally these are the rough-in, trim-out, and finish work.

One of the most used tools for creating the project schedule is Microsoft Project. MS Project is a very powerful software program that can help a project manager organize, prioritize, and track a project. While there are many tools and charts that a PM or GC can use, the chart used most often is the **Gantt chart**, an example of which is shown in Figure 14.9. This figure is a simple example, but you can see what a Gantt chart does with listing the tasks, their start and completion dates, and shows you their duration.

In Microsoft Project, the software allows you to do many things with the Gantt chart. When you first create the chart, you can save it as the **baseline**,

ID	Task Name	Start	Finish	Duration	Aug 2004
					9 10 11 12 13 14 15 16 17 18 19 20 21 22 23 24 25 26 27 28 29 30 31
1	Install Cable Supports in Ceiling	8/9/2004	8/10/2004	2d	
2	Rough In Cable	8/10/2004	8/13/2004	4d	
3	Dress Cable in Ceiling	8/16/2004	8/18/2004	3d	
4	Build MDF	8/18/2004	8/19/2004	2d	
5	Terminate Cables in MDF	8/20/2004	8/25/2004	4d	
6					
7					
8					
9					
1o					

Figure 14.9

Sample Gantt chart

which is what the software will use to compare any subsequent changes in the chart from the time you create it onward. By establishing a baseline schedule, you can see how far the schedule slips out or pulls in, if the project finishes ahead of schedule.

Within the Gantt chart, you can assign precedence to each of the tasks you list. There are four precedence categories: start-to-start, start-to-finish, finish-to-start, and finish-to-finish. A start-to-start setting means that the tasks tied to each other will all start at the same time. A finish-to-start assignment means that the task given this precedence will not start until the task tied to it finishes. A start-to-finish precedence means that the task assigned must start before the one tied to it finishes. A finish-to-finish precedence means that all tasks tied together must finish at the same time. In MS Project, the precedence typically shows up after the duration column. Knowing how these relationships work and how to assign them can help you prepare or understand a schedule that you are working with, and will tell you what tasks occur in what order. This is how a PM or GC will control the project's timeline.

Once the tasks, durations, start and finish dates, and precedence are set, you can quickly discern what the **critical path** of the project will be. The critical path is the longest series of tasks that, by adding up their durations, takes the longest time to complete. The critical path is dynamic, and will change as tasks are completed or extend in duration. It is useful in identifying the tasks that require prioritization and additional resources, since delaying a task that is on the critical path will delay completion of the entire project. By using a Gantt chart, you will know quickly whether you are on track, and with the baseline comparison, you can adjust your planning the next time you do a similar project.

14.4.3   Site Survey

The next step in the planning phase is the site survey. While a site survey may not be necessary in new construction where no building initially exists, it is always a good idea to perform one before you begin installation of the cabling. The reason for doing so is to verify that everything shown on the print is correct on the job site. Other factors often change once the project is started that affect the way construction progresses and these changes cause you to alter your planning. In a remodel or retrofit situation, a good site survey can identify potential issues on site that are not reflected in the print. As stated before, these changes will make you alter your plans, sometimes forcing you to be extremely creative in the methods you use to accomplish your objectives on time.

It is a good idea to make copies of the final set of drawings. This is so you can have one clean set to create additional professional quality copies as necessary, and multiple copies for notes for yourself, to give to your crew, and for documentation to give to the customer after project completion. One of these copied sets should accompany you during your site survey, so that you have a reference to compare what you see on site. You should also carry a pad of paper, a pen, pencil, and highlighters of many colors. These items are useful to make notes, to call attention to notes or sections of the drawing, and to give you a way to document questions that you have so that they remind you to ask someone else later. You may also keep a measuring tape with you to confirm placements of things like racks, equipment, and furniture.

While performing a site survey, take note of the proposed cable pathways on the drawings, if they exist. Following these routes allow you to see what your crew will encounter in the ceiling, verify that conduits or in-floor ducting are installed, and determine any other potential delay or issue that needs correcting prior to your crew starting work. This gives you the opportunity to make sure the site is ready for your crew to begin work rather than have them show up to find that they have nothing to do because the site is not ready for them. You will also want to note what other work is occurring simultaneously within the area so as to avoid conflicts and possible damage to your cables by other trades attempting to complete their own tasks.

You may want to have the foreman or lead technician for your crews accompany you on this survey if you are not the one who is on site at all times. By doing this, both of you can study the site and discuss possible solutions and alternatives to issues encountered, rather than trying to explain it to them later when they can easily mistake what and where you are talking about. Also, having an additional set of eyes during the visit may help identify other concerns that you may miss. Once the survey is complete, assemble all the notes, type them to assure they are organized and legible, and place them, along with the marked prints, in the project folder for future reference.

14.4.4   Initial Construction Meeting

Most GCs hold an initial construction meeting either prior to or right after construction begins. This meeting covers final planning questions, any concerns about the job site, and safety and job site rules. The architect, all representatives of the different trades, and sometimes the customer or an appointed representative attend this meeting, so that the right people are on hand to answer questions that arise and address issues. If any changes to the project occurred since the last meeting or update, distribution of the documentation takes place, ensuring that all the attendees receive the same information. An introduction of the GC and all of the onsite foremen for the trades may take place, so that everyone knows who is in charge of each crew.

14.4.5   Ordering and Staging Materials

Once you have acquired the project, planning for ordering, delivery, and staging of materials becomes the next part of the planning process. The size of the project and staging area as well as security concerns need to be taken in to consideration in your plan. Usually, the GC overseeing the project will let you know if and where you can store materials onsite. Additionally, the finances of the company may affect this process.

When ordering and staging materials, you would usually order what needs to be used first, namely, any cable support structures and the cable itself. Consumables such as pull string and electrical tape are part of this first order as well. Keep in mind that some items may be in a backorder status from the manufacturer and must be considered when placing the order. Sometimes the first order includes the racks for the equipment and cable terminations, depending on the project schedule and how far along construction is for the TRs and ERs.

When planning for delivery of materials, working with the GC becomes an important matter. The GC can tell you the best time for delivery, the correct address for the project provided you do not already have it, and where you will store the materials. The GC can also tell you about the security of the site, and may even be able to secure an unsecured area by adding a lock to a door or providing fencing. The GC is in charge of the project site as a whole, so working with him/her is a must.

Occasionally, you may be able to have all of the materials delivered at once. In cases such as these, the recommendation is that you ensure that they are stored in a secure area, as thievery is a major concern on site. Most of the time materials arrive at the site in accordance with the phases of the project. You should have an inventory list of ordered materials to verify the delivery before the driver leaves, because most distribution houses have the policy of once accepted, the delivery is yours whether it is all there or not, unless you have written exceptions on the receipt.

This goes for damage to the product as well. When having product delivered, you should always do some type of acceptance verification, be it a simple visual inspection to full acceptance testing of the cables. There are many ways of acceptance testing for both copper and fiber-optic cable, and it is highly advisable that you perform some type of acceptance testing on any cables you receive. Most cable tester manufacturers have an adapter that will perform tests on unterminated copper cable. For fiber-optic cable, the use of an Optical Time Domain Reflectometer (OTDR) with a bare fiber adapter can verify that there are no breaks within the cable. Whatever tests you perform, the results should be included in the project documentation so that they are retrievable later, if needed.

14.4.6   Testing, Labeling, and Documentation

The project documentation, testing requirements, and labeling requirements depend on the customer and what was within the project specification. Most often there will be some type of documentation required for acceptance of the project and again prior to the final payment. This package will most likely include copies of the test results on all installed cables, an **as-built drawing** of the site with all the locations labeled on it, and a copy of the warranty documentation showing that the installation is covered.

Acceptance testing for copper cables requires a device that can test to the standards for the category rating of the installed cable. While there are many manufacturers of copper testers, the most prominent ones are Fluke and Agilent (formerly Hewlett Packard). Because the standards change rapidly, you should ensure that the tester you use has the latest firmware for it installed and that it has been factory calibrated within the last twelve months. In addition, most units require field calibration prior to use to ensure that the two sides, called the primary and remote units, are set to the same requirements.

There are two types of field verifiable tests outlined by the TIA/EIA-568-B standards: the permanent link and channel tests. The permanent link test will test your cable from the jack to the patch panel, not including the patch cords or test cords used to perform the test. This test is the more stringent of the two and is the most used acceptance test set up. The Channel test includes the patch cords and the horizontal cross-connects that will eventually run from equipment to equipment. This test, once completed, has the stipulation that the patch cords and cross-connects stay in place. Moving or changing out the patch cords renders the test results null and void. This is one reason that the channel test is not requested by customers very often, preferring instead to use the permanent link test for acceptance.

As far as labeling is concerned, most customers will require a specific labeling scheme so that they can identify and troubleshoot their own system once installed. The desire of the cabling industry is that this scheme follows the TIA/EIA-606-A standard, which covers labeling; however, this is not always the desire of the customer. When all else fails, the customer is always right. As long as the cabling scheme makes sense to the people who are going to do the support and it is printed with an electronic printing device rather than handwritten so it is easily read, the actual schema does not matter.

The as-built drawing is the final piece that is common in the documentation given to the customer upon completion of the project. The as-built is a drawing that has the final labeling of the jacks next to their locations along with any other notes or diagrams necessary. There may even be representations of the installed cable trays and other cable pathways for future reference if the circumstances require it. Some customers ask that this drawing, or at least a copy of it, remain inside the TR or ER of the area that it represents for use by service personnel. Any writing on the as-built should be printed with an electronic printing device so that it can be read as easily as the labels.

When turning over the documentation, keep in mind that these items will portray your company image, so the more professional you can make them the better. With the upgraded cable testers and with the ability to create

CD-ROMs, when the test results are downloaded to a computer, you have the ability to immediately place them into a digital media format like a CD for your customer. Additionally, using software such as Microsoft Visio or Autodesk AutoCAD, you could provide digital copies of the as-built as well. By offering these services, you provide your customer with options, and by giving them an easily updated format, they can keep their records up-to-date as changes occur.

14.5   Chapter Summary

- Consulting with the end user is the first step in the planning and design process. By doing so, you can gain valuable insight into what they want, which is invaluable in planning a system that is right for them.

- Keep in mind that the pathway method you choose has advantages as well as disadvantages. By knowing these, providing a flexible solution that fits the project is much easier to do.

- Your design should provide the performance requirements for today and for the future. This will give your customer the advantage of being able to utilize the installed cable plant for many years.

- Remember that an installed cable system needs to conform to more than just the TIA standards. The NEC and any additional local codes need to be followed as well.

- Being familiar with the type, size, and space requirements for the equipment that the customer wants for the ER will make designing a much easier prospect.

- Commercial projects use a standard layout for specifications called the Master Format as set forth by the Construction Specifications Institute. Being able to read and understand these specifications allow you to respond to them correctly.

- When writing specifications, using a standard layout such as the MasterFormat gives your respondents a clear understanding of what you want; clear communication is the essence of written specifications.

- Drawings assist all involved with a construction project by providing a graphical representation of the construction site. They provide information that written specifications cannot.

- To make them easily readable, drawings are separated into different sheets, with each sheet displaying only one or two aspects of the construction project. Often there will be coinciding information that resides on multiple sheets.

- By paying attention to the drawings, you can discern whether you will be in conflict with other trades on the construction site. This will allow you to avoid conflict once the project begins.

- Make multiple copies of the drawings that have important information on them. This way, you can make notes and mark them as needed while still maintaining a clean set for future use.

- Working with retrofits, remodels, and historical sites have their own individual challenges. Using methods such as phased cutovers and floating the current system can make the installation process an easier prospect.

- Residential projects are similar to commercial ones and are usually on a much smaller scale. Flexibility and expandability are key points; however, unlike commercial projects, most residences have hard walls and ceilings, making access to the cable difficult if not impossible. To avoid this becoming an issue, always prewire everything, even if you do not use it later.

- When writing RFPs and RFQs, the more detailed you write them, the more accurate the responses and the better the chance of receiving exactly what you want.

- The contract is the legal and binding document that details all the aspects of the project, from the type of work performed and materials used to any penalties for late work, and subsequent warranties and service contracts at the project's completion.

- Managing the project can be done with many tools, the most common of which is the Gantt chart. A properly assembled Gantt chart can show you important information about your project, including the order of project tasks, the critical path of tasks, and the expected completion date based upon that critical path.

- By creating a baseline of your original Gantt chart prior to starting the project, you can compare your current chart to the original to track changes in the project schedule.

- Whenever possible, perform a site survey to identify potential issues and verify site conditions. Using a blank copy of the construction drawings to make notes during the survey can help you recall important information at a later date.

- Always attend the preconstruction meeting because important information regarding the site and project will be presented.

- Be certain to examine delivered products and perform acceptance testing of fiber-optic and copper cabling to ensure that no damage occurred during shipment.

- Provide project documentation to your customer including but not limited to preinstallation acceptance test results, as-built drawings, and final test results prior to project handoff. If at all possible, provide the documentation in the customer's requested format whether that format is paper or digital.

14.6   Key Terms

access floors: A system consisting of completely removable and interchangeable floor panels supported on adjustable pedestals or stringers (or both) to allow access to the areas beneath.

as-built drawing: A schematic of the cabling system representative of the system as installed.

baseline: In project management, a set of original tasks, start and finish dates, durations, work and cost estimates that you save after you have completely fine-tuned your project plan before the project begins. It is then used as the primary reference point to which you measure changes to your project.

cable tray: A support mechanism used to route and support telecommunications cable or power cable. It is typically equipped with sides allowing placement of cables within the sides over its entire length.

conduit: Rigid or flexible metallic or nonmetallic circular tubing used as an alternative pathway for cables.

critical path: In project management, a series of tasks that dictates the calculated finished date of the project.

equipment room: A centralized space that houses the telecommunications equipment that serves the occupants of a building. It is distinct from the

Telecommunications Room due to the nature and complexity of the equipment inside.

Gantt chart: A chart (named for Henry Laurence Gantt) that consists of a table of project task information and a bar chart that graphically displays the project schedule, depicting progress in relation to time; often used in planning and tracking a project.

junction box: A metal or plastic box used as an access for cable or wire that is typically seen in conduit or underfloor raceway systems.

MasterFormat: Created by the Construction Specifications Institute, a standard, organized list of numbers and titles for information about construction requirements, products, and activities.

National Electrical Code (NEC) Created by the National Fire Protection Association (NFPA), a series of documents covering electrical safety often adopted as the legal electrical code by many municipalities in the United States.

per-drop pricing: A method of charging a customer for cabling installation services given as a fixed amount for each location of cable installed.

phased cutover: A method of activating installed cables in small groups rather than all at once; useful for remodel/retrofit situations where the end user still requires their system to be active during construction.

Request for Proposal (RFP): A document used when a buyer is looking for costs on a product or service. The RFP is in-depth and detailed, while the RFQ is usually more exploratory and less defined.

Request for Quote (RFQ): A document used when a buyer is looking for costs on a product or service. Similar to an RFP, but usually more exploratory and less defined.

Scope of Work (SOW): A document outlining what a project will entail and can contain details on either products, services, or both.

telecommunications entrance facilities: The point of interconnection between the network demarcation point and/or the campus backbone and the intra-building wiring: sometimes referred to as the MPOE, or main point of entry. The entrance facility typically includes over-voltage protection and connecting hardware for the transition between outdoor and indoor cable.

telecommunications room: A room that houses telecommunications wiring and wiring equipment; usually contains one or more cross-connects.

underfloor ducts: A floor distribution method using a series of metal distribution channels, often embedded in concrete, for placing cables. Underfloor raceway is another term for these structures.

14.7 Challenge Questions

14.1 Explain the MasterFormat. What is it, what does it do, and who publishes it?

14.2 What is the most common tool used for planning a project schedule?

    a. PERT chart

    b. Gantt chart

    c. Work Breakdown Structure

    d. A notebook

14.3 What is the document published by the National Fire Protection Association, what is it used for, and what does it cover?

14.4 When should you use an RFP versus an RFQ?

14.5 Tests performed on cable that you have just received on the job site prior to installation are called _____.

    a. delivery tests

    b. pre-employment tests

    c. acceptance tests

    d. Permanent Link tests

14.6 When performing a site survey what should you bring?

    a. Your coffee

    b. A copy of the drawing

    c. Your punchdown tool

    d. A hard hat

14.7 List three possible items given to the customer at project completion.

14.8 List four types of cable distribution and support structures. Give a brief description of each.

14.9 Name the two types of field testing done at the end of the project. Explain the differences between them.

14.10 Describe each of the following terms.

a. Scope of Work

b. Terms and Conditions

c. Warranty and Service

d. Critical path

e. Baseline

f. As-built

14.11 What are the two types of bids and give a brief explanation of each.

14.12 In working with drawings, there are three types of views or sections that you may encounter. List them and provide a brief description or example of each.

14.13 Describe two retrofit/remodel situations discussed in the chapter.

14.14 The TIA/EIA standard covering labeling is _____?

a. TIA/EIA-569

b. TIA/EIA-607

c. TIA/EIA-568

d. TIA/EIA-606

14.15 The TIA/EIA standard covering residential installations is _____?

a. TIA/EIA-570

b. TIA/EIA-568

c. TIA/EIA-569

d. TIA/EIA-607

14.16 Explain the TIA/EIA residential standard recommendations for two levels or groups in residential wiring.

14.17 Are the TIA/EIA standards the only documents you need to follow when designing and installing cabling? Why?

14.18 Referring to question #14.17, what other documents should be followed when designing and installing cabling?

14.19 What are some similarities between using conduit and using underfloor ducts as cabling pathways?

14.20 Describe some advantages of using access flooring in an equipment room or telecommunications room.

14.8 Challenge Exercises

Challenge Exercise 14.1

In this exercise, you look at Chapter 14 of the *BICSI Telecommunications Design Methods Manual*, which covers design and project planning. For this exercise, you need a computer with a Web browser and an internet connection, preferably a high-speed connection.

14.1.1 Log on to your computer and open your Web browser.

14.1.2 In your Web browser, enter:
http://www.bicsi.org/Content/Files/PDF/9thchap14.pdf

14.1.3 Look through Chapter 14 and find tables that have symbols used on drawings.

14.1.4 List four different symbols and their meanings as listed in the BICSI chapter.

Challenge Exercise 14.2

In this exercise, you look at sample telecommunications specifications online. For this exercise, you need a computer with a Web browser and an internet connection, preferably a high-speed connection.

14.2.1 Log on to your computer and open your Web browser.

14.2.2 In your Web browser, enter: *http://www.google.com*.

14.2.3 Perform a search for **telecommunications specifications**. In the Google search text box, use quotations around the search term so that Google searches for that exact phrase. You can accomplish this by using the Advanced Search feature as well.

14.2.4 Look at two or three different sample specifications. Write down their URL locations and the organizations that host them. Note the differences and similarities between them and give at least three examples of each.

14.9   Challenge Scenarios

Challenge Scenario 14.1

You are the project manager on an Inside Plant Cabling Project for the XYZ Company of New York. The project is to cable a floor in their multi-story building with the floor consisting of several hard wall offices and open space for planned cubicles. You have been given a date of August 30 for the completion of your part in the project. Using this date as the finish date, create a list of tasks, estimate the time it will take to complete those tasks, and use this information to create a Gantt chart displaying your projected start date and the critical path for your project. Do not forget to include weekends and holidays when planning your work schedule, since you are not authorized to use any overtime on this project unless absolutely necessary.

Challenge Scenario 14.2

You have been asked to design the cabling infrastructure for a client who is looking for a network to run 1 Gigabit to the desktop areas and a minimum of 10 Gigabits in the backbone. Their requirements are to have two jacks at each location of a configuration of your choice, one expected to be used for their phone system and the other for their computers. They would like to use an overhead approach for their cabling distribution because the building is already built and the drop-tile ceiling is the most accessible area for this purpose. What products would you suggest and why? Would you use copper or fiber-optics? What type? What type of support system would you use? Why? Are there any other recommendations that you may have based on their current and future needs?

CHAPTER 15

Cabling Installation and Testing

Learning Objectives

After reading this chapter you will be able to:

- Set up a staging area and work team, and understand the basics of laying and pulling cabling, which is considered the rough-in phase of cable installation

- Perform a basic installation of horizontal and vertical cabling

- Perform cable splicing and termination, referred to as the trim-out phase of cable installation

- Perform cable testing and complete necessary documentation, which is part of the finish phase of cable installation

After learning the basics of electronic signaling—which include the details of what types of cable, connectors, terminators, and other supplies and equipment to use, and the necessary standards and safety procedures that must be adhered to in the cabling industry—the real work begins! This chapter covers the three main phases of cabling installation: laying and pulling cabling (also called the rough-in phase); splicing and terminating (trim-out phase); and testing and certification (finish phase). Throughout this chapter, all phases of cabling installation discussed are required to conform to the ANSI/TIA/EIA standards (568-B, 569-A, and 606-A in particular), which should be studied as a supplement to your learning. You should also be thoroughly familiar with the requirements described in the National Electrical Code (NEC), articles 770 and 800.52 in particular, and your local building and safety codes.

15.1   Phase 1: Laying and Pulling Cabling

Rough-in is one of the pivotal steps in cable installation. The **rough-in phase** includes: identifying the staging area; assembling work teams; installing conduit; and pulling cable from the telecommunications room (often referred to as the telecommunications or wiring closet) to the individual outlets throughout the location. These outlets are called the network jacks or plugs. It is at this stage in the cabling project that an installer may encounter **firestops**, which are barriers that prevent the spread of flames vertically or horizontally, when attempting to run cabling from one floor to another.

During the rough-in phase, you must lay all cables in appropriate places, such as in conduits or raceways, according to ANSI/TIA/EIA 569. You should take precautions to make sure that you do not bend the cables too tightly or pull with too much force, both of which may damage the cabling. A simple rule of thumb is to leave more cable than is necessary in each run, which builds in flexibility. If the project specifications do not indicate the exact overage to allow, leave at least 3 meters of additional cable in the telecommunications closet (TC) for both twisted-pair and fiber cable, and 30 centimeters for twisted-pair cable at the jack.

The installer must be able to locate potential hazards in the workplace such as electrical power sources or conduit. This can be accomplished by visually inspecting the location for potential problem areas, and using blueprints or by thoroughly documenting the building layout prior to installation. Lay cables, especially unshielded copper cables, as far as possible from sources of electromagnetic interference (EMI), such as ballasts (i.e., fluorescent light fixtures). Three feet is an acceptable distance. You must also make sure that cabling properly crosses firestops, which can include bricks, caulking, or pillows, and restore the fire rating of the firewall or firestop after the installation is complete.

A cabling technician should also protect the cabling during the installation process. Because a single damaged cable can cause interrupted, or the complete loss of, data transmission, your careful techniques during the cabling rough-in phase can save hours of troubleshooting and frustration to detect and repair a damaged cable during the finish phase or—worse—during client use. The best way to prevent postinstallation problems is to test the cabling during every step of the rough-in phase, and then perform the finish phase testing as described in Section 15.3, "Phase 3: Testing and Certifying the Cabling," of this chapter.

Another consideration for an installer is space, which applies to the raceways, TCs, and so forth. In a new installation, the designer is responsible for providing adequate space for all runs and hook-ups, and you should therefore not run into too many problems. However, in a project that involves a pre-existing cabling infrastructure or one that doesn't have the benefit of an architect or designer, you may need to make adjustments to fit the existing conditions. Any modifications must be made according to standards and codes. If you're faced with a space dilemma during the installation, consult the project specifications and the project manager to determine the most efficient and safest way to proceed. This consultation will undoubtedly involve a review of standards and codes.

NOTE

Although this chapter focuses on the phases of a new cabling installation project, many of the methods and techniques apply to retrofit projects as well. However, a retrofit project may involve many more challenges than a new cabling installation project.

When the installation of cabling is conducted in an environment that is already occupied by tenants, it is essential that the work be conducted in a manner that minimizes debris and that every attempt is made to not interfere with the client's daily business. Noise and debris should be kept to a minimum.

TIP Point out to tenants and other workers that cable is present. This will help avoid possible damage to the new cabling and ensure that there is a minimal chance that one of the occupants or workers could be injured. Additionally, precautions should be taken to protect office equipment, furniture, and property from potential damage from dust or debris.

Installation of cabling in a new or unoccupied area is not without its requirements, but it does provide more leeway when the positioning of hardware or mounting devices requires more space to work. The same precautions should be taken in this environment as in the occupied space. Identify electrical power lines and sources, and identify where fluorescent lighting will be located to avoid installing cables in high EMI areas. No matter what project environment is encountered, you must adequately plan the installation to reduce the chance of unexpected delays or problems. Planning includes, but is not limited to, identifying the installation team and delegating tasks to team members involved.

TIP If taking on a new or old project, make every attempt to gather documentation on the layout or previous work beforehand and always document everything you do, as it will save you time and resources later.

15.1.1   Cabling Installation Team

When the time comes to identify the number of workers needed, it is essential that the following criteria be followed to ensure adequate coverage of the work area. Whether using reels or boxes of cabling, there needs to be at least one individual who helps feed cable to the work area and another individual who guides the cabling as it passes into the ceiling or walls. These individuals observe the cabling to ensure that there are no snags and to avoid stretching of the cable. On the receiving end of the cable, one or two individuals must be available to pull the cable. Using two individuals is the preferred method, because one can pull and the other can guide the cable to its destination. Installers should also have some type of communications equipment, such as cell phones or radios, to talk to one another and other contractors while installing cabling.

Adequate eye protection, hearing protection, and gloves should be provided to protect the members of the ▣ **NOTE**
installation team from injury.

15.1.2   Work Area Security

The security of the work area is very important because the equipment and supplies used are usually the responsibility of the installer. You should store materials in a locked room when not being used to avoid theft. During work hours, the work area should be barricaded to prevent access by onlookers.

Personnel who are not directly involved in the project should be made aware of the cabling and work being conducted to prevent accidents. Use marking tape or some other device to lead people away from the area as an additional precaution. You should also keep the cabling as close to the wall and out of walkways as possible. Security and safety should be your primary concerns throughout the project.

15.1.3   Staging Area

The staging area should be considered your base of operations because it is where the majority of work will be conducted. The amount of equipment and supplies may change depending on the type of cable being installed, but the purpose of the staging area does not. It is generally located within the telecommunications room and is where one end of the cable run(s) will be terminated. The installation of network distribution cabling may require numerous boxes or reels of cable, unlike backbone cabling, which generally only requires one reel.

15.1.4   Rough-In Support Tools and Equipment

If using network distribution cable or backbone cable, tools are available that can make the process easier. Distribution cable is generally sold in a box that unrolls easily and contains an average of 1,000 feet of cable. The box allows for easy labeling, handling, and storage of cable.

Backbone cable is usually distributed on a reel (Figure 15.1) because of the count, or number of pairs, that are associated with backbone cables. The counts can be as few as 25 pairs or as many as 2,700 pairs. Because cable reels

Figure 15.1

Cable reel

can be heavy, the use of cable trees, jack stands, and reel dollies or rollers can make the process easier and much safer for the installation team. A jack stand holds a reel of cable; a reel dolly is essentially a jack stand with wheels.

In addition, for all types of cables, use the tools described in Chapter 13, "Building Your Cabling Toolkit," such as cable casters, fish tape, push/pull rods, and gopher poles. You should also assemble your crimping and splicing tools, wire cutters, labeling equipment and supplies (including a permanent ink marker), fasteners, cable ties, D-rings for supporting cables and cross-connects, connectors and wall plates, and other jacks, as necessary. Acquire punch-down blocks, patch panels, backboards, cable trays, cross-connects, and conduit, per the project specifications.

Finally, another useful "tool" is a detailed checklist of all of the previously mentioned items that are needed to install the cabling. The checklist should be created before the actual installation begins, and consulted and updated at the beginning and end of each workday throughout the project.

TIP Keep notes, either on a pad of paper or an electronic device (such as a personal digital assistant [PDA]) of cables installed, their locations, and other identifying information throughout all phases of cabling installation. Alternatively, if your project specifications are properly created and contain a great amount of detail, you can use them as a checklist and to jot down notes throughout the project. Either way, these notes will be used to complete necessary documentation after the installation is completed.

15.1.5   Backbone Installation

The point at which an outside service provider's (SP's) equipment/cabling, such as that provided by a telecommunications utility, meets the entrance facility of the premises is called the **point of demarcation**. The **backbone**

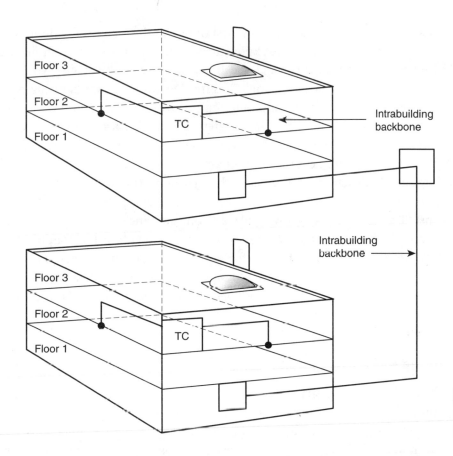

Figure 15.2

Intrabuilding and inter-building backbones

interconnects the major features in a structured cabling environment—entrance facilities, TCs, and equipment rooms—and includes the raceways, intermediate and main cross-connects, patch panels, cable supports, and the actual cables themselves such as coaxial, twisted-pair, and/or fiber-optic. Depending on the size of the area in which the cabling is being installed, backbone cabling can span from floor to floor and between buildings. Intrabuilding backbone cabling is contained within a single building, whereas interbuilding backbone cabling links two or more buildings. Figure 15.2 illustrates an example of intrabuilding and interbuilding backbones.

Conduits/Raceways

Install conduits/raceways in a professional manner. Generally, conduits should be installed at right angles and parallel to building grids. When installing a raceway under a floor, first draw out the plan on the floor with a pencil or in chalk. You must know the length of all runs, which should be indicated in the blueprints. If the blueprints are not available, you can measure the runs ahead of time. Be sure that your raceway housing does not bump up against existing building features that might interfere with data transmissions or pose a fire or safety hazard. You also need to consider conduit bends in this stage. After you have the basic conduit runs laid out in pencil (or chalk), cut the conduit and bends to the proper length, making sure that each length fits well within its connecting segment. If the conduit material is stiff in nature, smooth any sharp burrs or edges. If installing conduit along a wall, place a conduit hanger on the conduit and screw the hanger to the wall, repeating for the length of the conduit. You should expect to use a hanger for every 12 inches of conduit installed, unless otherwise specified in the design plans or local codes. Throughout this process, check that the conduit is installed plumb to the appropriate edge. An elbow should not be used unless you're connecting conduit that differs in diameter size.

Punch-Down Blocks and Patch Panels

The cable runs in a structured cabling environment terminate in a punch-down block, which is usually a 66-block or a 110-block, or BIX- or Krone-style blocks, as described in Chapter 8, "Cabling System Connections and Termination." The 110-block is most commonly used for voice and data cabling termination, although you will find many installations that use a 110-block for terminating voice systems and patch panels for terminating data systems. Punch-down block termination provides a cross-connect from one cable set to another, allowing for easier moves, adds, and changes (MACs) as the need arises.

A punch-down block is mounted to a backboard, which is usually made of plywood and secured to the wall of a TC. If you install cabling on more than one floor, each floor must have a separate punch-down block with terminations for the cable drops from the higher floors. Backbone cables should be installed with 10-foot service coils at the termination points, which are commonly located on the backboard in the closet. Figure 15.3 illustrates a typical TC.

Install patch cables from the punch-down block to a patch panel. The purpose of the patch panel is to connect the backbone system to networking

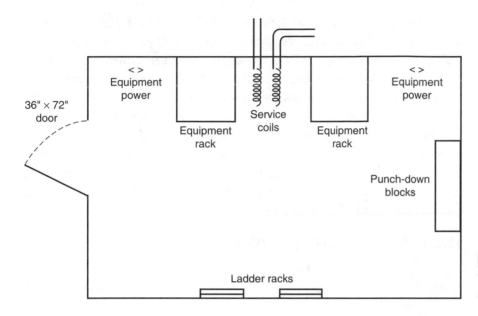

Figure 15.3
TC with punch-down block and service coils

equipment such as a hub or router. End-user equipment, which includes workstations, network printers and scanners, and other shared electronic equipment, generally connect to a hub (also called a concentrator) or router via RJ-45 cable jacks or outlets.

> **NOTE**
>
> For complete details about local area network (LAN) and wide area network (WAN) configuration and maintenance, refer to *Networking Illuminated*, which is part of the Jones and Bartlett Illuminated series.

Pulling the Cable

Set up your jack stands or reel dolly near or in the TC. If you're installing backbone cabling to connect several floors, it's easier to pull cable down than to push it up, although you might not have a choice but to push cable if you are unable to move the reels to an upper floor. Carefully pull the cable from the bottom of the reel, and try to avoid crimping or kinking the cable if at all possible. Pull more cable than you need (approximately an additional 3 meters), cut the cable, and label it according to the project specifications or work order. For details about labeling, see the "Cable Labeling and Management" section 15.2.3. It is important to support and secure the cabling and equipment throughout the pulling process, and to avoid over-bending the cable or excessive cinching when fastening the cable or cable bundles to hangers. The maximum allowable bend radius is 20 times the outside diameter of the cable that's being pulled, and 10 times the diameter of the cable after it's been placed. You should avoid using conduits, pipes, or electrical fixtures as support structures during the pulling process. Finally, when pulling cable to the termination point, be careful not to exceed the maximum allowable amount of cable pair untwist, per the following specifications: (1) no more than 3 inches of conductor untwist for the termination of Category (Cat) 3; (2) no more than 1/2 inches of conductor untwist when terminating Cat 5 or 5e; and (3) no more than 1/4 inches of conductor untwist when terminating Cat 6. Chapter 12, "Structured Cabling," details additional ANSI/TIA/EIA-568-B design requirements for backbone cabling installation.

15.1.6   Horizontal Cable Installation

Horizontal cabling basically runs from the equipment room to each TC, tying all cabling to the backbone cable. This configuration is often referred to as a **home run** (Figure 15.4). One of the most important aspects in planning horizontal cable installation is to build in scalability so that future MACs don't require installation of new horizontal cabling or the relocation of existing runs, if possible. You can achieve this goal through the use of a horizontal cross-connect methodology, as described in Chapter 12.

Horizontal cable is generally four-pair unshielded twisted-pair (UTP) cable or two-pair shielded twisted-pair (STP) cable. Total cable lengths should not exceed ANSI/TIA/EIA-568 standards, such as 90 meters from the TC to the work area jacks.

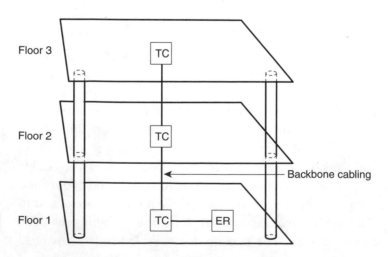

Figure 15.4

Home run

The process of installing horizontal cable is quite similar in many aspects to backbone cabling installation, with some notable differences. For horizontal cabling, you use conduit made of polyvinyl chloride (PVC), fiberglass, or rigid electrometallic tubing (EMT) as the raceway, although you can use flexible metallic tubing if the preferred conduit materials do not lend themselves to the cabling environment, and only if local codes allow for the use of flexible conduit. Horizontal cabling, and therefore conduits, should be hidden from view in areas such as drop ceilings, wire raceways, cable trays, raised access floors, and so forth.

To lay horizontal cable, pull the cable from the box in which it is shipped through a conduit with a pull string or fish tape (which is where the term "fishing a conduit" originates). If you have the necessary equipment, you can pull the cable through the conduit by means of a vacuum or push the cable with some type of compressed-air device or blower. However, you must use a pull string when installing cable in an open ceiling. A gopher pole is handy for pulling the cable through the ceiling space, or you can attach a ball to the end of the string and toss it through the ceiling space. Leave the pull string in place after the job is completed in case you need to make changes in the future. Care should be taken at all times not to pinch, squeeze, or crimp the cable. Per ANSI/TIA/EIA-568-A specifications, you can use Cat 5 horizontal wire in distances up to 90 meters for "in-the-wall" cabling, whereas a Cat 5 patch wire, which has a more flexible core, can be used for distances up to 10 meters in areas that may require some bending of the cable.

15.1.7   Considerations for Special Cable Types

Most of the procedures we've covered thus far pertain to typical backbone, horizontal, and vertical copper cabling installations, with some exceptions. The following three sections address special cabling situations, such as fiber-optic, aerial, and underground cabling.

Fiber-Optic Cabling

Fiber-optic is an ideal cable choice for connecting two or more buildings on a campus, in addition to connecting several floors within a building, because of its reliability and data transmission speed. However, because fiber-optic cabling is still more expensive to install and maintain than copper cabling, you should treat the fiber cabling with even greater care than copper cabling during the installation process. For example, be careful not to nick or scratch the fiber when pulling, and do not exceed the allowable pulling tension as recommended by the manufacturer. You should also strive to create cable runs without splices. Splicing introduces some level of data loss, no matter how sound the splice. Fiber-optic runs should be as straight as possible, and preferably installed in innerducts located within conduits. Project specifications should indicate whether the innerducts must be plenum or non-plenum rated.

Fiber can be pulled, as described previously, or air-blown. Air-blown fiber (ABF) is installed by pushing a thin fiber-optic bundle through a duct by using compressed nitrogen gas. The main benefit of ABF is the flexibility to easily add additional conduits, or modules, as the need arises.

Aerial Cabling

Aerial cabling is installed via some overhead route, such as on buildings, poles, or some other aboveground structure. The cable may have a *messenger* built into it, which is a strand that supports it over the length of the cable run. A cable that has a built-in messenger is often referred to as Figure 8 cable, because the cross-section of the cable is shaped like the number 8. A cable without a built-in messenger may be secured to some other form of support. Aerial cabling is generally lower in cost and easier to install than other forms of cabling (such as in-building raceway systems) but is more susceptible to damage from pedestrians, vehicles, lightning strikes, and harsh weather conditions. Refer to ANSI/TIA/EIA-758-A for guidance on aerial cabling and outside plant installations.

Underground Cabling

Underground cabling, which is generally preferred over aerial cabling because it is hidden and therefore more secure from vandalism and weather conditions, comes with a higher cost. Underground cabling requires careful planning, especially in regard to pulling points, maintenance points (called *manholes*), location of nearby utility line runs, type of soil and near-surface environment, and depth.

You should use fiberglass, galvanized steel, or PVC types B, C, or D ducts. Duct trenches should be at least 24 inches in depth, unless your local code specifies otherwise. Pay close attention to the bend radius of cabling in the duct, which should also be in accordance with code. Ducts should be capped at both ends to prevent seepage of water or other materials.

TIP

When installing fiber-optic cabling underground, include a copper cable with the fiber-optic cable to more easily locate the buried fiber cable after the installation is complete. Do so only if local code permits this step.

Now that we've become familiar with the various steps involved in the rough-in phase, we move to the intermediate major part of cable installation: the trim-out phase.

15.2   Phase 2: Trimming and Terminating the Cabling

The **trim-out** phase of cabling involves cutting the cable to the desired length, stripping the outer sheathing, crimping on a connector, and terminating the cable. This phase also includes cable labeling and management, using such items as ladder systems, J-hooks, and cable ties. Because cable creation, crimping, and termination were covered in depth in Chapters 8 and 13, we summarize these topics in the following two sections as a refresher, and offer additional tips and considerations.

15.2.1   Cutting and Splicing Cables

In this section, we cover how to create the most commonly used data communications cable: a Cat 5 cable with an RJ-45 connector—and then discuss cable splicing. To create a Cat 5 cable, pull a length of copper cable (always allowing for more than is required, which gives you some "wiggle" room) and cut the cable with a cable knife or a sharp pair of heavy-duty scissors. Use a stripping tool to strip back approximately 1/1 inch of the sheathing. Do not remove more sheathing than is absolutely necessary. Carefully untwist the pairs no more than 1/2 inch from the end of the cable, and separate the pairs according to their appropriate color combinations. Insert the individual wires into the RJ-45 connector, making sure that each wire is well seated within the connector. Finally, using a crimping tool, press the connector firmly onto the wires.

Cable splicing is simply connecting two copper cable conductors together, or two fiber strands in the case of fiber-optic cabling, and enclosing them in a permanent housing to prevent disconnection. You may splice copper and fiber cables used in backbone runs, but you should not use spliced copper cables in horizontal runs for safety reasons. Spliced backbone cables must adhere to NEC standards, consist of a single, fire-retardant sheath, and be categorized as Communications Riser Rated.

For copper, you commonly use one of two splicing methods: foldback or in-line. A foldback splice results in the conductors being "folded" into the splice, whereas an in-line splice results in a straight-across configuration. You use splice enclosures called "B" connectors, Scotchloks, or Picabonds to keep the cables spliced together. Fiber-optic cabling splice methods include fusion and mechanical. A mechanical splice results in two perfectly aligned fiber cables, housed in a splice enclosure that maintains the alignment. A fusion splice involves using heat or electricity to fuse the two glass ends together.

15.2.2   Terminating Cables Using a Punch-Down Block

For twisted-pair cabling, pull the cabling to the punch-down block, making sure to maintain the twists on the cable. Failing to do so can result in crosstalk. You can use a flat screwdriver to gently push the cable wires into the punch-down block receptacle, or use an impact tool to make the job easier on you. Orient the tool in the correct direction with the cutting edge facing the end of the cable; reversing the tool can result in accidentally cutting the wire itself. If this occurs, remove the cable from the punch-down block and reterminate it.

15.2.3   Cable Labeling and Management

Throughout the cable creating, splicing, and termination processes, you should label both ends of every cable. As described in Chapter 13, you can use a variety of methods, such as a permanent ink marker to write identifying text on the cable itself, or a more sophisticated labeling device or software that outputs labels that you apply to both ends of each cable. (Generally, labels should appear approximately 1/2 inch from each end.) Labeling devices and software usually include preconfigured descriptions of cables in addition to symbols and bar codes. The project specification's documentation should help you determine the best method to use.

In addition to cables, clearly label all jacks, punch-down blocks, patch panels, and so forth. Label punch-down blocks and patch panels by mounting a label on the backboard.

TABLE 15.1   Example of Structured Cabling Labeling Codes

Component	Code	Description
Building and floor	XXXYY	XXX represents the building number and YY represents the floor number. Example: 01005 is building 10, floor 5
TC	TCYYZ	TC stands for telecommunications closet, YY represents the floor number, and Z is an alphanumeric code. Example: TC051 is the first TC on the fifth floor
Cable	DDDYYZ	DDD represents the cable number and YYZ represents the TC code (omitting the code "TC" for brevity). Example: 005051 is cable number 5 in the first TC on the fifth floor

You should develop and use a labeling scheme, if one is not already detailed in the project specifications. This scheme consists of identifying codes for each component of the structured cabling environment, such as the buildings, floors, TCs, cables, outlets, jacks, and equipment. An example of a few labeling codes are outlined in Table 15.1.

To help keep your cables orderly and well managed, use raceway systems and cable trays, as previously discussed. In addition, you may use a ladder rack, which is a support structure that closely resembles a section of ladder, to support data and power cables (although you must abide by the separation guidelines for the two different types of cables). Ladder racks are usually made of steel, come in various dimensions, and are installed in ceiling spaces, equipment rooms, and TCs. You can mount J-hooks in ceiling spaces to hold cables. Space the J-hooks no more than 4 feet apart. Use cable ties or Velcro ties to attach cables to the cable management equipment, or to keep cables bundled together.

15.2.4   Consolidation Points and Multi-User Telecommunications Outlet Assemblies (MUTOAs)

Consolidation points (CPs) and MUTOAs, as discussed in Chapter 12, are installed where modular furniture is going to serve as work areas. You can install only one CP in a single horizontal run, and you must attach the MUTOA to a permanent structure of the building. The MUTOA must be labeled according to the computers it serves and the maximum allowable length of the patch cord per computer must be indicated.

The next section covers the final phase of the cable installation process, the finish phase.

15.3   Phase 3: Testing and Certifying the Cabling

The **finish phase** of cable installation includes testing, certifying, and documenting the cabling. All of these procedures must be completed to the customer's satisfaction before the project can be "signed off."

15.3.1   Testing Before Certification

A cabling installation must be inspected and tested, proving that it performs satisfactorily within the quality control standards set forth in the project specifications. Test the cable on the reel prior to installation with a time domain reflectometer (TDR) and record the baseline readings. Individual cables should be field tested after creation, splicing (if applicable), and termination using handheld cable testers.

To certify that a structured cabling system is working according to specifications and standards, you must perform tests on the cabling and connecting hardware according to TIA/EIA-568-B. Category 5 cabling tests were originally specified in TIA/EIA TSB 67, "Transmission Performance Specifications for Field Testing of Twisted Pair Cabling Systems" and TIA/EIA TSB 95, "Additional Transmission Performance Guidelines for 4-Pair 100 Ohm Category 5 Cabling." Both of those documents are now part of TIA/EIA-568-B; however, they are still widely referred to.

TIA/EIA-568-B.1, TIA/EIA-568-B.2, and TIA/EIA-568-B.3 address testing of structured cabling systems. The 568-B.1 document addresses design, installation, and field testing for a generic structured cabling system. The 568-B.2 and 568-B.3 documents address manufacturing and component tests for cables, patch cords, and connecting hardware. UTP and screened balanced twisted-pair cabling (or Category 3, 5e, and 6) are covered in 568-B.2, whereas 568-B.3 covers fiber-optic cabling.

All cabling must be free from kinks and nicks. The tests measure for the following measurement points, using parameters specified in the TIA/EIA-568-B standards:

- **Wire map:** A detection of wiring errors, such as shorts, crossed pairs, reversed pairs, or split pairs. (The results include the continuity of the shield connection, if present.) Each wire in a cable must match the termination point (such as the patch panel) pin to pin. That is, pin 1 on the cable must match pin 1 on the patch panel, etc.

- **Length:** Identification of the length of a cable. A cable tester is placed on one end of a twisted-pair cable, and a signal is generated from the cable tester to the other side where another device is connected that will reflect back the signal. By performing a calculation based on the amount of time it takes to receive the signal back, the distance of the cable can be determined.

- **Insertion loss (attenuation):** A measurement of end-to-end signal loss of each pair of wires within a cable. The value of the highest signal loss reading must be lower than TIA/EIA-568-B standards for the cable to pass the test. Insertion loss is typically tested across a range of frequencies, from 1 MHz through 100 MHz, in 1 MHz steps.

- **Propagation delay:** A measurement of the travel time of a signal from Point A to Point B. You must report the longest delay. Measurements are made across a range of frequencies, which is typically 1 to 100 MHz.

- **Near-end crosstalk (NEXT):** A measurement of the crosstalk that occurs at the transmitting end of a cable when the pairs on one end of a cable are bleeding over to another pair (Figure 15.5). This is caused by insufficient twisting of the pairs, or twisting the wrong pairs together.

- **Power sum near-end crosstalk (PSNEXT):** A calculation of the sum of all pair combinations for crosstalk (that is, the NEXT effect on a particular pair by the other three pairs), measured at the near end to the transmitter (Figure 15.6). Because PSNEXT varies significantly with frequency, you

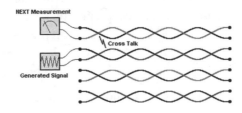

Figure 15.5

NEXT. Courtesy of Ian Patrick, © www.datacottage.com 2005

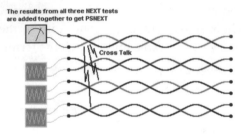

Figure 15.6

PSNEXT. Courtesy of Ian Patrick, © www.datacottage. com 2005

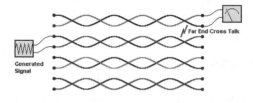

Figure 15.7

FEXT. Courtesy of Ian Patrick, © www.datacottage.com 2005

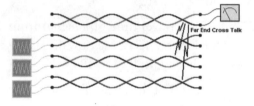

Figure 15.8

PSELFEXT. Courtesy of Ian Patrick, © www.datacottage. com 2005

must make measurements across a range of frequencies, which is typically 1 to 100 MHz.

- **Far-end crosstalk (FEXT):** Much like NEXT, a measurement of crosstalk at the opposite end of the cable from which the signal originated (see Figure 15.7).

- **Equal level far-end crosstalk (ELFEXT):** A calculation of the difference in attenuation of a *disturbing* pair and the FEXT that the disturbing pair causes in an adjacent pair. The results are normalized for the entire length of the cable. ELFEXT measurements are taken in steps in the frequency range of 1 through 100 MHz. Results and documentation include identification of the wire pair combination that exhibits worst value for ELFEXT.

- **Power sum equal level far-end crosstalk (PSELFEXT):** A calculation of the sum of the individual ELFEXT effects on each wire pair by the other three pairs. Because there are four wire pairs in a cable, there are four

PSELFEXT results for each end. Each of wire pair is tested in a range of frequencies from 1 MHz through 100 Mhz.

- **Return loss:** A measurement of the total energy that is reflected on each wire pair in a cable due to slight manufacturing imperfections. Return loss is measured across a frequency range of 1 MHz to 100 Mhz. (Return loss is not specified for Category 3.)

- **Delay skew:** A measurement of the difference in propagation delay between the fastest and slowest pairs. Basically, when a signal is split into four and transmitted down each pair of wires, the signals are recombined into one signal at the far end. A high delay skew value can result in transmission errors, which is of great concern on high-speed networks. The delay skew parameters are covered in EIA/TIA-568-B.

NOTE

Other tests than those specified in TIA/EIA-568-B may be necessary to obtain a warranty from the premise wiring vendor(s).

Refer to the most recently released TIA/EIA-568-B documents for acceptable values (that is, minimum or maximum allowable values) of the various tests. Tests results fall into Pass or Fail categories. Any link that fails must be corrected, repaired, or replaced, and re-tested.

WARNING

Do not move the cabling or any of the newly installed components while performing the tests. If you make any changes to the configuration after the tests are completed, you must retest the system.

The previously described tests are performed across several categories, mainly permanent link and channel. Permanent link tests detect open faults, shorts, crosses, and splits in wall jacks, horizontal wiring, and main cross-connects. These tests are generally performed during installation and before patch cords and networking equipment, such as hubs, routers, etc., are installed. Channel tests measure end-to-end cable runs in a more comprehensive and thorough level than basic tests, and are performed after patch cords and networking equipment are installed.

TSB-67 and TSB-95 specify three levels of accuracy—Level I, Level II (specifically Level II-E), and Level III. Level I accuracy reflects performance when testing through a Figure 8 connection, which is prone to crosstalk.

Level II-E represents a much higher level of accuracy than Level I and is used when testing Category 5e cabling. Level III accuracy includes all of the TSB 95 tests for Category 5e, and is geared toward Category 6 (and eventually Category 7) 200 MHz to 250 MHz traffic.

15.3.2   Certification and Documentation

Passing the previously discussed tests indicates compliance with TIA/EIA standards. You should provide printouts of all tests, or detailed notes taken during the testing processes, along with an up-to-date network cabling scheme and diagram to the client. Local building and fire safety inspectors also will inspect the work and provide independent reports, all of which must be part of the client's final documentation.

15.4   Chapter Summary

- The first phase of cable installation is called rough-in, during which the staging area is identified, work team is assembled, conduit is installed, and cable is pulled from the telecommunications room (or wiring closet) to the individual network jacks or plugs throughout the location. During this phase, backbone and horizontal cabling are installed. Other types of cabling, such as aerial and underground, are also installed during this phase if the project specifications call for them.

- The second phase of cable installation is the trim-out phase, during which cables are cut, stripped, crimped onto a connector, and terminated. Cable and component labeling also takes place during this phase, which involves using a preplanned naming scheme, or code, to differentiate each specific item in a structured cabling system. Specific items on the labels include buildings, floors, TCs, cables, jacks, and others. It is during the trim-out phase when cable management equipment is employed, such as ladder systems, J-hooks, and cable ties.

- The third and final phase of cable installation is the finish phase, in which the cabling and components are tested for adherence to quality standards, certified, and documented. The project is not considered complete and eligible for customer sign-off until all of the parts of the finish phase are complete.

15.5   Key Terms

backbone: Cabling configuration that interconnects the major features in a structured cabling environment—entrance facilities, TCs, and equipment rooms—and includes the raceways, intermediate and main cross-connects, patch panels, cable supports, and the actual cables themselves such as coaxial, twisted-pair, and/or fiber-optic.

finish phase: The phase of cable installation that includes testing and certifying the cabling.

firestops: Barriers in a structure designed to prevent the vertical or horizontal spread of flames.

home run: Horizontal cabling that runs from the equipment room to each TC, tying all cabling to the backbone cable.

point of demarcation: The point at which an outside service provider's (SP's) equipment/cabling, such as that provided by a telecommunications utility, meets the entrance facility of the premises.

rough-in phase: The phase of cable installation that includes identifying the staging area and work teams, installing conduit, and pulling cable from the telecommunications room to the individual outlets throughout the location.

trim-out phase: The phase of cable installation that involves cutting the cable to the desired length, stripping the outer sheathing, crimping on a connector, and terminating the cable.

15.6   Challenge Questions

15.1   There are two main types of backbone cabling that pertain to location; _____ cabling is contained within a single building and _____ connects two or more buildings.

15.2   The backbone includes which of the following features? (Choose all that apply.)

    a. Patch panel

    b. Fiber-optic cabling

    c. Messenger

      d. Category 5 cabling

      e. Raceway

15.3 When should an elbow be used when installing conduit?

15.4 When pulling cable, how much overage should you leave for flexibility in the telecommunications closet?

      a. 30 centimeters

      b. 70 centimeters

      c. 3 meters

      d. 6 meters

15.5 When pulling cable, how much overage should you leave for flexibility at the jack?

      a. 10 centimeters

      b. 20 centimeters

      c. 30 centimeters

      d. 40 centimeters

15.6 What is the type of support feature that is sometimes built into aerial cabling?

      a. Messenger

      b. Hanger

      c. Gopher pole

      d. Vertical strand

15.7 Underground cabling conduits should consist of which of the following materials? (Choose all that apply.)

      a. PVC types B, C, or D

      b. Fiberglass

      c. Galvanized steel ducts

      d. EMT

15.8 You should never splice Category 5 cabling used in a _____ run.

15.9   Horizontal cabling conduits should consist of which of the following materials? (Choose all that apply.)

a. PVC types B, C, or D

b. Fiberglass

c. Galvanized steel ducts

d. EMT

15.10   In a horizontal cable run, what is the maximum total cable length from the TC to the work area jacks allowed by TIA/EIA-568 standards?

a. 10 meters

b. 70 meters

c. 90 meters

d. 105 meters

15.11   ABF is the abbreviation for _____ . Describe one of its benefits.

15.12   A cable that has a built-in messenger is often referred to as a _____ cable.

15.13   What are the two main splicing methods for copper cables?

a. Foldback

b. In-line

c. Picabond

d. Mechanical

15.14   What are the two main splicing methods for fiber-optic cables?

a. Foldback

b. Fusion

c. Picabond

d. Mechanical

15.15   What is a point of demarcation?

15.16   Name five field tests that Category 5 cabling must pass.

15.17   What is the purpose of a ladder rack?

15.18 CPs and MUTOAs are used primarily in work areas that include which of the following:

a. Underfloor conduit

b. 66-blocks

c. Perimeter raceways

d. Modular furniture

15.19 _____ tests detect open faults, shorts, crosses, and splits in wall jacks, horizontal wiring, and main cross-connects.

15.20 What is the TIA/EIA specification that describes permanent link and channel testing?

a. 568-B (formerly TSB-67)

b. TSB-76

c. Article 700

d. NEXT

15.21 With which of the following is a bend radius a significant installation factor?

a. Fiber-optic cabling

b. Copper cabling

c. B connectors

d. PVC type C

15.22 What is a labeling scheme?

15.7   Challenge Exercises

Challenge Exercise 15.1

In this exercise, you create a pre-installation checklist of the equipment and supplies needed to install backbone cabling. Your checklist should be detailed and include cabling, connectors, supports, tools (from standard hand tools such as wire cutters to cable reels), labeling equipment, and miscellaneous supplies. Include specific cabling types and specifications, where appropriate. To complete this exercise, you need a pen and paper, or a computer with a word processing application and a printer. *Optional:* Consider searching the Internet for "data cable installation checklists" or "telecommunications installation checklist" to compare a variety of checklists already in use by campus-type organizations.

Challenge Exercise 15.2

In this exercise, you tour a campus that includes most or all of the features discussed in this chapter. If you are participating in an instructor-led classroom setting, your instructor will lead the tour. If you are studying remotely, arrange for a tour at a local college or multi-building corporate complex. During the tour, take notes and draw diagrams of the equipment room, TC, underfloor conduit and other accessible raceway systems, and various equipment in the TC (such as a 110-block or BIX). Be prepared to describe the purpose and function of each component.

Challenge Exercise 15.3

The purpose of this exercise is to have the students perform in-depth and meaningful research about Category 5 and 6 field tests and TIA/EIA-568-B testing requirements.

Some Web sites that the students might find helpful are:

- TIA Online: *http://www.tiaonline.org/*
- Cablingdb.com: *http://cablingdb.com/Standards*
- Siemon: *http://www.siemon.com*
- ECCO Computers: *http://www.eccocomputers.com*

15.8 Challenge Scenarios

Challenge Scenario 15.1

The Prentice city council wants to construct a three-story municipal complex to house various city department offices. The new complex will incorporate a structured cabling system, integrating voice, data, and video. The council has asked you to help them create specifications for the complex, which will become a request for bid (RFB). Your task is to provide details for the following outline topics. The first outline section—work area subsystem—has been completed:

I. Work Area Subsystem

    A. Provide work area patch cable

    B. Install work area patch cable

II.   Horizontal and Vertical Subsystems

III. Equipment Subsystem

IV.   Engineering and Project Management

V.   Training

VI.   Documentation

APPENDIX A

Obtaining and Keeping Employment

Simply put, today's job market is rather more grim than cheery. In a world of rising unemployment, decreasing demand for IT and technical jobs, and fiscal conservatism—which translates into fewer opportunities for those seeking new positions and less likelihood of promotion or job changes for those already employed—how can savvy IT professionals find jobs or keep the jobs they already have? Those are precisely the topics that this Appendix addresses, in that order.

A.1 Finding Employment

A new generation of IT employees is emerging that is more accustomed to using the Internet than any preceding generation. Our findings on the most commonly practiced job-hunting technique—namely, posting your resume on the Internet—may therefore be somewhat shocking. According to recent media reports, only 5% of all jobs are filled through Internet postings, resume services, and other online avenues. Stated differently, 95% of all such positions are filled through more conventional means. This helps to explain why understanding and making the most of those "more conventional means" remains absolutely key to finding a job, if only because the odds are 19-to-1 that those means ultimately will result in finding work.

To begin, it is essential to understand the job search and placement processes. We review these in their normal order, and provide what we hope are helpful tips and suggestions about how to make them work best for you as you undergo the process. Along the way, we also remind you of some sound rules about how, where, and when to go looking for work.

A.2   Ten Steps to the Workplace, Plus One

Our recommended sequence of steps may be shorter or longer than what you find in the real world. It all depends on what work you have already done, materials you've already assembled, or additional tasks or avenues you may discover between your departure down the job search trail and your ultimate arrival in the workplace.

Nevertheless, this sequence will prepare you for the journey to employment.

1. **Do an inventory: Skills, experiences, education, interests, and more.** Before you can start looking for work, you must get organized. To begin, you need to perform a lengthy and completely honest personal inventory to determine what you can bring to any job you might pursue. You should reflect on your past and your experiences so you can take stock of all the things you know how to do (skills), all the potentially work-related things you have done (experiences), your schooling and any other training you may have completed (education), the subjects and types of work activities that interest you most (interests), and anything else you can think of that might be relevant to your life on the job. Make a list, keep a journal, jot down some notes. Do whatever works for you to help create an inventory, and keep adding to it as you work your way through the rest of the process (or as you discover new interests, discard old ones, and so forth).

TIP   Self-assessments are great tools. On the Web, in your reading, through job placement services or career counseling, and so forth, you'll find all kinds of tests and surveys you can conduct to help perform various parts of the inventory process. Use them not just to document your inventory, but to discover and explore possibilities you might otherwise overlook.

2. **Set employment goals: What kind of job, what kind of work, and what kind of pay.** Ask yourself what you want to do for a living, what kind of company you want to work for, and how much money you want to make. It is important that you set goals for

yourself to drive the process of looking for work: searches are most successful when you know what you are looking for.

 TIP

When searching for anything, it is always easier to find something in particular than it is to deal with the results of "anything will do." Although desperation may lead to broader searches rather than narrower ones, it is still worthwhile to spend time thinking about and researching what kind of work you want, what kind of organization you want to work for, and what kind of pay you want (and need) to make it worth your while.

3. **Do a sanity-check: Does your inventory match your job goals?**
    Look at some classified ads for the kinds of jobs you think you are interested in. Talk to a friend, a colleague, a recruiter, a hiring manager at a company of the type that interests you, or other job placement professional about your goals. Does what you bring to the situation match what you are looking for? If so, keep following the sequence; if not, back up as far as you must (this may require adding to your inventory or simply resetting your goals) and repeat until a reasonable match occurs.

4. **Sell yourself: Put your best foot forward.** Much of what is involved in finding work depends on your ability to stand out in a potential crowd of other applicants. You have to be able to explain what you know in terms that work for your listeners or interlocutors, whoever they may be. The same is true for setting a value on your experiences, education, skills, and so forth. In fact, your success will often depend on how attractive you can make yourself to a person with input on hiring decisions. This means you not only must know your history, interests, abilities, skills, passions, strengths, and weaknesses really well, it also means you need to be able to communicate them in the most positive, appealing way possible.

 TIP

There is a benefit to this seemingly difficult and onerous task. It forces you to evaluate yourself and the ingredients you bring to the workplace. It also helps you clarify your goals, especially in terms of what you want to do with yourself at work. Nowhere else in the process are the returns better matched to the effort you expend: the more you put in, the more you will get back, in many different ways.

5. **Prepare application materials: Resume, cover letter, work history, and more.** You'll need a resume that describes your prior work history, training, education, and more (see the

"Recommended Resources" section at the end of this Appendix). You'll want to create a basic cover letter that introduces yourself, states what job you are applying for (and why you are qualified to do it), and promotes your skills and experience that make you uniquely qualified for the position. At this point, it is also a good idea to compile a work history with names, places, dates, and contact information for prior employers. This will come in handy when you find yourself filling out an application form in the future, and will help you make sure you have left nothing important out of your resume or cover letter. Be sure to double-check all contact information. In an age where employers routinely check references and prior employment, it is potentially damaging to your chances if they can not use the information you provide to make contact when it is needed.

TIP Ask several people to look over your materials. Make sure everything is perfect: spelling, grammar, layout and design, and tone. Remember also to put that "best foot forward" whenever you can!

---

6. **Begin the search process: Get the word out!** Remember that 95% statistic we cited in the introduction to this Appendix? Although you'll want to post your resume to job sites online, you should spend the bulk of your time and effort communicating with real people. This means old-fashioned networking, where you list all of the people you know (not just those who work in your chosen field) and contact them with ruthless dedication: friends, family, former co-workers, teachers, and anyone else you can think of. Everybody knows somebody, and somewhere out there are opportunities waiting to be found and taken.

TIP Selling yourself applies during the networking process. Even for family members (who presumably care about your situation and want to help you out), it is necessary to be pleasant, positive, and grateful for their time and attention—even if they can not help you out immediately. Make sure they know what sort of job you are looking for, and how to contact you. Many initial contacts that produce no immediate results do produce results later on for those who know how to ask for help in finding work.

Of course, you will also want to follow the "usual routes" to finding jobs as well. This means reading classified advertisements and job postings, and calling companies or organizations where you might want to work to inquire about opportunities. Recruiters report that roughly half of all jobs are never advertised, so you must count on personal contacts and inquiries to turn things up about which you otherwise might never know.

7. **Follow up 1: Check on reception, progress, timing, and opportunities.** As you begin to submit materials to prospective employers, you will want to also start an ongoing process of follow-up to make sure that your materials have been received, and help you understand the internal processes that might lead to an interview or to a rejection. Remember that rejection is the more likely outcome, so do not set your hopes too high at this point. Always ask for as much information about the way things work: how are applications handled, how long does it take to get a response, what form(s) will responses take, and so forth. Be sure to ask if other jobs might be available in or around the workplace roles you have decided to pursue.

 TIP

Many job applications lead to rejection in the form of no response (especially online applications or posting replies). Keep track of what you have inquired about and when you did it. Assume that if you hear nothing in a month, you need to talk to somebody about your materials, start the process over, or give it up.

8. **Follow up 2: Schedule, research, and deal for an interview.** If you are approached by a prospective employer, do not get too excited unless they want to schedule an immediate interview. Even then, although a small thrill is natural, remember that this is the beginning of a process in which many are called but few are chosen. This is when you want to find out where and when you might be interviewed, and to start asking about the people who might interview you. If travel is required, it is also entirely appropriate to ask about the kinds of costs the interviewer will cover and what costs you yourself must bear. Be conservative with what you must spend to apply for any job, because it may not be repaid unless the interviewing organization agrees in writing to cover such costs.

9. **Do the interview: Prepare, anticipate, and show yourself well.** Assuming you are granted an interview, temper your excitement with the knowledge that you have got some real work to do to get ready. One of the biggest complaints that interviewers levy against

applicants is lack of knowledge. Refrains such as, "they didn't know anything about our products" or "they knew nothing about our sales strategy" are common and fatal.

To begin, learn as much as you can about your prospective employer. The Internet can come in handy for this task. You should look for information about the employer's business, financial condition, products, key partners and customers, sales and promotions, and recent mentions in the news media.

You should also learn as much as you can about your prospective position: what tasks does it entail? What responsibilities? What tools and technologies? What kind of training or certification may be required or involved? This will not only help you understand what you are getting yourself into, but will also help you anticipate what the employer is looking for and what kinds of questions they are likely to ask in an interview.

During the interview, balance trepidation and excitement by concentrating on showing yourself to the best possible advantage, assuming, of course, you do want the job. In that case, look for ways to make what the employer needs and wants relevant to what you know and can do, experiences you have had, the kind of problems you have solved, and situations you have handled. Think and speak in terms that make sense to your interviewer as much as you possibly can; this facilitates communication and understanding.

TIP When preparing for an interview, prepare a list of likely questions (you will find plenty of them in the interview-related resources at the end of this Appendix) and get somebody to role-play various types of interviews with you. If you can find someone who is or has been a hiring manager or who has conducted interviews on the job in the past, so much the better!

---

10. **Follow up 3: Write thank-yous, perform status checks, and deal with the consequences.** Once the interview is over, don't forget to say thank you. You will probably do this verbally as you leave, but do not forget to follow up with a thank-you note. Written communication weighs strong in this situation, according to many experts and recruiters. Even if you do not get that particular job, a

request in a thank-you note for future consideration may lead to exactly that.

During the interview, try to get a sense of what is going to happen next and when. Take notes on the spot or as soon as you can, and then schedule status checks and other follow-ups based on the outcome. Polite, friendly, and active ongoing indications of interest sometimes can make a difference.

If you get an offer, ask yourself, "Do I really want the job?" If that answer is yes, review the terms of the offer carefully: Is it the position you want? Is the pay acceptable? What are the benefits? Is there room for future advancement? You will be given some time (usually two weeks or less) to respond to the offer, so take time to think things over and decide what to do next. If you are fielding multiple offers, weigh each one in terms of its pros and cons, and rank them accordingly. You must often respond in writing to an offer, so be prepared to write a letter and inform the company of your decision.

TIP

Responding to an offer is your last opportunity to negotiate terms. Pay, start date, vacation days, benefits, and so forth are all part of this picture. If a job comes close to meeting your requirements but doesn't quite get there, you can respond to an offer with a counteroffer. As long as you don't ask for too much more than the original offer, you may get what you request—but be prepared to get a "take it or leave it" response.

---

11.   **If nothing happens, start over.** If you go through this process as far as Follow up 1 (or further) and have found no opportunities, you should start the process over. Because things are not working for you in their present form, do whatever you can to further refine your inventory, sanity-check your goals, prepare your materials, and get the word out.

Notice how many steps occur in this recommended process before a job search ever begins. You can do some of these steps in parallel (such as preparing materials and spreading the word that you are looking for work). Be aware that skipping preliminary steps can cause problems with later parts of the process. Likewise, doing things in parallel can cause problems of their own if you get unexpectedly lucky and are asked to provide materials before they are ready. Whatever you choose to do, be equally prepared for failure and success, and you will be ready for anything.

A.3   Staying Employed

In an era of consolidation, downsizing, and layoffs, even those of us lucky enough to have jobs today may not be lucky enough to have them in the near future. Although outright paranoia is probably not warranted in most circumstances, proactive measures to protect your position are always important to keeping your job. What does this mean?

When employers have to cut employees, they do so in a variety of ways. Sometimes, entire divisions or operating units are dismantled and everybody within them is asked to leave. In that situation, there is not much you can do to keep your job. However, the steps we recommend in the preceding section will help make you a more attractive job candidate and help you find other work. In many cases, managers are asked to rank their employees on the basis of performance, contribution to the bottom line, value of skills and knowledge, and other criteria that can help to separate the keepers from those who will be let go. Simply put, your second job is to place yourself into the "keeper" category.

Notice the criteria that were mentioned in the ranking process:

- Performance
- Bottom line contribution
- Value of skills and knowledge

All of these provide opportunities to protect your position, if you know how to use them to your benefit. Each of these topics is the focus for a short discussion in the sections that follow.

A.3.1   Taking Advantage of Performance Boosters

In general, performance is a topic that comes up on the job in an informal way all of the time, and in a formal context during annual or other scheduled reviews. To manage your performance, you must do what you can to make it the subject of positive feedback whenever possible. If you do a good job for a customer or internal users, for example, and can ask for support without damaging your working relationship, it is smart to solicit e-mails or letters thanking you for a job well done. Likewise, if performance-related awards are given in your organization, make it your goal to earn some. Prior to your next formal review, provide your reviewers with copies of these acknowledgments so they will be sure to consider them when rating your performance.

In general, anything you can do to shed positive light or put a positive spin on your job performance will help move you to the top of the rankings in that area. Also, if the unthinkable happens, future employers will also be interested in your accolades and achievements, not to mention your stellar performance appraisals.

A.3.2 Adding to the Bottom Line

The key here is to look for ways to contribute to the bottom line and to make sure that others hear about it. If you work in a cost center, for example, this often means talking about how you saved the organization money, helped the organization realize various efficiencies, or eliminated unnecessary costs. If you work in a revenue center, you have to figure out how what you do helps to contribute to whatever money your part of the organization is making. (This can be difficult when your company is losing money, but as long as some cash is flowing in, you should be able to make some kind of case for your contribution to that flow.)

Mercenary though it may seem, it is equally important to be able to put some kind of monetary value to your contributions. The bottom line is all about numbers, so you must supply some numbers, too. This will often force you not only to be creative in assessing and evaluating your contributions, but it will often help you better understand your organization's business model and practices as well as how your job fits into the bigger picture of financial success or failure. This kind of information is of great interest to prospective and current employers.

A.3.3 Increasing Skills and Knowledge

In an important way, we saved the best way to protect your job for last, because we move away from the realms of perception as to how well you perform and how much you contribute to realms that you can control. In short, the more you do to keep your skills and knowledge current, the more valuable those skills and knowledge will be to your employer. Even better, the more you learn about new topics, tools, and technologies, the better you will be able to participate in tracking new directions and developments within your organization.

In this case, it is important to be able to document what you know and can do. Although obtaining IT certifications and keeping them current is a traditional way to demonstrate measurable skills and knowledge, as long as you can talk intelligently about what you have learned and how you have used resulting skills and knowledge, you can demonstrate value to your organization. Activity and learning above and beyond the call of immediate duty is usually a sign of initiative, interest, and motivation—characteristics of great interest to current and prospective employers. Thus, if you concentrate on keeping up with the fields in which you work, and learning about new topics, tools, and technologies, you will be able to take a place at the head of the rankings for employees based on these criteria.

A.4   Recommended Resources for Job Seekers and Changers

Although there are countless sources of information for finding jobs and developing your career, we have found the following to be particularly useful. We hope you will try them and feel the same way:

- **Your favorite search engine**: Rather than play favorites when there are so many good search engines to choose from, we instead recommend that you get to know one or more of these excellent tools and concentrate your efforts on using them to help you find the plethora of "good stuff" that is out there. To begin, you will find a wealth of additional resources simply by using the headings for the remaining items in this very list as search terms, with thousands of responses to each and every one.

- **Resume resources**: Online and in print, hundreds of models and tools are available to help you create or improve your resume and other traditional job application materials such as the cover letter. The big job posting sites routinely offer resume builders; most are worth investigating. In this area, our favorite resource is Paula Moreira's excellent book, *ACE the IT Resume: Resumes and Cover Letters to Get You Hired* (Osborne/McGraw-Hill), because it not only explains the development process and provides good examples, it also addresses how to customize your materials to help you stand out from the crowd.

- **Job interview resources**: As with other resources mentioned here, a lot of information and advice is available to help you prepare for and do your best during the interview. Search for "career advice," "job interview advice," or "job interview resources," and choose what is most useful to you. Here, our favorite resource is Paula Moreira's companion to the previously cited book, *ACE the IT Interview* (Osborne/McGraw-Hill, 2002), because it provides many practice questions, interview tips, research guidelines, and much more to help you get and survive the interview with confidence and grace.

- **Job search tips and tricks**: This is an area in which it is difficult to make specific recommendations, except to tell you that you must be as precise as possible about your chosen field of work and the job you are seeking to find the most focused and useful advice. Be prepared to spend hours online looking for just the right kind of advice. For those looking for good places to start, Richard Bolles (author of the famous job search book, *What Color Is Your Parachute?*) has a terrific Web site at *http://www. jobhuntersbible.com.*

- **Training and certification information**: For those seeking to demonstrate or add value for technical knowledge or skills, training and certification may be powerful additions to your arsenal. We strongly recommend visiting key certification and training portals online, such as *http://www.certcities.com, http://www.certmag.com, http://www.gocertify.com, http://www.cramsession.com, http://www.examcram2.com*, and more. They provide the most up-to-date information and resources for the ever-changing training and certification arena.

APPENDIX B

The Future of Cabling

To discuss the future of the communications cabling industry, we should spend a moment looking at the past and present. Prior to the 1984 divestiture of AT&T, the topic of communication wiring systems, as they were referred to back then, were virtually nonexistent, unless you were within the industry itself working for a utility company. The primary application of this early time period was voice applications. For the few who were even involved with data networks, the architectures and speeds were considered to be state-of-the-art if the networks were designed with such topologies as two-pair STP-A/Type 2 IBM cabling, Thicknet, Thinnet, or ARCnet, with blazing speeds of 4 Mbps transfer rates. Unlike today's vendor-neutral network designs, systems of the past were vendor-specific, and the installations and maintenance of the physical cable plant itself was solely provided by the equipment's vendor or specifically chosen contractors.

The data side of the telecommunications industry has seen tremendous evolution over the past 25 years. What was once referred to as communications wiring systems became ISP (inside plant) wiring, which evolved into local area networks (LAN), which became premise networks, and then evolved into campus networks, and today are now referred to as enterprise networks. Ironically, the early term of "communications wiring system" was an across-the-board generic description of the physical cable plant within the industry at the time. Today's terminology refers only to a small section of what the data side of the industry has evolved into and does not cross over universally to include other sectors such as metropolitan area networks (MANs), wide area networks (WANs), service area networks (SANs), or global area networks (GANs), just to name a few.

The telecommunications industry has also seen massive regulatory changes over the past 25 years, in the form of standards and minimal performance guidelines and/or recommendations. After the 1984 divestiture of AT&T, it was suddenly realized that end users had been left to fend for themselves. In an effort to resolve confusion, two groups, the TIA and EIA, under the guidance of ANSI, formed the standards bodies of today and created generic cabling standards to serve as minimal performance criteria for the next generation of industry designers, installers, troubleshooters, and end users.

Since the introduction of the original TIA/EIA-568 commercial document in 1991, these standards have provided the guidance to successfully deploy such mediums as Category 3 and Category 4 UTP cable. Once again in 1995, the TIA/EIA-568-A revised document played a key role in the deployment of Category 5 UTP cable, which vaulted the telecommunications industry from the 16 MHz performance of Category 3 and the 20 MHz performance of Category 4 to an extraordinary performance level of 100 MHz per second of Category 5. This technological achievement has served as a defining point within the telecommunications industry.

Although the industry's achievements between 1977 to 1995 are extraordinary, the advancements starting in 1997 with the proliferation of Category 5e/E UTP, Category 6 UTP, Category 7 SSTP, SM/MM/LO fiber-optic, and wireless technologies have literally reshaped today's world. (Note that fiber optics played an enormous role in long-distance/long-haul technologies before its integration within today's data communication backbones.)

The TIA and EIA continue to provide the industry with comprehensive revisions including the current TIA/EIA-568-B document. This document has served as a minimum guideline to assist in the successful deployment of Category 5e/E cable for Gigabit technologies and will be referenced for the successful deployment of Category 6 cable for pending deployment of 10 Gigabit transmission technology.

B.1   The Future

Although the past and present of the communications industry has been predominantly based on copper technologies such as coaxial and UTP, the future centers around fiber optics, tunable laser systems, wireless topologies, broadband over power, and Voice over Internet Protocol (VoIP). To be fair, it should be pointed out that with the evolution of Categories 6 and 7 providing us with speeds of 250 MHz and 600 MHz, respectively, copper mediums will continue to be a viable option within the Physical layer choices of the future.

Looking into the future of the telecommunications industry as far as one can, the bandwidth demands and the transmission speeds of today's network protocols show absolutely no signs of slowing down. To that end, the end users and consumers will likely see the continued proliferation of network topologies to evolve from current gigabit transmission speeds to 10 Gigabit to 40 Gigabit, and to what many are referring to as *super gig*, or 100 Gigabits per second network speeds by the year 2015. It will be the future transmissions of information packets that consist of voice, video, and data within a single packet, multidimensional illustration-based data bases that also include traditional text, 3-dimensional scalable molding by the engineering sector, and conference room-based as well as stand-alone PC-based video conferencing. It would be irresponsible if the ever-evolving telecommuting sector was left out the overall picture. Additionally, the residential side of the telecommunications industry will continue to explode with the evolution of structured cabling and wireless topologies in the home. Let's take a look at the Physical layer options in the past, present, and beyond.

B.1.1   Coaxial Cable

Coaxial cable has played a vital role in the telecommunications industry, ranging from its usage with stereos and TVs to its most widely known role as the medium of choice for the implementation/deployment of the 10Base5 and 10Base2 Ethernet platforms of the past. However, there is room for argument that even though the TIA/EIA no longer recommends coax as a cabling option, in fact it is a superior choice to UTP.

The downfall of coaxial cabling was caused by two major factors: space consumption and installation time. However, looking at the amount of research and development that has been spent on UTP component-based networks over the years, one has to wonder if the choice had been to go with coax rather than UTP, and had the industry spent most of the available dollars and time on coaxial research and development, what the outcome would have been.

Over the years, comparisons have indicated that coaxial-based networks were capable of delivering higher-speed networks than their UTP counterparts, and that the cost of coaxial networks would have been cheaper than UTP-based networks. However, if you are currently in the process of deciding what type of cable to deploy in your network, it would be wise to consider the following factors as vital to the longevity, performance, and potential expansion of your network: component integration; level of support within the market; staff knowledge; available resources; cost of repairs, replacements, and upgrades; and the projected future availability.

B.1.2   Unshielded Twisted Pair (UTP)

UTP has gained by far the greatest acceptance of all of the cable mediums. The industry has evolved through the Category 3/16 MHz solution to the Category 5e/E 1000 MHz/gigabit solution. Currently, the industry is beginning to enter the Category 6 era, what many refer to as the "10 gig era." The reasons for UTP's wide acceptance are twofold: One is its ease of installation based on a simplistic color code and pair placement system, and the second is the familiarity of electron-based technology and how they travel over conductive materials.

Although end users and consumers have already experienced the benefits of past and present technologies, the twisted-pair industry shows no signs of slowing down or giving way to other options. In the near future, there will be discussions, development, and the deployments of Category 6E/625 MHz UTP technology, and Category 7 stainless-steel shielded twisted-pair (SSTP) cabling with an original targeted bandwidth of 600 MHz, which is now being reviewed based on other advancements. A Category 8 cable technology is beginning to be discussed, however, with no word on the performance criteria to date.

B.1.3   Shielded Twisted Pair (STP), Screened Twisted Pair (ScTP), and Foil Twisted Pair (FTP)

This technology has received most of its attention with two specific cables over the years: in the 1970s, IBM type II/2-pair 300 MHz cable and, recently, Category 7/4-pair 600 MHz technology in a SSTP. These media types have not gained anywhere near the type of acceptance that UTP has gained for two primary reasons: cost of the cable itself and the additionally required installation time that an STP cable type requires to be installed properly. The advantages of shielded cable designs are well documented and have always been the overwhelming choice in Europe and other areas that have to deal with excessive RFI. Despite the advantages of these mediums and their superior transmission performance characteristics, it is doubtful that many areas of the world will embrace them, including North America.

However, do not sell Category 7 short. Ironically, it addresses several challenging issues that other twisted-pair solutions have struggled with, such as alien crosstalk, return loss, and EMI over long transmission distances as well as in unusual applications because of its shielded design.

B.1.4 Wireless

One of the most popular topics in the telecommunications world over the past five years has been the 802.11 or wireless world. It has evolved at a torrid pace and achieved transmission rates that were at one time considered impossible to reach in a wireless network environment. In a very short period of time, the wireless industry has already brought to market the following protocols: 802.11a (5 GHz/54 MB/s), 802.11b (2.4 GHz/11 MB/s), and 802.11g (2.4 GHz/54 MB/s), and current discussions are about an 802.11n option reaching 100 MB/s speeds. Although the evolution of these products have hit retail in an almost overnight fashion, the market has not been as quick to embrace them at the same rate.

The coverage and exposure that these technologies have received over the past five years would have you believing that all fixed/physical mediums are doomed. This has not even remotely been the case. Although advantages such as portability, flexibility, and minimal installation time have been well documented, there is a major concern—security—and no one to this point has really been able to completely address that concern.

B.1.5 Fiber Optics

The advantages of fiber-optic cable are unequalled and well documented: resistance to EMI and RFI, greater transmission distances, significantly easier to troubleshoot, and significantly less data loss over the length of the cable. It is regarded as the best solution for future-proofing one's network. However, fiber optics is not without its drawbacks and, contrary to popular belief, these drawbacks do not always include that it is difficult to install or that it is fragile. The main drawback has been cost. Fiber optics, like all other media types, requires an acquired skill set. However, until recently, it has been a more expensive solution.

As the deployment of fiber optics has grown over the years to virtually take over the backbone area of today's networks, the price of the technology has fallen and now is actually cheaper than some of the other options. Some of the cost reduction can be directly attributed to a reduced cable cost. However, the majority of the cost reduction has occurred with the required electronics.

Advancements within the industry have paved the way to being able to produce much cheaper LEDs. It has spurred the introduction of VCSEL technology, and electronics have continued to reduce in price because of heavy competition and automation processes. The future of fiber optics is extremely bright. Recent achievements such as tunable laser systems, WDM/DWDM/DTWDM/CWDM, laser-optimized fiber, EDFA technology, as well as PONs technology and many other medical uses have reignited the fiber industry. In case you are still contemplating what would be the best solution to future-proof your network needs, consider a laser optimized solution with incorporated CDWM.

B.2   Conclusion

When was the last time anyone can remember a technology going backwards or failing to continue to evolve? The UTP market will continue to prosper, based on continued refinement of the cable manufacturing processes and the advancements of the associated electronics. The STP/ScTP and FTP markets will continue to find their way into environments that are being saturated by EMI/RFI. The wireless market will continue to stretch the limitations of the mediums and continue to gain a strong acceptance into existing residential structures. New residential construction will be hard wired with what's known as structured cabling. Finally, top it all off with the fiber-optics industry, which is already firmly entrenched in the long haul/WAN/access network and enterprise backbone areas of the industry. Fiber optics will continue to migrate into the remaining horizontal areas of the enterprise networks by continuing to lower component costs, and via the use of media converters and the introduction of optical network interface cards. In addition, fiber optics will continue to evolve in the last mile markets of FTTP, FTTB, and FTTC by deploying the PONs technology.

APPENDIX C

Cabling Resources and Information

This appendix is a compendium of resources found throughout the book in addition to other links and print sources that will help you learn more about various cabling topics.

C.1 General Resources

Cabling Installation and Maintenance Magazine: http://cim.pennnet.com/home.cfm

Timely, regularly updated information about standards, tools, technologies, and techniques. A great resource for anybody involved in designing, installing, or maintaining cable plants.

RF Café: *http://www.rfcafe.com*

A menagerie of radio frequency-related articles, forums, software, utilities, and Web links.

Mobile Electronics Glossary: *http://www.the12volt.com/glossary/glossary.asp*

A complementary glossary provided by the12volt.com.

C.2   Cable Testers and Other Test Equipment

Agilent Technologies: *http://we.home.agilent.com*

Test and measurement, automated test equipment and communications products, and services.

Electro Rent Corporation: *http://www.electrorent.com*

Leading supplier of test and other equipment for purchase, lease, or rent for power, telecommunications, fiber-optic cable, and general test and measurement purposes.

Extech Instruments: *http://www.extech.com*

Manufacturer of handheld test equipment for electricity, power, networking, fiber-optic metering, and more.

Fluke Electronics: *http://www.fluke.com*

Manufacturer of hand-held and portable test equipment for cables, networking, power, and digital and analog signals.

C.3   Cabling, Structured Wiring, and Component Companies

3M: *http://www.3m.com*

Search on "access network" for cabling, diagnostic equipment, and more.

Accu-Tech Corporation: *http://www.accu-tech.com*

Network cabling, premises wiring supplies, and components.

ACS Industries: *http://www.acsindustries.com*

Manufacturing company with wire, fiber-optic, and specialty wire divisions.

AFL Telecommunications (an ALCOA company): *http://www.alcoa.com/afl_tele/*

Fiber-optic products, engineering, and services for telecommunications.

America Cable Systems: *http://www.afcweb.com/acs/*

Designer and manufacturer of integrated modular wiring systems for commercial and industrial applications.

Anixter: *http://www.anixter.com*

Cabling, wire, electrical products, networking infrastructure, connectors, and components.

Archtech Electronics Corporation (AEC): *http://www.archtech.net*

Manufacturer and importer of networking products, cables, switches, racks and cabinets, conduit and raceways, fiber-optic cables and assemblies, bulk cable, and tools and testing equipment.

Arlington Industries: *http://www.aifittings.com*

Electrical, conduit, and networking components, assemblies, fittings, and grounding products.

Atlantic Cable: *http://acicable.com/*

Distributor of telephone and electrical wire and cable products.

Belden: *http://www.belden.com*

One of the world's largest manufacturers of cable, connectors, components, and related equipment, Belden also operates a "cable college" for general training and product education (also includes a terrific online reference, with cable finder, glossary, color code list, and product frequently asked questions [FAQs]).

Berk-Tek: *http://www.berktek.com*

Manufacturer of copper, telecommunications, and fiber-optic cables plus various stuctured cabling systems and related solutions and services.

Bogen Communications International, Inc.: *http://www.bogen.com*

Manufacturer of commercial audio and telecommunications products, special purpose training, remote learning, networked clocks, voice mail, and messaging systems.

Bud Industries: *http://www.budind.com*

Manufacturer of enclosures: cabinet racks, equipment chassis, open racks, and card racks.

Cable Design Technologies (CDT)

- Mohawk: *http://www.mohawk-cdt.com*

  The Mohawk division of CDT specializes in wire and cable technology, including copper, fiber-optic, and hybrids. The company stresses standards compliance in all products and solutions.

- NORDX: *http://www.nordx.com*

  The NORDX division of CDT specializes in cabling manufacture, including central office cabling, coaxial, drop cables, fiber-optics, and broadband solutions. The company also provides design and installation of large-scale corporate or organizational networking infrastructures and solutions.

- Red Hawk: *http://www.red-hawk.com*

  The Red Hawk division of CDT specializes in high-speed, high-bandwidth fiber-optic products for use in voice/data networks, including assemblies, components, test and verification enclosures, and custom products.

- X-Mark: *http://www.metalenclosures.com*

  The X-Mark division of CDT specializes in the design and manufacture of metal enclosures for cabling and networking equipment, and related cable management racks, trays, and so forth.

Cablofil: *http://www.cablofil.com*

Manufacturer of cable trays, troughs, supports, brackets, and related accessories for cable routing and management.

Chatsworth Products Inc.: *http://www.chatsworth.com*

Manufacturer of customer premises mounting products, including relay, server, and equipment racks, and cable management products.

Clauss Company: *http://www.claussco.com*

Manufacturer of hand tools for fiber-optics and cable installation.

CommScope: *http://www.comscope.com*

Manufacturer of high-performance, high-bandwidth cables for telecomm and networking applications, including fiber-optic, coaxial, and twisted-pair.

Damac Products, Inc.: *http://www.damac.com*

Manufacturer of enclosures, cabinets, premises systems, racks, cable management, network furniture, power strips, shelves, and related accessories.

Draka Comteq USA: *http://www.drakausa.com*

Manufacturer of copper cable and copper fiber composites for networking, data communications, and more, plus the Harmony structured cabling system.

Endot Industries: *http://www.endot.com*

Manufacturer of pipe, tubing, riser, raceway, duct, and inner duct products for cable routing and management.

Erico Inc.: *http://www.erico.com*

Manufacturer of fasteners, lighting protection, grounding, surge protection, and power protection products and components.

Geist Manufacturing: *http://www.geistmfg.com*

Manufacturer of power distribution, cord, and cable management products.

General Cable: *http://www.generalcable.com*

Manufacturer of wire and cable products, including data communications, telecommunications, and fiber-optic cables for voice, data, and other high-speed networking applications.

Graybar Electric: *http:// www.graybar.com*

Large international distributor of electrical and communications products and services, including cables, connectors, tools, and test equipment.

Greenlee: *http://www.greenlee.com*

Source of professional grade wiring and cabling tools and test equipment.

Harbour Industries: *http://www.harbourind.com*

Manufacturer of high-performance wire and cable for aerospace, commercial, industrial, LAN, and other markets, including premises networking cabling and coaxial cables.

Harger Lighting and Grounding: *http:// www.harger.com*

Equipment for lightning protection, grounding and bonding, transient voltage and surge suppression, and ground testing.

Harris Corporation: *http://www.harris.com*

Large multinational corporation that offers various communications and networking products and services, including network management, and test and measurement services and equipment.

Hellerman Tyton: *http://www.tyton.com*

Manufacturer and OEM supplier of systems and solutions for connecting, fastening, identifying, organizing, and routing wire and cable, including RSCS and Cat 6 information and technologies.

Hitachi Cable Manchester Inc.: *http://www.hcm.hitachi.com*

A manufacturer of copper, fiber-optic, and electronics cable, including a broad range of premises cable types: Cat 1, Cat 2, Cat 3, Cat 5e, Cat 6, and Cat 6e.

HOLOCOM Networks: *http://www.holocomnetworks.com*

Manufacturer of cabling management solutions including telecommunications enclosures, furniture and office networking applications, and secure networking solutions.

HOMACO: *http://www.homaco.com*

Manufacturer of wiring frames and racks, equipment racks, and cable organizers.

Hubble Premise Wiring: *http:// www.hubbell-premise.com*

Manufacturer and distributor of cross-connect, cable management, and optical fiber products and solutions, along with tools for complete installations, test equipment, and components, mounts, and so forth.

IDEAL Industries: *http://www.idealindustries.com*

Manufacturer of tools, test equipment, connectors, wire management supplies, and installation tools and equipment for cable installers and maintainers.

iNNOdata, LTD: *http://www.ezmt.com*

Manufacturer of compact enclosures and wiring managment solutions for small LAN/telecom network installations for home, SOHO, and small businesses.

Jensen Tools: *http://www.jensentools.com*

Leading distributor of tools, test equipment, and tool kits for computer, electrical, electronic, cabling, networking, and other technical specialties.

KGP Telecommunications: *http://www.kgptel.com*

Distributor of networking and communications tools, cables, supplies, and equipment; also manufactures protection and termination systems for high-speed voice, data, and video applications.

Kitco Fiber Optics: *http://www.kitcofo.com*

Provider of fiber-optic connectorization products and certified training, plus field and consulting services to government and industry.

Klein Tools: *http://www.Kleintools.com*

Producer of professional-quality hand tools for the electrical and telecommunications industry, in addition to construction and mining.

Krone: *http://www.krone.com*

International supplier of copper- and fiber-based cabling systems for telecommunications and data networks, along with network planning and design, installation, implementation, and testing services.

Labor Saving Devices Inc.: *http://lsdinc.com*

Leading designer and manufacturer of advanced specialty tools for wiring and cabling.

L-Com Connectivity Products: *http://www.l-com.com*

Cabling and networking products of all kinds, including Ethernet, wireless, home networking, media converters, power products, and racks and cabinets.

Lemco Tool Corporation: *http://www.lemco-tool.com*

Manufacturer of tools for construction and maintenance of fiber-optic and coaxial cable systems, including aerial components, splicing and installation tools, cable preparation tools, and ground kit tools.

Leviton Voice & Data: *http://www.levitonvoicedata.com*

Distributor of voice, data, and video systems for residential, industrial, and SOHO applications, including performance systems, connectors, wall plates, copper and fiber components, structured media centers, cable management, and tools.

MilesTek Corporation: *http://www.milestek.com*

Supplier of custom-designed telecom products, and distributor of products and custom cable assemblies for home automation, alarm and security systems, and professional audio and video.

Molex Premise Networks: *http://www.molexpn.com*

Manufacturer of twisted-pair and fiber-optic cables, with patching systems, outlet solutions, cable management, tools, and testers as well.

Nelson Firestop Products: *http://www.nelsonfirestop.com*

Manufacturer of firestop products of all kinds including cable management, sealants, compounds, pillows, coatings, composite sheets, putty, wrap strips, and pipe chokes.

Optical Cable Corporation: *http://www.occfiber.com*

Manufacturer of numerous types of fiber-optic cables for high bandwidth transmission of data, video, and audio communications over moderate distances.

Optimum Fiberoptics, Inc.: *http://www.optimumfiberoptics.com*

Fiber-optic cables, cable assemblies, consumables, test equipment, installation tools and supplies, adhesives, components, termination kits, and distribution equipment.

Ortronics: *http://www.ortronics.com*

Supplier of custom, modular networking products including patch panels, data adapters, 110 cross-connect systems, network cabling systems, racks, cabinets, and cable management.

Paladin Tools: *http://www.paladin-tools.com*

Supplier of coaxial, twisted-pair, fiber-optic, and flat telephone cable; hand tools; tool kits; and wire/cable pulling tools. Offices worldwide.

Panduit Corporation: *http://www.panduit.com*

International manufacturer of wiring and communciations products including network cabling systems, cable, and wiring accessories.

PI Manufacturing Corporation: *http://www.pimfg.com*

Manufacturer of networking cabling, connectors, equipment, cable management, power distribution, and voice and data components.

Porta Systems Corporation: *http://www.portasystems.com*

Designer, manufacturer, and marketer of products that connect, protect, test, manage, and administer public and private telecommunications lines and networks, with a special emphasis on network protection and central office products.

Quabbin Wire & Cable: *http://www.quabbin.com*

Manufacturer of networking cables of all kinds; this site offers a library of interesting technology briefs and a cable finder tool.

Rip-Tie, Inc.: *http://www.riptie.com*

Manufacturer of cable ties, straps, tape, and other cable management products.

Roxtec: *http://www.roxtec.com*

Manufacturer of cable seals for all kinds of networking needs from access networks to backbone networks as well as antenna cables, radio link cables, and wave guides for wireless applications.

Seatek Company Inc.: *http://www.seatekco.com*

Manufacturer of cable cutting and stripping equipment and tools.

Siemon: *http://www.siemon.com*

Originally a rubber products company that began in 1903, The Siemon Company has become a popular network cabling solutions provider.

Snake Tray: *http://www.snaketray.com*

Manufacturer of flexible steel cable trays designed to bend easily in any direction, for use overhead, direct wall mount, or below raised floors.

Specified Technologies, Inc. (STI): *http://www.stifirestop.com*

Manufacturer of various firestop products, including sprays, pillows, foam, sealants, mortar, and putty.

Speer Fiber Optics, Inc.: *http://www.speerfiberoptics.com*

Supplier of testing and maintenance equipment for telecom and datacom fiber-optic applications.

Sumitomo Electric Lightwave: *http://www.sumitomoelectric.com*

Manufacturer of fiber-optic cables, interconnects, assemblies, components, management systems, splicers, and cabling systems.

Superior Essex: *http://www.superioressex.com*

Large-scale manufacturer of wire and cable products, including datacom and telecom applications. Customers include OEMs, RBOCs, major independent telcos, and datacom/telecom distributors.

Telect, Inc.: *http://www.telect.com*

Manufacturer of fiber-optic, digital, and analog connectivity products and power distribution panels for data and telecom industries.

Tennet Technologies: *http://www.tennet.com.sg/network_solution/infraplus.htm*

Manufacturer of structured cabling systems for voice/data/video networks and power cabling.

Toner Cable Equipment, Inc.: *http://www.tonercable.com*

Manufacturer of broadband RF systems, modulators, splitters, and connectors, under its own Toner line and for numerous other well-known OEMs (Force Inc. and Sony).

Trompeter Electronic: *http://www.trompeter.com*

Manufacturer of RF interconnect products, cross-connects, tools, and patch panel insertion-controlled interconnection modules, plus custom assemblies for fiber-optic and RF cables.

Tyco Electronics: *http://www.tycoelectronics.com*

Large-scale distributor of interconnects, terminal blocks, relays, circuit protect and wireless devices, cable identification systems, and connectors.

Unique Firestop Products: *http://www.uniquefirestop.com*

Manufacturer of various firestop products, including bulkhead plates and sealants designed for retrofits without reinstallation.

U.S. Conec, Ltd.: *http://www.usconec.com*

Manufacturer of fiber-optic cable and connection hardware for telecommunications and business premises telecommunications.

Wilcom Inc.: *http://www.wilcominc.com*

Manufacturer of test equipment, splitters, repeaters, and other telecom products aimed at fiber-optic, ADSL, and telecom users.

C.4   Electricity and Energy Basics

The following Web sites provide fundamental information on electricity and related topics.

Alpha Wire Co.: *http://www.alphawire.com/*

Analog-to-digital and digital-to-analog conversion: *http://www.cs.tut.fi/~ypsilon/80545/ADDA.html*

Digital Encoding: *http://www.rad.com/networks/1994/digi_enc/main.htm*

Forms of energy: *http://easyweb.easynet.co.uk/~jesus.heals/java/basic/basic.htm*

Fundamentals of electricity: *http://zebu.uoregon.edu/1997/ph161/l2.html*

Tests on conducted and electrical noise: *http://epics.aps.anl.gov/techpub/lsnotes/ls232/ls232.html*

Voltage and static electricity: *http://www.amasci.com/emotor/voltmeas.html*

C.5   Forums

Visit the following forums for an exchange of information on the topics of ATM and Frame Relay:

- ATM Forum: http://www.atmforum.com
- Frame Relay Forum: http://www.frforum.com

C.6   History of Cabling

The following links provide detailed information about the origin and timeline of cabling and related technologies.

- An Introduction to Network Cabling: *http://www.firewall.cx/ cabling_intro.php*
- Analog Telephony: *http://www.privateline.com/TelephoneHistory/ History1.htm*
- AT&T History: *http://www.att.com/corporate/restructure/history. html*
- Emerging cabling technologies: *http://www.corningcablesystems. com/web/college/*
- History of Communications: *http://www.ciolek.com/PAPERS/ milestones.html*
- History of Computer Cabling: *http://ftp.arl.mil/ftp/historic- computers/*
- PC History: *http://www.pc-history.org/*
- The History of Data Cabling: *http://www.datacottage.com/nch/ cablinghist.htm*

C.7   Home Cabling

The following Web sites offer information on cabling systems geared toward SOHO environments.

- HomePNA organization: *http://www.homepna.org*
- HomePNA vendor: *http://www.homepna.com/*

- Powerline Carrier (PLC) systems vendor and home automation: *http://www.smarthome.com*

- X-10 in detail: *http://www.smarthome.com/about_x10.html*

C.8   IEEE 1394: FireWire

The following Web sites offer information on FireWire.

- Overview: *http://developer.apple.com/firewire/overview.html*

- Plastic Optical Fiber (POF) Trade Organization: *http://www.pofto.com/*

C.9   Personal Safety

Ladder Safety Guidelines: *http://www.osha.gov/Publications/osha3124.pdf*

C.10   Planning, Documentation, and Drawings

Division17.Net: *http://www.division17.net*

An initiative aimed at supporting integration of telecommunication systems during all phases of building design and construction, with access to downloads, drawing symbols, and formatting tools and conventions.

C.11   Standards Groups and Bodies

American National Standards Institute (ANSI): *http://www.ansi.org*

An organization involved in the development of technology standards in the United States, ANSI focuses primarily on networking technologies, programming languages, and communications techniques and methods.

Canadian Standards Association (CSA): *http://www.csa.ca*

A not-for-profit membership-based association that serves business, industry, government, and consumers in Canada (and the global marketplace).

Electronic Industries Alliance (EIA): *http://www.eia.org*

A national trade organization that represents over 2,500 members who are mostly manufacturers working in the electronics industry in the United States.

Environmental Protection Agency: *http://www.epa.gov/*

A U.S. federal agency that regulates environmental standards, helping to protect natural resources.

European Commission (EC), CE marking: *http://europa.eu.int/comm/ enterprise/faq/ce-mark.htm*

A description of how to reproduce the CE mark.

European Committee for Electrotechnical Standardization (CENELEC): *http://www.cenelec.org/Cenelec/Homepage.htm*

A nonprofit standards body made up of National Electrotechnical Committees from 28 European countries.

Federal Communications Commission (FCC): *http://www.fcc.gov/*

A U.S. federal agency that regulates interstate and international radio, television, wire, satellite, and cable communications.

International Code Council (ICC): *http://www.iccsafe.org/*

A nonprofit organization that offers, among other services and items, a listing of all of the basic building codes in effect in the United States and internationally.

International Electrotechnical Commission (IEC): *http://www.iec.ch*

A Swiss-based, global organization that creates and publishes international standards for electronic, electrical, and related technologies.

International Organization for Standardization (aka ISO): *http://www. iso.ch*

An organization that defines, maintains, and promotes global standards for the worldwide networking community.

International Telecommunications Union-Telecommunication Standardization Sector (ITU-T): *http:// www.itu.org*

An organization that manages standards that relate to various networking topics and technologies, including communications, telecommunications, and outright networking services.

Institute of Electrical and Electronics Engineers (IEEE): *http://www.ieee.org*

A nonprofit, professional association that acts as a technical authority on a broad range of topics and technologies that range from computer engineering and telecommunications to biomedicine. As part of its focus, the IEEE produces and maintains a large body of networking-related specifications and standards. Work that originates in the IEEE is often shared with ANSI, and in turn may be shared with ISO. This explains why many important IEEE networking standards are ANSI and ISO standards as well. The most important IEEE networking project is the collection of standards known as the 802 project.

Japanese Industrial Standards Committee (JISC): *http://www.jisc.go.jp/eng/*

An organization that plays a major role in helping to form industrial standards in Japan.

Japanese Standards Association (JSA): *http://www.jsa.or.jp/default_english.asp*

An organization that strives to "educate the public regarding the standardization and unification of industrial standards, and thereby to contribute to the improvement of technology and the enhancement of production efficiency."

National Electrical Code (NEC): *http://www.nfpa.org/categoryList.asp?categoryID=192&URL=Publications/necdigest*

Hosted by *NEC Magazine*, a publication of NFPA, the most current NEC is available for purchase from this Web site. You can download the previous edition for free.

National Fire Protection Association (NFPA): *http://www.nfpa.org/*

A nonprofit organization whose mission is to reduce fires and other hazards worldwide via consensus codes and standards, training, education, and research activities.

National Institute for Occupational Safety and Health (NIOSH): *http://www.cdc.gov/niosh/homepage.html*

An arm of the Centers for Disease Control, the NIOSH is a U.S. federal agency that conducts research focused on preventing workplace injuries and illnesses.

Occupational Safety and Health Administration (OSHA): *http:// www.osha.gov*

An official U.S. government body that regulates and monitors health and safety issues as they relate to the workplace.

Telecommunications Industry Association (TIA): *http:// www.tiaonline.org*

A leading, U.S.-based nonprofit trade association that serves communications and information technology industries.

Underwriters Laboratories (UL): *http://www.ul.com*

An independent, not-for-profit product safety testing and certification organization that aims to warrant electrical and electronic equipment safe for use in the home or in the workplace.

- Standards Listing: *http://ulstandardsinfonet.ul.com/*
- Information Technology: *http://www.ul.com/ite/*
- UL marks: *http://www.ul.com/mark/*
- Power-Limited Circuit Cables (UL 13): *http://ulstandardsinfonet.ul.com/scopes/0013.html*
- Communications Cables (UL 444): *http://ulstandardsinfonet.ul.com/scopes/0444.html*

C.12   Structured Cabling Guidelines

BICSI Telecommunications Dictionary (1999), BICSI: A Telecommunications Association, Tampa, FL.

Construction Specifications Institute (CSI): *http://www.csinet.org*

Provides MasterFormat definition and information.

Newton, Harry (1999), *Newton's Telecom Dictionary: 15th Expanded Edition*, Miller Freeman, Inc. New York, NY.

The Telecommunications Distribution Methods Manual: 10th Edition (2003), BICSI: A Telecommunications Association, Tampa, FL.

C.13   Training Providers

Anixter: *http://www.anixter.com*

Preparation for BICSI Installer Levels 1 and 2, and Technician.

CET Networking Education: *http://www.cetweb.com*

Preparation for Registered Communications Distribution Designer (RCDD), LAN, and outside plant (OSP) exams.

Communications Supply Corporation (CSC): *http://www.gocsc.com/training.html*

Data communication products distributor.

EffecTec: *http://www.effectec.com*

Preparation for BICSI.

Fiber Network Training: *http://www.f-n-t.com*

Fiber-optic and communications training.

Light Brigade: *http://www.lightbrigade.com*

Fiber-optic training.

New Jersey Institute of Technology: *http://cpe.njit.edu/cable*

Cabling e-learning program.

Servamatic: *http://www.servamatic.net*

Residential cabling theory and hands-on learning.

C.14   Wireless Systems

802.1x: *http://www.ieee802.org/1/pages/802.1x.html*

How Wireless Works: *http://computer.howstuffworks.com/wireless-network.htm*

APPENDIX D

Glossary

0–9

802: A standard series for network architectures and data formats.

A

access floors: A system consisting of completely removable and interchangeable floor panels supported on adjustable pedestals or stringers (or both) to allow access to the areas beneath.

alternating current (AC): A type of current that switches back and forth at a rate of 50 to 60 Hz (cycles per second).

American National Standards Institute (ANSI): The North American standards body for most software and hardware areas of the computer industry.

American Telephone and Telegraph (AT&T) Corporation: Also known as "Ma Bell," it was the telecommunications provider that dominated the telephony industry until the 1980s.

American Wire Gauge (AWG): The standard for cabling dealing with cable diameter and electrical resistance.

amps: A unit of measurement for electrical current.

analog: A type of signal that changes shape continuously even though it has a steady source.

as-built drawing: A schematic of the cabling system representative of the system as installed.

Asynchronous Transfer Mode (ATM): A type of service used for high-speed network switching using 53-byte cells.

attachment unit interface (AUI): A type of coaxial cable connector used in Thicknet Ethernet (10Base5) networks.

attenuation: The reduction of signal strength over the length of a cable. Attenuation is cumulative over the length of a cable run, and is measured in decibels.

B

backbone: Cabling configuration that interconnects the major features in a structured cabling environment—entrance facilities, telecommunications rooms, and equipment rooms—and includes the raceways, intermediate and main cross-connects, patch panels, cable supports, and the actual cables themselves such as coaxial, twisted-pair, and/or fiber-optic. Also referred to as *backbone cabling*.

bandwidth: The data capacity of a link; specifically, the width of a band of electromagnetic frequencies.

baseband: A single, unmultiplexed channel dedicated to sending a single signal.

baseline: In project management, a set of original tasks, start and finish dates, durations, and work and cost estimates that you save after you have completely fine-tuned your project plan before the project begins. It is then used as the primary reference point to which you measure changes to your project.

basic service set (BSS): Uses one SSID (network name). An access point acts as the central point for all devices wanting to participate on the network.

Bluetooth: A wireless protocol used to communicate from one device to another in a small location where the devices being connected are in the same general proximity from each other. Bluetooth uses the 2.4 GHz band to communicate.

bonding: The creation of a relationship between two devices that need to seek ground, and the method used to produce a good electrical contact between multiple metallic parts for the purpose of sharing ground.

British Naval Connector (BNC): A type of coaxial cable connector used in video and Thinnet Ethernet (10Base2) networks.

broadband: A multiplexed channel that can send more than one signal at a time over the same media.

buffer coating: A durable outer coating that protects the core and cladding as well as adds durability.

building distributor (BD): A secondary level to backbone cabling; also called intermediate cross-connect (ICC).

C

cable caster: A device similar in operation to a fishing rod and reel that is used to pull cable over a long open span like a drop ceiling.

cable jacket: The surrounding insulator of a cable; the first line of a cable's defense from the outer elements.

cable plant: The collection of bundled cabling inside a commercial building.

cable tester: A multi-function cable testing device that can provide information regarding the length of a cable, and faults in a cable (short circuits or open circuits) as well as the distance to a fault.

cable tray: A support mechanism used to route and support telecommunications cable or power cable. It is typically trough-shaped, allowing placement of cables within the sides over its entire length.

campus distributor (CD): The main communications center in a large cabling infrastructure; also called the main data frame/facility, main closet, or main cross-connect.

Canadian Standards Association (CSA): A standards body responsible for developing the Canadian electric standard.

cancellation effect: The process by which the magnetic fields of one wire are crossed with the fields of another, limiting the exposure of any one wire to the other.

capacitance: The limit of the amount of signal the cable can carry; often depends on the proximity of the connectors.

Category 1: Twisted-pair cabling used exclusively for voice-based communications; can support transmission speeds up to 1 Mbps.

Category 2: A type of twisted-pair cabling used in older IBM networks; can support transmission speeds up to 4 Mbps.

Category 3: Formerly the minimum standard for voice and data networks; a type of twisted-pair cabling that supports transmission speeds up to 10 Mbps.

Category 4: A type of twisted-pair cabling used to support the 16-Mbps Token Ring networks.

Category 5: The most commonly installed medium; a type of twisted-pair cabling designed to support Fast Ethernet and other networks.

Category 5e: Enhanced Category 5; a recent standard designed to support all high-speed voice and data networks.

Category 6: A standard designed to support Gigabit Ethernet and other technologies over greater distances.

central network architecture (CNA): A cabling concept that deploys fiber optics for data transmission, and then either Category 3 copper cable, if an actual physical cable for voice transmission, or Voice over IP (VoIP) technology for voice transmission. Like DNA, this architecture provides connectivity between intrabuilding and interbuilding designs as well as vertical connections between floors and connections between closets/distributors.

certification meter: A device used to measure compliance with published cabling standards.

cladding: A type of shielding used in cables.

cladding: Material surrounding the core of the fiber-optic cable composed of a material with a different refractive index to provide signal reflection.

closed-circuit TVs (CCTV): Does not broadcast TV signals, but transmits them over a closed circuit through a cable or wireless transmitter and receiver; used often in security solutions and home security systems.

coax stripper: A tool that is used to expose the core, grounding, and protective sheathing of a coaxial cable.

coaxial cable: A type of electrical cabling, often called "coax," consisting of a single solid or stranded core surrounded by shielding and inner and outer insulation; found in both voice and data networks.

community antenna TV (CATV): A technology used for pay TV services and high-speed data services using a hybrid fiber-coaxial technology.

composite video: A type of video where the chrominance, luminance, and synching information are all carried across the same wire.

concentricity: Signal loss in a fiber-optic cable due to the core not being perfectly centered in the cladding.

conductor: A material through which electricity can flow easily; metals (such as copper and aluminum) are excellent conductors, as are humans due to high water content.

conduit: Rigid or flexible metallic or non-metallic circular tubing used as an alternative pathway for cables.

connector: A small component, usually made of plastic metal, located at one end of a cable, allowing you to plug it into a NIC, jack, etc.

consolidation point (CP): Used in zone cabling, the location of the inter-connection between horizontal cables that extend from building pathways and horizontal cables that extend into work area pathways.

core: The inner conductive element of a cable over which data are transmitted; the fiber-optic cable core is made of high-quality glass and is very small in diameter.

critical angle: The angle at which internal reflection will not occur.

critical path: In project management, a series of tasks that dictates the calculated completion date of the project.

cross-connect: A location within an infrastructure that allows the connection of new or existing cables. This is inclusive of, but not limited to, optical jumpers, copper patch cords, and termination blocks such as 110-style devices.

crosstalk: The coupling of transmission circuits (e.g., wire pairs in cable) that results in undesired signals, impeding the flow of data across a cable.

current: The speed or flow rate of electricity on a circuit.

D

data communications: The transmission of data signals over a transmission medium, which requires interfaced devices to exchange data.

Data Over Cable Service Interface Specification (DOCSIS): The standard for CATV data networks. DOCSIS versions include 1.1 and 2.0.

de facto: A standard that originated privately and emerged on a widespread basis through market acceptance.

de jure: A legislated or official standard, usually the product of an independent standards body or organization.

decibels (dB): In regard to signals, a measurement of attenuation.

Deutsche Industrie Norm (DIN): A type of coaxial connector used for European land mobile radio applications and analog cellular systems.

dielectric material: Foam or plastic insulation that spaces between the central core and the shielding inside of coaxial cabling.

digital subscriber line (DSL): A technology that offers speed improvements to other network access methods using the same physical media available to legacy voice services.

digital: A signal type that is discrete (changes from one state to another [on/off]).

direct broadcast satellite (DBS): An alternative to most current cable-based television systems. DBS provides television programming directly from satellites using small, home-mounted satellite dishes.

direct current (DC): Current that flows only in one direction (negative to positive).

dispersion: The scattering of light and signal on a fiber cable.

distributed network architecture (DNA): The traditional cabling concept that deploys both voice and data cable from the campus distributor to the floor distributor over various cable types. Like CNA, this architecture provides connectivity between intrabuilding and interbuilding designs, as well as vertical connections between floors and connections between closets/distributors.

drawing: A process of wire creation in which the metal is pulled under tension to form a wire.

dual-shielded coaxial cable: A type of coaxial cable that differs from standard coaxial cable in that it has two shields covering the dielectric insulator.

duplex: In terms of communications (also called bidirectional communication), transmissions that can move in both directions along the same communication path. Two types of duplex transmissions are half duplex and full duplex.

E

earthing: Another term for grounding, the creation of a conducting connection to the earth or a body that serves as the earth. Earthing is a commonly used term outside of the United States.

elastomers: Insulation materials that are plastic and very flexible.

electromagnetic interference (EMI): Any electromagnetic disturbance that disrupts, impedes, or degrades the electromagnetic field of another device by being in close proximity to it.

electrostatic discharge (ESD): The release of static electricity when two objects come in contact; ESD can damage electronic equipment and impair electrical circuitry, resulting in complete or intermittent failures of that equipment.

end gap: A space between two fiber-optic cables that are spliced or between a fiber-optic cable and a terminating connector. End gap forces light to jump across the gap in the space. In addition, the air gap between the fiber ends causes a difference in refractive index, which can cause some of the signal to reflect back to the source.

entrance facility: The point of interconnection between the network demarcation point and/or the campus backbone and the intrabuilding wiring. The entrance facility typically includes overvoltage protection and connecting hardware for the transition between outdoor and indoor cable.

equipment room (ER): A centralized space that houses telecommunications equipment that serves the occupants of a building; distinct from the telecommunications room (TR) due to the nature and complexity of the equipment inside.

exposure: A telecommunications cable's susceptibility to introduced electrical currents not used to transmit the telecommunications signal itself.

extended service set (ESS): In an ESS, the BSS access points communicate with each other to forward traffic from one BSS to another, which allows for devices to communicate with other devices via the ESS.

extrusion: A process of wire spinning in which the wire is stretched to form a thin strand through pushing.

F

Federal Communications Commission (FCC): An independent U.S. government agency that answers directly to Congress, and is responsible for regulating interstate and international radio, television, wire, satellite, and cable communications.

ferrule: A metal or fiberglass sleeve with a barrel-like appearance that is used when terminating coaxial and some fiber-optic cables. A ferrule adds strength and stability to a cable connector.

Fiber Distributed Data Interface (FDDI): A dual-ring fiber physical topology using token-passing logical topology.

fiber-optic cabling: A type of cabling designed to carry coherent light as either pulses or a continuous modulated stream; all fiber is made of the same inner core of pure glass.

fiber optics: A technology that uses glass or plastic fibers (also called threads or optical waveguides) to transmit data, instead of metal-cored cables. Fiber-optic cables have more bandwidth than metal cables and can transmit data digitally, but they are also much more expensive and fragile. Most telcos, however, are gradually replacing their regular telephone lines with fiber-optic cables.

finish phase: The phase of cable installation that includes testing and certifying the cabling.

firestops: Barriers in a structure designed to prevent the vertical or horizontal spread of flames.

fish tape: A tool that is used to pull cable through a hollow wall, floor, or ceiling.

floor distributor (FD): The horizontal area that houses the horizontal cable, its applicable connecting hardware, and other specific equipment.

fluoropolymers: Plenum-rated materials that are required for fire-rated insulators.

frequency: (1) The number of complete cycles per second in alternating current direction for an oscillating or varying electrical current; hertz is the unit of measurement for frequency. (2) A particular range of the radio spectrum.

frequency division multiplexing (FDM): A technique used to allow for the transmission of data across the available bandwidth of a circuit to be divided by frequency into narrower bands, each used for a separate voice or data transmission channel. FDM allows for more than one conversation to be carried on a single circuit.

full-duplex: Communications in which data can travel in both directions simultaneously, with the sender and receiver each using half of the communication path.

G

Gantt chart: A type of chart (named for Henry Laurence Gantt) often used in planning and tracking a project; consists of a table of project task information and a bar chart that graphically displays a project schedule, depicting progress in relation to time.

gigabit interface converter (GBIC): A transceiver that converts one type of connection and signaling to another.

gopher pole: A lightweight, telescoping pole with a hook on one end, used for pulling or moving cabling in hard-to-reach places.

ground: A conductive path or connection, which is typically a grounding rod, that carries an electrical charge to the earth or some other body that absorbs the charge.

grounding: The procedure used to carry an electrical charge to ground (earth ground) through a conductive path, which is typically a grounding rod. Electricity is always looking for a path to travel and the most common path to disperse unwanted or unneeded electricity is ground.

grounding equalizer (GE): Formerly known as the TBB interconnecting bonding conductor; connects multiple TBBs together in a multi-story building.

H

half-duplex: Communications in which data can move in both directions, but in only one direction at a time; senders on each end are able to use the entire communication path.

hertz (Hz): A unit of measurement that represents cycles per second.

Home Phoneline Networking Alliance (HomePNA): The standard for using copper phone lines within your home as a way to connect network devices; is based on a set of standards that enables voice and data transmissions over a home's existing telephone cabling.

Home Radio Frequency (HomeRF): A WLAN standard that supports up to 10 Mbps in the 2.4 GHz band. HomeRf is not interoperable with 802.11 or Bluetooth.

home run: Horizontal cabling that runs from the equipment room to each TC, tying all cabling to the backbone cable; describes the central point (usually in a centralized closet) where all of the wiring in the SOHO terminates.

horizontal cabling: The connecting station cable that links the floor distributor to an individual workstation.

I

impedance: A type of resistance that is intrinsic to the cable itself, and is the sum of inductive reactance and capacitive reactance.

induction: Stray voltage introduced onto telecommunications cables from higher voltage lines that run in parallel.

infrared: A radiated technology that uses infrared light; for point-to-point connections.

Institute of Electrical and Electronics Engineers (IEEE): A nonprofit standards union that is a leading authority in technical areas ranging from computer engineering, biomedical technologies, and communications, to electrical power, aerospace, and consumer electronics.

insulation displacement contact (IDC): The design characteristics of an RJ-45 data jack.

insulation: The outer material that makes up the cable jacket.

insulator: A material, such as special rubber, plastic, or glass, that does not conduct electricity well, but covers a material that does conduct electricity. In cabling, insulators keep electricity from leaving a wire that is in a voice or data cable.

Integrated Services Digital Network (ISDN): A type of service that provides circuit-switched video, voice, and data access.

interference: The external elements that can impede the transmission of information across a cable.

intermediate data frame/facility (IDF): Another name for horizontal closet or cabling, horizontal cross-connect, telecommunications room, and floor distributor; refers to the horizontal area that houses the horizontal cable and its applicable connecting hardware and other specific equipment.

International Electrotechnical Committee (IEC): Founded in 1906, an organization that publishes international standards associated with electrical, electronic, and other related topologies.

International Organization for Standardization (ISO): An organization responsible for developing standards and reference models for member nations and organizations for communication and other commercial areas.

International Telecommunications Union (ITU-T): A U.N.-governed organization that is responsible for international communications standards.

ion: An atom with more or fewer electrons than normal.

J

jack: The female component in a wall plate, patch panel, or other device.

junction box: A metal or plastic box used as an access for cable or wire typically seen in conduit or underfloor raceway systems.

L

Lightguide Interconnection Unit (LIU): A chassis for enclosing fiber-optic patch panels used to maintain cable integrity.

loose tube: A fiber-optic cable type that encloses many individual strands of fiber-optic cable (core and cladding) in a single sheath.

M

MasterFormat: Created by the Construction Specifications Institute, a list of numbers and titles for organizing information about construction requirements, products, and activities into a standard sequence.

medium: The manner in which information is sent within a communication system.

micron: One millionth of a meter (10^{-6}), equivalent to one millimeter.

MM fiber: Short for multimode optical fiber.

mode: The path light takes as it travels the length of a fiber-optic cable.

modulation: The process of translating information into a continuous signal.

Morse code: The first communications protocol, or language, to be used across a cabled network.

moves, adds, and changes (MACs): Upgrades or modifications made to an existing cabling infrastructure.

multimeter: An inexpensive device that measures resistance and voltages on a cable.

multipair cable: Twenty-five pair cable.

multi-user telecommunication outlet assembly (MUTOA): A device that can service 12 or more work area locations, equalizing the cabling distances to each work area and uniformly locating each unit for troubleshooting and overall system documentation purposes.

N

National Electrical Code (NEC): A code established to protect persons from the risks of using and working with electricity, which if improperly handled, could result in the damage of property and the loss of life; NEC is sponsored by the NFPA.

National Fire Protection Association (NFPA): The association responsible for the establishment of guidelines to prevent fire.

National Institute for Occupational Safety and Health (NIOSH): The U.S. agency responsible for attempting to prevent workplace injury through research and prevention. NIOSH is a division of the Centers for Disease Control and Prevention (CDC).

near-end cross talk (NEXT): The leaking of an electrical signal between two pairs of copper cabling.

network interface card (NIC): A personal computer (PC)-based card installed to create an interface to the network. Combo cards allow you to use the card with either an RJ-45 or coaxial BNC connection. NICs allow you to insert the physical connection (the cable itself) into the NIC installed in the PC.

noise: Unwanted electrical or radio signals on a cable.

O

Occupational Safety and Health Administration (OSHA): A U.S. Department of Labor organization responsible for regulating workplace safety.

ohm: A unit of measurement for electrical resistance.

Ohm's law: A property that states the mathematical relationship between electrical voltage, resistance, and current: $V = IR$.

Open Systems Interconnection (OSI) reference model: An international protocol stack and reference model used in education and protocol development.

optical free-space: The atmosphere or the vacuum of space where wireless signals cross.

optical time domain reflectometer (OTDR): A device that measures the length of a fiber-optic cable as well as any faults in the cable and the distance to the fault.

P

per-drop pricing: A method of charging a customer for cabling installation services given as a fixed amount for each location of cable installed.

phased cutover: A method of activating installed cables in small groups rather than all at once; useful for remodel/retrofit situations where the end user requires the system to be active during construction.

photodiode: An electronic component that is sensitive to light. Light signals hit the photodiode and are converted to electrical pulses.

photon: A particle of light.

pins: The areas on a connector into which cable wires are fed.

plain old telephone service (POTS): The legacy phone service designed to carry primarily voice traffic.

plastic optical fiber (POF): A type of optical fiber that uses polymethylmethacrylate (PMMA)—a general-purpose resin—as the core material and fluorinated polymers for the clad material.

plenum-rated cable: Contains a wire coating that burns at a much higher temperature and emits fewer fumes than non-plenum–rated cabling. The NEC requires that plenum-rated cabling be used in plenums (the space above the ceiling tiles and below the next story into which two or more air ducts run).

plug: The male component crimped onto the end of a cable.

point of demarcation: The point at which an outside service provider's equipment/cabling, such as that provided by a telecommunications utility, meets the entrance facility of the premises.

polyethylene: The most common insulation material used in coaxial cable insulation jackets.

polyvinyl chloride (PVC): A common thermoplastic insulator.

power: The measurement of rate at which work can get done using the electricity available.

Powerline Carrier (PLC) systems: Another way of connecting devices in a residential home using the existing high-voltage power wiring (60 Hz/120 Volt AC). PLC systems normally use the X-10 protocol to communicate.

premises wiring (or cabling): Customer-owned communications and power transmission media, including copper and fiber-optic cabling, installed within or between buildings. Premises wiring begins at the load end of a service drop, may consist of backbone, horizontal, and vertical wiring, and extends to a user's computer.

push/pull rod: A series of narrow lightweight rods in which cable is threaded and then pushed/pulled through hard-to-access areas such as wall and ceiling drops. Rods are attached to a base pole, which is held by the cable technician.

R

radio frequency (RF): 1. Regarding radio wave propogation, refers to any frequency within the electromagnetic spectrum. Applying an RF current to an antenna creates an electromagnetic field through which communications can take place. 2. Another term for coaxial cabling; RF cable designations usually apply to coaxial cabling used in RF applications.

radio frequency interference (RFI): Electromagnetic interference at radio frequencies, often occurring at a specific frequency or within a specific range of frequencies.

Radio frequency Government (RG): The designations used for coaxial cabling that were derived from older military standards. Formerly referred to as *Radio Guide*.

radio spectrum: The complete spectrum of electromagnetic frequencies used for communications.

reactance: Opposition to the flow of alternating current due to capacitance or inductance.

reflection: Return of light or sound waves from a surface (i.e., image in a mirror).

refraction: A ray of light or energy that is deflected from a straight path when passing from one medium (air) through a different medium (water).

refractive index: The amount of refraction a given element will allow.

refractive index: The property of optical materials that relates to the velocity of light in the material.

regional Bell operating carriers (RBOCs): The seven companies that were formed as the result of the 1983 breakup of AT&T.

registered jack (RJ): A connector type used by twisted-pair cabling.

Request for Proposal (RFP): A document used when a buyer is looking for costs on a product or service. An RFP is in-depth and detailed, while an RFQ is usually more exploratory and less defined.

Request for Quote (RFQ): A document used when a buyer is looking for costs on a product or service; similar to an RFP, but usually more exploratory and less defined.

resistance: The opposition of the flow of electricity; represented as R.

RJ-11 connector: Abbreviation for Registered Jack-11, which is a four- or six-wire connector used primarily to connect telephone equipment in the United States, but can also be used (although not recommended) for data.

RJ-45 connector: Short for Registered Jack-45, which is an eight-wire connector used commonly to connect computers on a data network.

rough-in phase: The phase of cable installation that includes identifying the staging area and work teams, installing conduit, and pulling cable from the telecommunications room to the individual outlets throughout the location.

S

Scope of Work (SOW): A document outlining what a project will entail; can include details on either products, services, or both.

screened twisted-pair (ScTP): A type of twisted-pair cabling that contains an outer screen or shield.

service set identifier (SSID): Specifies which 802.11b network you are attempting to join. This is how you participate in a wireless network.

shielded twisted-pair (STP): A type of twisted-pair cabling that uses inner and outer shielding for protection against electromagnetic interference.

signal: The force of energy transmitted across a communication medium; it represents the information being communicated.

simplex: Transmissions that travel in only one direction, allowing the sender to use the whole communication path for the transmission; examples of simplex transmissions include broadcast television and stock ticker-tape machines.

skin effect: The phenomenon that occurs when the current migrates out to the insulation (skin) of the conductor, rather than traveling through the core.

SM fiber: Abbreviation for single-mode optical fiber.

Small Office Home Office (SOHO): A home office used for doing work; connectivity is achieved through wired and wireless systems.

sneak current: Voltages that are too low for the primary protectors to respond to; alternatively, a current drawn by faulty equipment that overheats the cabling (can be a fire threat and/or disable data and voice transmissions).

solid core: A core consisting of a dense single wire.

spread spectrum: A type of wireless communication in which the frequency of the transmitted signal is varied, resulting in a greater bandwidth than if the signal was not varied; spread spectrum signals are usually specified in MHz or GHz.

standard: Guidelines representing the layout, operation, and/or the physical characteristics of a particular technology or the best practices for implementing a technology.

standards bodies: Organizations formed to establish and maintain rules and guidelines.

stranded core: Tiny fibers of copper packed together to form a single conductive core.

structured cabling system: Guidelines for such things as wiring, connections, terminations, supporting equipment, insulation, distribution centers, and cable installation methods.

structured cabling: The methodology used for cabling in the commercial and residential cabling industries.

stud finder: A tool used to identify where wooden framing (studs) are located in a hollow wall.

Subminiature A (SMA): A type of coaxial connector designed for semi-rigid, small diameter metal jacketed cable.

Synchronous Optical Network (SONET): A subscriber WAN service that aggregates multiple signaling types into a single large pipe. Developed in the late 1980s, SONET represents the North American Public Optical network.

T

T-1: A high-speed digital circuit equal to 24 DS-0 (64-kbps) channels, or 1.544 Mbps.

T-3: A high-speed digital circuit equal to 672 DS-0 channels (44.736 Mbps), or 28 T-1s.

T568-A: The first of two ANSI/TIA/EIA standards for twisted-pair color-coding.

T568-B: The second of two ANSI/TIA/EIA standards for twisted-pair color-coding. T568-B is most commonly used today.

telecommunications bonding backbone (TBB): Connects the TMGB to all the TGBs. If multiple TBBs are included by design, they are connected via GEs. The TBBs in a multi-story building connect with a GE at the top floor and at every third floor in-between, at a minimum.

telecommunications entrance facility: See *entrance facility*.

telecommunications grounding busbar (TGB): A miniature version of the TMGB.

Telecommunications Industry Association/Electronic Industries Alliance (TIA/EIA): North American organizations that establish and govern most of the standards for voice and data cabling.

telecommunications main grounding busbar (TMGB): Represents the separation between the electrical grounding system and the telecommunications grounding system. The TMGB is a pre-drilled, copper busbar, ideally located in a building's telecommunications entrance facility or placed to minimize the length of the grounding conductor. The TMGB serves as the grounding point for any telecommunications equipment within the room and as a central attachment for the TBB.

telecommunications room (TR): A room that houses telecommunications wiring and wiring equipment usually containing one or more cross-connects.

telegraph: The first communications device, patented in 1837; allowed the transmission of messages over electronic cables.

telephone test set: A tool used by telecomm engineers to test phone circuits.

telephony: The process of converting sound or data into an electrical signal and transmitting the signal over electrical cabling.

termination: The process of property terminating a cable to provide stable communications.

thermoplastics: The most common type of insulation material.

Thicknet: The type of coaxial cabling used in 10Base5 Ethernet networks, and once was the recommended cabling for network backbones. Thicknet is easily known by its bright yellow insulation jacket.

Thinnet: The type of coaxial cabling used in 10Base2 Ethernet networks. Thinnet cabling is more flexible than Thicknet but signals do not travel as great a distance.

Threaded Neill Concelman (TNC): A type of coaxial cable connector that is simply an enhanced version of the BNC connector; instead of a twist-crimp-lock, the interface is threaded.

TIA/EIA-568: Commercial building standards for telecommunications wiring.

TIA/EIA-RS-232: An interface standard for serial connections.

tight buffered: A fiber-optic cable type that encloses a single strand of fiber-optic cable (core and cladding) in a sheath.

time domain reflectometer (TDR): A device that measures the length of a copper cable as well as any faults in the cable and the distance to the fault.

tone and probe set: A test tool that is used to identify a pair of copper wires; usually used in telecom circuits.

total internal reflection: The principal that light will bounce off the junction between core and cladding with no signal loss.

transformer: A mechanism that converts and reduces power on a power line.

Transmission Control Protocol/Internet Protocol (TCP/IP): The protocol suite used for the Internet and for internal intranets.

transmission mode: The specific order in which data are transferred across whatever physical media is being used, whether the data communication is simplex or duplex. Transmission modes are designed for digital communications, but can be transferred to and from analog signals at any point, depending upon the underlying architecture.

triaxial: A type of coaxial cabling that has a single core like standard coaxial cabling and has two shields like dual-shielded coaxial; uses both the inner conductor and the inner shield to transmit information while the outer shield provides ground potential.

trim-out phase: The phase of cable installation that involves cutting the cable to the desired length, stripping the outer sheathing, crimping on a connector, and terminating the cable.

turnaround time: In half-duplex communications, the time it takes to hand control over to the other side. Turnaround time can be crucial in certain transmissions such as CB radio transmissions.

twinaxial: A type of coaxial cabling that contains two insulated conductors either running in parallel or twisted together having a common shield and insulator jacket.

twisted-pair cable: The predominating cable type in voice and baseband data networks; composed of pairs of twisted cabling.

twisted-pair cabling: The most widely used type of network cabling, using pairs of twisted wires. This twisting eliminates cross-talk and other signal degradation.

Type F: A common coaxial connector used in video, radio-frequency, and audio-visual entertainment systems.

Type N: One of the oldest coaxial cable connectors in use; type N connectors are waterproof and have good signal stability.

U

UL 13: The safety standard for power-limited circuit cable.

UL 444: The safety standard for communications cable.

Ultra High Frequency (UHF): A very common type of coaxial connector developed for RF communications above 50 MHz.

underfloor ducts: A floor distribution method using a series of metal distribution channels, often embedded in concrete, for placing cables. Underfloor raceway is another term for these structures.

Underwriters Laboratories (UL), Inc.: An independent, not-for-profit organization that is responsible for the testing and rating of electrical and electronic products for safety.

unshielded twisted-pair (UTP): The most common type of twisted-pair cabling used for voice and data networks; UTP relies solely on the cancellation effect.

V

vendor: A provider or manufacturer of a technical product or service.

vendor neutral: A design not related to a specific vendor; unilateral interoperability of a product between an industry's vendors.

vendor proprietary: A design related to a specific vendor; product is usually incompatible with another vendor's products.

volt: A unit of measurement for voltage.

voltage: The force used to create a flow of current when a closed circuit is connected between any two points; also, electromagnetic force or pressure, represented as V or E.

W

watt: A unit of measurement for power.

wire finder: A tool used to locate the path a cable follows through walls, floors, ceilings, and underground.

wire spacing: The carefully controlled spacing between conductors and shields.

wired media: Also known as cabled media, in which information travels through a contained physical path or circuit.

wireless media: Also known as radiated media, in which information travels freely through the air or the vacuum of space.

wireless signals: Electromagnetic signals that travel over free space and do not require a cable.

X

X-10: the leading home automation protocol used today. With X-10 controllers, which send digital signals over home wiring systems to receivers that are plugged into the SOHO wall outlets, X-10 is transmitted over the home wiring at 60 bps (bits per second).

Z

zone of protection: An area that is under or nearly under a lightning protection system.

Index

Outstanding New Titles:

Computer Science Illuminated, Second Edition
Nell Dale and John Lewis
ISBN: 0-7637-2626-5
©2004

Introduction to Programming with Visual Basic .NET
Gary J. Bronson and David Rosenthal
ISBN: 0-7637-2478-5
©2005

Information Security Illuminated
Michael G. Solomon and Mike Chapple
ISBN: 0-7637-2677-X
©2005

The Tao of Computing
Henry Walker
ISBN: 0-7637-2552-8
©2005

Databases Illuminated
Catherine Ricardo
ISBN: 0 7637 3314 8
©2004

Foundations of Algorithms Using Java Pseudocode
Richard Neapolitan and Kumarss Naimipour
ISBN: 0-7637-2129-8
©2004

Artificial Intelligence Illuminated
Ben Coppin
ISBN: 0-7637-3230-3
©2004

Programming and Problem Solving with C++, Fourth Edition
Nell Dale and Chip Weems
ISBN: 0-7637-0798-8
©2004

Java 5 Illuminated
Julie Anderson and Herve Franceschi
ISBN: 0-7637-1667-7
©2005

Programming in C++, Third Edition
Nell Dale and Chip Weems
ISBN: 0-7637-3234-6
©2005

Computer Networking Illuminated
Diane Barrett and Todd King
ISBN: 0-7637-2676-1
©2005

Computer Systems, Third Edition
J. Stanley Warford
ISBN: 0-7637-3239-7
©2005

C#.NET Illuminated
Art Gittleman
ISBN: 0-7637-2593-5
©2005

Computer Graphics: Theory into Practice
Jeffrey McConnell
ISBN: 0-7637-2250-2
©2006

Ethics, Computing, and Genomics
Herman Tavani
ISBN: 0-7637-3620-1
©2006

http://www.jbpub.com/

JONES AND BARTLETT
PUBLISHERS
BOSTON   TORONTO   LONDON   SINGAPORE

1.800.832.0034

Take Your Courses to the Next Level

Turn the page to preview new and forthcoming titles in Computer Science and Math from Jones and Bartlett…

Providing solutions for students and educators in the following disciplines:

- Introductory Computer Science
- Java
- C++
- Databases
- C#
- Data Structures

- Algorithms
- Network Security
- Software Engineering
- Discrete Mathematics
- Engineering Mathematics
- Complex Analysis

Please visit http://computerscience.jbpub.com/ and http://math.jbpub.com/ to learn more about our exciting publishing programs in these disciplines.

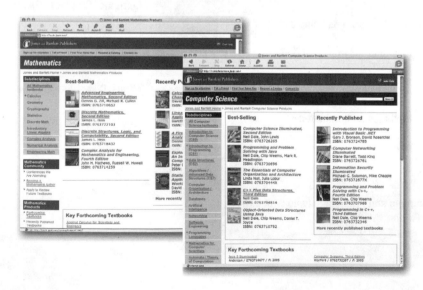